Benchmark Papers
on Energy

Series Editors:
R. Bruce Lindsay, Brown University
Mones E. Hawley, Jack Faucett Associates

Volume

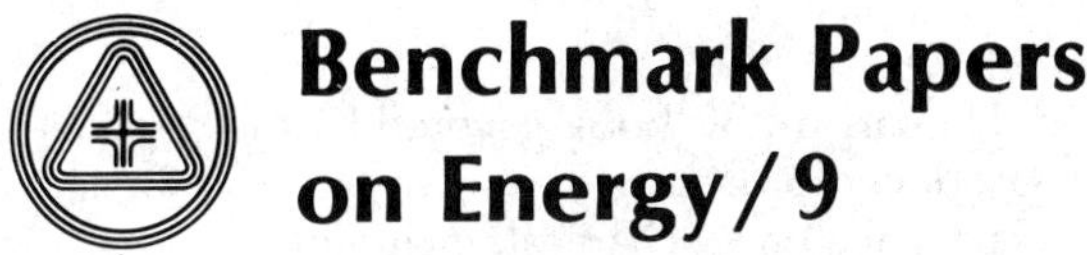

Benchmark Papers on Energy/9

A BENCHMARK® Books Series

MECHANICAL, THERMAL, AND CHEMICAL STORAGE OF ENERGY

Edited by

W. V. HASSENZAHL
University of California

Hutchinson Ross Publishing Company

Stroudsburg, Pennsylvania

LIBRARY OF CONGRESS CATALOGING IN PUBLICATION DATA
Main entry under title:
Mechanical, thermal, and chemical storage of energy.
  (Benchmark papers on energy; 9)
  Includes bibliographies and indexes.
  1. Energy storage—Addresses, essays, lectures.
I. Hassenzahl, W. V. (William V.), 1940–   . II. Series.
TJ165.M4   621.042   81-6485
ISBN 0-87933-392-8   AACR2

Distributed world wide by Academic Press,
a subsidiary of Harcourt Brace Jovanovich,
Publishers.

CONTENTS

Contents

SERIES EDITOR'S FOREWORD

The Benchmark papers on Energy constitute a series of volumes that makes available to the reader in carefully organized form important and seminal articles on the concept of energy, including its historical development, its applications in all fields of science and technology, and its role in civilization in general. This concept is generally admitted to be the most far-reaching idea that the human mind has developed to date, and its fundamental significance for human life and society is everywhere evident.

One group of volumes of the series contains papers bearing primarily on the evolution of the energy concept and its current applications in the various branches of science. Another group of volumes concentrates on the technological and industrial applications of the concept and its socioeconomic implications.

Each volume has been organized and edited by an authority in the area to which it pertains and offers the editor's careful selection of the appropriate seminal papers, that is, those articles that have significantly influenced further development of that phase of the whole subject. In this way every aspect of the concept of energy is placed in proper perspective, and each volume represents an introduction and guide to further work.

Each volume includes an editorial introduction by the volume editor, summarizing the significance of the field being covered. Every article or group of articles is accompanied by editorial commentary, with explanatory notes where necessary. Both an author index and a subject index are provided for ready reference. Articles in languages other than English are either translated or summarized in English. It is the hope of the publisher and editor that these volumes will serve as a working library of the most important scientific, technological, and social literature connected with the idea of energy.

The present volume is one of two edited by Dr. William V. Hassenzahl of the University of California. Their subject is the storage of energy. The purpose of storage is to make energy available at some future time. The concept of serial time, of thinking about time as a series of events as opposed to the unity of "for all time," is an evolutionary step and a basic characteristic in the civilization of mankind. Artifacts for use in the future world are a unique characteristic of human burials. Early examples of cognitive storage of energy include heated rocks, elevated reservoirs, potters' wheels, and animal snares that use bent branches.

The present volumes deal with a wide range of more modern forms of energy storage and are distinguished by the physical sciences with which the forms of energy storage are most closely associated—electrochemistry, electricity, magnetism, chemistry, mechanics, and heat. The emphasis of each volume is on current topics of great concern: use of renewable energy sources that are not available continuously, such as solar; conservation of petroleum energy; and reduction of the cost of providing electricity to meet peak power demands. For readers whose basic science needs refreshing, Dr. Hassenzahl has provided an introduction that also facilitates comparisons of the various forms of energy on a common basis. His comments on the reprinted papers and the bibliographies lead the reader to greater depth of specialization. The tables of contents indicate clearly the coverage of each volume.

MONES E. HAWLEY

PREFACE

Effective energy utilization is the basis of today's advanced industrial society. Usually energy is used neither where, nor when, nor in the form in which it is found, thus creating a need for its conversion, transmission, and storage. Perhaps the largest artificial storage systems are the dams and reservoirs that contain enough water to produce electricity sufficient to power the world's largest cities for months. At the other extreme, both in magnitude and kind, almost every living organism stores energy in the form of hydrocarbons to satisfy its own particular variable energy requirements.

Tens of thousands of papers have been written on energy storage. To reduce this topic to a level that is tractable in two volumes, *Electrochemical, Electrical, and Magnetic Storage of Energy* and *Mechanical, Thermal, and Chemical Storage of Energy*, the perspective has been narrowed to the specific and willful storage of energy by use of special devices. This limitation leaves an almost overwhelming subject, which includes batteries, hydroelectric pumped reservoirs, capacitors, magnets, flywheels, hydrogen and certain specially produced chemicals, and thermal energy storage. Fuels such as gasoline are a special topic and not included here.

In selecting articles for inclusion, emphasis was placed on electric utilities' energy storage, which requires far more storage in terms of both capital investment and total stored energy than any other application. A few articles of historical significance set the stage for present developments; some articles outline the basic laws or rules that govern the developments of a given technology, and others describe the present status of the technologies or the materials required for and the material problems associated with these technologies.

The Introduction describes in simple terms the various forms in which energy exists and the mechanisms available for its storage. It is not intended to be a basic science course but should supply a vocabulary for readers who have not had introductory courses in chemistry and physics; should provide a consistent picture of the interrelationships among the various forms of energy; and, in addition, should give readers some feeling for the capabilities, capacities, costs, and sizes of the various energy storage methods.

Bibliographies are provided at the end of each commentary to aid researchers in obtaining more information on each particular energy storage technology.

W. V. HASSENZAHL

CONTENTS BY AUTHOR

MECHANICAL, THERMAL, AND CHEMICAL STORAGE OF ENERGY

INTRODUCTION

ENERGY AND ENERGY CONVERSION

The developments and changes of society over the past few millennia, and in particular over the past two hundred years, have been characterized by the substitution of animals and, more recently, mechanical devices for human muscle power. We have witnessed an ever-increasing utilization of energy and diversification of energy sources. These changes have occurred throughout the world but have been most profound in the West.

The developments in agriculture, mining, construction, transportation, and other areas have generated the need for various types of energy that must be deliverable at a high rate or equivalently at a high power level. Frequently the available energy is neither in a convenient form for the planned use nor in a convenient location. For example, petroleum is not naturally found on highways, and chemical energy must be converted into mechanical energy to propel a vehicle. Thus it is often necessary to transform energy from one form to another. Further, because the processes of energy transformation are generally expensive, it has become common to attempt to produce some forms of energy at a constant rate and on a large scale.

Electric utility companies frequently produce excess electrical energy at night and store it in a form that can be converted back into electrical energy for use during the peak load periods, which typically occur during the daylight hours. Of course, larger, more expensive generating plants could be installed and their output cycled, but the use of storage facilities reduces cost and in some cases improves overall system performances. To understand why one type of energy storage is selected over other possiblities, it is necessary to consider the forms in which energy can exist and the relationship among the various forms.

Energy, defined as "the capacity to do work,"* appears in a variety of natural forms, such as the *potential* energy of water in a high lake relative to its final resting place in the ocean, the kinetic energy of the water as it flows from the lake to the sea, the electromagnetic energy in light, and the thermal energy or heat that exists in all matter.

In our society usually we do not use energy directly in these forms; rather it is converted into a more convenient form. For example, the kinetic energy of flowing water can be converted into the rotational kinetic energy of a generator, which in turn is converted into electrical energy. The electrical energy might then be changed back into mechanical energy or used to produce heat, and so forth. People rarely use electrical energy directly, yet nearly 20 percent of all the energy used in the United States is electrical.

Energy is always a positive quantity. One system may have less energy than another, but it can never have zero energy. Thus, it is a change in the energy state or level of one system that may be utilized to effect some change in the energy state or level of another· system. Clearly the energy of the driving system must decrease and that of the driven system must increase. Energy is never lost, never destroyed, and never created, yet it can be changed. One of the powerful fundamental laws of physics is the *Law of Energy Conservation*, which states that the total energy of a closed system cannot be changed without some outside influence.

Energy is available for our use in five basic forms:

> Mechanical energy,
> Thermal energy or heat,
> Electrical or electromagnetic energy,
> Chemical energy, and
> Nuclear energy or mass.

Each of these types of energy will be described either in this volume or in the companion volume, *Electrochemical, Electrical, and Magnetic Storage of Energy.* In almost all cases one form may be transformed into another, although frequently an intermediate step is required.

Although energy may exist in many forms, all may be expressed in the same units. In the new Standard International (SI) set of units the preferred unit of energy is the joule (J). The SI unit

Webster's Third New International Dictionary (Springfield, Mass.: G. C. Merriam Co., 1965).

of distance is the meter (m), 39.37 inches; the unit of time is the second (s); and the unit of mass is the kilogram (kg). The joule is a derived unit given by

$$1 \text{ J} = 1 \text{ kg m}^2/\text{s}^2. \tag{1}$$

Other units of energy are used for some special applications. The electric utilities use the kilowatt hour (kWh) and the megawatt hour (MWh). These may be expressed in terms of joules:

$$1 \text{ kWh} = 3,600,000 \text{ J} = 3.6 \times 10^6 \text{ J} \tag{1a}$$
$$1 \text{ MWh} = 3,600,000,000 \text{ J} = 3.6 \times 10^9 \text{ J}.$$

The watt (W), a unit of power, is the rate of using energy:

$$1 \text{ watt} = 1 \text{ joule/second} = 1 \text{ J/s}. \tag{1b}$$

If the power required for a process, or available from a generator, is known, then the total energy required, or available, is equal to the power multiplied by the time of operation. Heat is a form of energy that is usually expressed in calories or kilocalories. One calorie (cal) is defined as the energy required to elevate the temperature of 1g of water 1 Celsius degree and equivalently 1 kilocalorie will raise the temperature of 1 kg of water by 1 Celsius degree. The calorie is now defined in relation to the joule (several types of calories are defined but all are within 1 percent of 4.18 J):

$$1 \text{ cal} = 4.18 \text{ J}. \tag{1c}$$

Another unit of thermal energy is the British thermal unit (Btu), which was originally defined as the heat required to raise the temperature of 1 pound (lb) of water by one Fahrenheit degree.

$$1 \text{ Btu} \approx 252 \text{ cal} \approx 1050 \text{ J}. \tag{1d}$$

Energy can be produced by a force (F) being exerted through a distance (d).

$$E = Fd. \tag{2}$$

In SI units the force is expressed in newtons (N):

$$1 \text{ N} = 1 \text{ J/m} = 1 \text{ kg m/s}^2. \tag{3}$$

The unit of voltage in SI is the volt (V) and the unit of current is the ampere (A). Frequently the electric field is used in calculations. The unit is volts/meter (V/m).

Mechanical Energy

Perhaps the energy form that is easiest to comprehend is mechanical energy. Mechanical energy can be the result of a force acting through a distance and may appear as kinetic, potential, strain, or, for a gas, compressional energy.

Kinetic energy is the energy associated with the motion of an object. Linear kinetic energy is given by

$$E = Mv^2/2, \tag{4}$$

where M is the mass (kg) of the object and v is its velocity (m/s). Rotational kinetic energy is given by

$$E = I\omega^2/2, \tag{5}$$

where I is the moment of inertia (kg · m^2) and ω is the angular velocity (s^{-1}).

Potential energy is the energy associated with the position of matter in a gravitational field. (This concept may be extended to other more general fields.) Usually the quantity of interest is the relative potential energy of an object as it moves from one height (h) to another in the gravitational field of the earth:

$$\Delta E = Mg\,\Delta h, \tag{6}$$

where ΔE (J) is the energy available from or required by a change in elevation Δh (m) of the mass M (kg), and g is the gravitational constant at the earth's surface ($g = 9.8$ m/s^2).

The gravitational constant at the earth's surface is equal to the acceleration of an unsupported object toward the earth. This acceleration will increase the object's velocity by 9.8 m/s (neglecting air resistance) during each second of free fall.

Elastic or strain energy is associated with the stretching or compressing of a solid substance. In the typical one-dimensional case of a spring, the energy is given by

$$E = k\,(\Delta x)^2/2, \tag{7}$$

where k (kg/s^2) is the "spring constant" and Δx the change in length of the material. This expression may be generalized to two or three dimensions for solids, and in a modified form it applies to liquids.

Compressional energy of a gas is the energy gained by a gas as it is compressed from one pressure P_i and volume V_i to a higher pressure P_f and a smaller volume V_f. The change in energy depends on the characteristics of the gas being compressed. No simple expression adequately describes this dependence. When most gases are compressed, a significant fraction of the energy goes into heating the gas. If this heat is removed during or after compression, then the relative amount of energy stored in the gas is considerably reduced.

The transformation of one form of mechanical energy into another is generally simple and relatively efficient. For example, potential energy is converted into kinetic energy when water from a reservoir flows downhill.

Thermal Energy

Thermal energy and heat are identical, but they are different from temperature. We are all familiar with temperatures and can distinguish a rather small temperature difference. As the temperature of a substance increases, the energy content also increases. The energy required to heat a volume V of a substance from a temperature T_1 to a temperature T_2 is given by

$$E = C(T_2 - T_1) \cdot V, \tag{8}$$

where C is the specific heat of a unit volume of that particular substance. A given amount of energy may heat the same weight or volume of other substances to temperatures greater or lower than T_2. The value of C may vary from about 1.0 cal/gm °C (or cal/cm^3 °C) for water to 0.0001 cal/gm °C for some materials at very low temperatures.

The energy given off by a material as its temperature is reduced, or taken up by a material as its temperature is increased, or equivalently the energy required to raise the temperature, is called the sensible heat.

Another form of thermal energy, the latent heat, is associated with the changes of state, or phase changes, of a material. For example, energy is required to convert ice into water, to change water into steam, and to melt paraffin wax. The energy required to cause these changes is called the *heat of fusion* at the melting point and the *heat of vaporization* at the boiling point.

Using our example of water, suppose one wishes to evaporate 1 g of ice by converting it to liquid and then heating it until it boils away. First, 80 cal will be required to change ice at 0°C to water at 0°C; then it will take about 100 cal to raise the temperature of the water to 100°C; finally 540 cal will be needed to boil the water, giving a total of 720 cal.

Just as the sensible heat varies from one material to another, so the latent heat also varies a great deal.

It is straightforward to determine the value of the sensible heat for solids and liquids, but the situation is more complicated for gases. If a gas restricted to a certain volume is heated, both the temperature and the pressure will increase. The specific heat observed in this case is called the specific heat at constant volume, C_v. If instead the volume is allowed to vary and the pressure is fixed, the specific heat at constant pressure, C_p, is obtained. The ratio C_p/C_v and the fraction of the heat produced during compression that can be saved significantly affect the storage efficiency.

Electromagnetic Energy

Energy is a characteristic of both electric and magnetic fields and may be transmitted through free space from one place to another through an oscillating combination of the two types of fields. This type of energy transmission is known as radiant energy. In addition to light, it includes radiowaves, microwaves, X rays, and γ radiation.

The storage of electromagnetic energy requires either a capacitor, a magnet, or a microwave cavity. Because it is quite difficult to store large quantities of microwave energy, in which the electric and magnetic fields oscillate, only magnets and capacitors, where the fields are essentially constant, will be considered here.

The energy in a magnetic field is given by

$$E = B^2/2\mu, \tag{9}$$

where B is in tesla (1 tesla = 1 volt s/m² = 1 weber/m² = 10,000 gauss) and μ is the permeability of the medium. The permeability of a vacuum is $4\pi \times 10^{-7}$ henrys/meter (H/m). The permeabilities of most nonferromagnetic materials (those not attracted to a magnet) are almost identical to the permeability of free space. The energy stored in a magnetic field of 10 tesla (T) is about 40×10^6 J/m^3.

The energy of an electric field is given by

$$E = \epsilon \cdot \frac{(\Delta V / \Delta x)^2}{2} , \qquad (10)$$

where ϵ is the permittivity of the medium and $\Delta V / \Delta x$ is the field gradient (V/m). In free space, $\epsilon = 8.8 \times 10^{-12}$ farads/meter (F/m). The energy density possible depends on ϵ and the maximum allowable field gradient. Air, which has a permittivity almost equal to that of free space, can withstand a field of about 20×10^6 V/m over short distances. Thus the maximum stored energy in an air capacitor is about 2000 J/m^3.

Magnetic fields are produced by the motion of electric charges—for example, the ionized particles in a plasma or the electrons in a current-carrying wire. Electric fields occur when positive and negative charges are separated.

Electromagnetic radiation, including light, X rays, and γ radiation, is produced by the acceleration and deceleration of electric charges, usually electrons.

The most familiar form of man-made electromagnetic energy is that which we receive by power transmission lines from a generator some tens to hundreds of kilometers distant. This form of energy is associated with a varying current in the conductors of the power lines, which is approximately in-phase with the varying voltages between the lines. To transmit the maximum possible power, the current and voltage in an ac power system must increase and decrease together. The energy delivered to a load during a given period (t) is

$$E = V_{av} \cdot I_{av} \cdot t, \qquad (11)$$

where V_{av} and I_{av} are the average voltage and current, respectively. One ampere and 1 volt delivered for one second amount to 1 joule. Electric utility companies try to provide a constant voltage, V_{av}, whereas the current, I_{av}, is controlled by the customers' demand for energy.

Energy is transmitted from the generator to the load at the speed of light, c, 300,000 km/s or 186,000 miles per second. The energy stored in the line at any given time is very small. For a 1,000 MW capacity, 300 km long transmission line, the stored energy (disregarding the inductance and capacitance of the line) is

$$1000 \text{ MW} \cdot \frac{300 \text{ km}}{300{,}000 \text{ km/s}} = 1 \text{ MJ}. \tag{12}$$

Chemical Energy

Chemical energy is the energy associated with the changing of one or more chemical compounds to form other compounds. The burning or oxidation of fuels is the most familiar use of chemical energy. Fuels, which are composed mainly of hydrogen and carbon, combine with the oxygen in the atmosphere to form various compounds and to give off heat. Perhaps the simplest example of this is the formation of water from hydrogen (H) and oxygen (O). Both hydrogen and oxygen exist as diatomic molecules (two atoms forming a single molecule). This is shown symbolically as H_2 and O_2. Oxygen occurs freely in the atmosphere, and hydrogen can be produced by a variety of chemical processes. At elevated temperatures or in the presence of certain catalysts, these two elements combine to form steam (H_2O) through the process

$$2H_2 + O_2 \rightarrow 2H_2O + \text{Energy}. \tag{13}$$

The energy released for each 18 g of water produced is 57,700 cal; equivalently, 1.21×10^8 J is released for each kg of hydrogen consumed. At 25°C the total energy released in this process is 1.34×10^7 J per kg of water produced. This energy is called the heat of formation of water. A similar energetic relationship exists for every other chemical reaction. In some reactions, a single compound decomposes and energy is released. For example, nitroglycerin [really trinitroglycerol $C_3H_5(NO_3)_3$] releases 6.8×10^6 J/kg when it decomposes. The energy available in the formation of 1 kg of water is greater than the energy released when 1 kg of nitroglycerin decomposes, but the nitroglycerin decomposition is much faster. Very rapid decomposition is the characteristic that defines explosives.

The energy associated with chemical reactions is another manifestation of electromagnetic energy. Chemical bonds are the result of the electric fields that exist between negatively charged electrons and the positive nuclei of atoms. It is the changing of these fields during chemical reactions that leads to the release of energy or requires external energy input.

We have seen that chemical energy may be available in the

form of fuels such as hydrogen, alcohol, gasoline, or fuel oil or in compounds such as nitroglycerin. It may also be in the form of two or more compounds stored in a container, such as a battery. In a lead acid storage battery, the charging reaction is given by

$$2PbSO_4 + Energy + 2H_2O \rightarrow PbO_2 + Pb + 2H_2SO_4. \qquad (14)$$

During charging, lead, lead oxide, and sulfuric acid, which have higher heats of formation, are produced from lead sulfate and water. The energy stored in a battery is typically referred to as electrochemical energy because in a battery cell, the electrical energy induces chemical reactions at the plates and is thereby converted back to electrical energy. Similarly, during discharge, the chemical energy is converted back to electrical energy.

Nuclear Energy

The energy associated with nuclear reactions is perhaps the most fundamental form of energy. It is in fact the simplest in concept, but it is not well understood because it is so remote from our day-to-day experiences. This form of energy is associated with the different mass of the original and final nuclei involved in a nuclear reaction. Thus, when a uranium atom breaks up in the process called fission, the neutrons and atoms that are produced have a total mass that is less than the original mass of the uranium atom. The excess mass has gone into the energy of motion of the resulting neutrons and atoms and possibly γ rays and neutrinos, particles produced during a certain type of nuclear reaction that are difficult to detect because they interact very weakly with other forms of matter and energy. Thus:

$$U^{235} \rightarrow \text{fission products} + \text{neutrons} + \gamma \text{ rays} + \text{neutrinos} + \text{energy}.$$

The energy available in any reaction can be determined by weighing the original and final particles and then applying Einstein's mass-energy relation to the mass difference:

$$E = \Delta M c^2. \qquad (15)$$

The fission of 1 kg of uranium will yield 8×10^{13} J, which is equal to the heat from burning 2 to 3×10^6 kg of coal. Although uranium may be a very valuable material as a fuel, it is not suitable for energy storage because it cannot be created from other nuclei except at the temperatures and pressures typical of the centers of some of the more massive stars.

Nuclear energy is also available from fusion, the opposite of fission. In fusion two nuclei combine to form one or more different nuclei, of which at least one is larger than the larger of the original nuclei. Thus, deuterium (D) plus tritium (T) combine to form helium (He) and a neutron (n) that subsequently decays into a hydrogen atom:

$$D^2 + T^3 \rightarrow He^4 + n \quad + Energy. \tag{16}$$
$$\llcorner H^1$$

This process is a source of energy in some stars. It is also being considered as a source of energy for electrical power reactors. One kg of a mixture of 2/5 deuterium and 3/5 tritium will yield 3.38×10^{14} J, which is four times as much as the energy available through the fission of 1 kg of U^{235} and is equal to the energy available from burning about 10^7 kg of coal.

Unfortunately, for the fusion reaction to occur, the initial elements, deuterium and tritium, must be raised to very high temperatures so that thermal motion will overcome the electrostatic repulsion between the two nuclei and allow them to meet and fuse. There are great hopes for fusion as a power source of the future and research is being carried out in this area, but the technical difficulties are great.

Although this form of energy is available, it is almost impossible to reverse the reaction and produce tritium and deuterium from He^4 and a neutron, or two deuterium atoms from He^4. Thus, because the conversion to other forms is difficult, nuclear energy does not appear to be suitable for energy storage.

It is useful to look for a thread that brings all of these energy forms together. It is in this perspective that the elegance of Einstein's mass-energy relationship may be understood. Through very careful weighing of the constituents of any reaction, we would find that the mass of the products is less than that of the original components when energy is produced in a reaction and that the final mass is greater than the initial mass when energy is required to make a reaction proceed. Thus, equation 15 applies not only to nuclear reactions but also to mechanical, electrical, and chemical processes. Although it would be impossible to measure with conventional techniques, the mass of a storage battery is greater in the charged state than in the discharged state. The difference for a typical automotive ignition battery would be about 10^{-9} kg out of a total mass of about 20 kg.

ENERGY STORAGE

Energy storage devices can be classified and categorized in several different ways. Each of the parts in this book and in the companion book *Electrochemical, Electrical, and Magnetic Storage of Energy,* describes the storage of one particular kind of energy. Although articles are included on small-scale applications of energy storage in the parts on specific technologies, one major, large-scale application for energy storage—meeting the diurnal load variations of electric power systems—is the primary application addressed in both volumes.

To understand the magnitude of this storage requirement, we can compare the stored energy in all the automotive starting, lighting, and ignition (SLI) batteries in the United States to the energy stored in one large pumped hydroelectric plant. (A pumped-hydro plant is a hydroelectric power plant that can be filled by operating electrically driven pumps. The reservoir may or may not have a natural water source in addition to the pumps.) A standard automotive SLI battery stores 1 to 2 MJ when new and typically discharges 0.1 to 10 percent during a starting operation. In round numbers, there are about 10^8 of these batteries in the United States today. If the average storage capacity is 1 MJ, then the total capacity is about 10^{14} J. In comparison, the Ludington, Michigan, pumped-hydro facility has a capacity of about 4×10^{13} J and can be discharged in six to eight hours. That one large pumped-hydro storage plant has 40 percent of the storage capacity of all the automotive SLI batteries in service gives an idea of the magnitude of the need for energy storage by the electric utility companies.

This storage requirement comes about because, on a yearly basis, the U.S. electric utility industry transmits power to its customers at a rate equivalent to only about 60 percent of its generating capacity. Aside from some downtime necessary for plant maintenance and standby equipment for emergencies, the unused 40 percent of capacity represents essentially wasted capital investment. This situation arises because the demand for power varies periodically on a daily, weekly, and seasonal basis and varies randomly during time periods of seconds to tens of minutes. Load variations are such that on a yearly basis, the minimum and peak power requirements for most electric utilities differ by a factor of two or more, and even on a daily basis the minimum load averages only 60 to 70 percent of the peak load.

The electric power industry must be prepared to meet the peak power demands, and generally this is done through a combination of three (or more) power sources: the base-load generation (typically about 45 percent of the peak power) is furnished by relatively efficient, large fossil-fueled or nuclear generators that operate around the clock at or near full throttle; the intermediate load (typically about 40 percent of peak power) is met by older, less efficient plants that are cycled in and out of service as the power demand ranges from 45 to 85 percent of the peak demands; and finally, the peak 15 percent of the power is usually delivered by gas turbines. The base-load units supply about 70 percent of the total system energy at the lowest delivered cost, while the intermediate cycling plants furnish about 25 percent of the energy at significantly higher cost. The remaining 5 percent is derived from the peaking units, which, although relatively inexpensive in terms of capital investment, require costly and scarce special fuels (such as JP-4 or number 2 fuel oil), operate at low thermal efficiencies, and necessitate high expenditures for maintenance.

Delays in deliveries of base-load plants have forced utility companies to substitute gas turbine peaking units for the generators on order. These turbines may deliver up to 15 percent of the total energy of some systems. To add to the power industry's problems, the partially successful drive to reduce overall power consumption between 1974 and 1979 often has not simultaneously reduced peak demand. The net result is that an even larger percentage of the total system power must be derived from the more expensive sources.

If we assume that it is not possible to control the variable load, then an ideal solution to this problem would be to have base-load plants provide all of the system's energy requirements by operating at full capacity (corresponding to the annual average load) and to have energy-storage systems that could absorb excess energy generated during periods of less-than-average demand and deliver it back to the system during periods of greater-than-average demand. These energy-storage systems should be highly efficient, nonpolluting, easily sited, reliable, long-lived, and easily and inexpensively maintained, and they should have a capital cost ($/kW) lower than the peak- or intermediate-load generating plants they replace. To date only pumped-hydro installations such as the one at Ludington have been used for this application.

One area where energy storage will aid the utilities is in capital investment. Partly because of the load variations, the electric utilities are the most capital intensive of all major industries. This would not be a serious problem if the total, yearly demand for electric power were not increasing, because the capital investment would then be relatively constant, with a few old or worn-out generation facilities being replaced each year. However, electric energy usage doubles every decade or so. Thus, the electric utility industry is expected to install about 400,000 MW of generating capacity at a total cost of $120 billion between 1977 and 1987. If inexpensive energy storage were available today, however, part of this investment in generating capacity could be postponed.

The potential savings associated with the utilization of energy storage systems on electric power systems are in fact much greater than just those in the area of capital investments. Let us assume that 10 percent of all generating capacity is in the form of gas turbines and that 5 percent of the total electrical energy used in the United States is generated by these units. Then about 8×10^8 barrels of number 2 fuel oil are consumed each year by these devices on utility systems. Since it should be possible for energy-storage systems to replace a large fraction of the gas turbine capacity (50 percent substitution should be possible) the potential saving is 4×10^8 barrels of oil per year. Of course some other fuel such as coal or uranium would have to be consumed to generate the electrical energy that is to be stored. However, coal and nuclear fuels are considerably more abundant than oil and natural gas.

Another potentially large-scale application of energy storage is for vehicular propulsion. It has been estimated that hydrogen, ethanol, batteries, and flywheels could replace 70 to 80 percent of the petroleum, including oil derivatives, now used for transportation. Three major applications of these storage technologies in the transportation sector are: battery-, hydrogen-, and ethanol-powered automobiles, buses, and trucks; flywheels for power augmentation and regenerative braking; and liquid hydrogen-powered aircraft. The rapid integration of storage technologies into the electric utility and transportation sectors could extend for many years the availability to humanity of the earth's very limited petroleum supplies.

The types of energy storage available are determined by the forms that energy can assume. The value of a storage technology for a particular application depends on the ease with which the stored energy may be used or converted into another useful form.

The energy in a small dry-cell flashlight battery is converted to heat and then to light quite effectively, and, though low efficiency and high cost remove these cells from consideration for electric power system applications, they have been of great value for a variety of uses. Thus the type of storage must fit the end use of the energy.

Continuing with this example of a dry-cell, primary battery, it must have a long shelf life; that is, it must retain a charge for an extended period, it must be relatively inexpensive, and it must deliver energy for a period of hours once discharge has started. Though rechargeable batteries are now widely available, this capability does not appear to be a primary consideration for most consumers because several hundred million dry-cell flashlight batteries are sold in the United States each year. For some applications, efficiency (energy withdrawn divided by energy deposited) is a primary criterion. In particular, storage units for electric utility systems must have an efficiency of about 60 percent or greater.

There are several characteristics of an energy-storage device that make it effective for a particular application. Cost (\$/J) is perhaps the most important factor. However, for some applications, such as satellites and space probes, reliability and lifetime are the primary considerations, and cost is of little concern because the cost of the system is certain to be only a small fraction of the total mission cost.

Energy storage density (J/m^3) and specific energy storage (J/kg) may be important considerations for other applications. For example, most car owners require a vehicle that will travel 300 km or more without refueling. A quantity of gasoline, hydrogen, or alcohol sufficient to supply the energy required to move a 2,000-kg vehicle this distance occupies a modest volume, (less than $0.1 \ m^3$) and has a modest weight (less than 100 kg). But lead acid batteries or steel flywheels for the same purpose would be considerably larger and more massive.

Another characteristic of the storage device is the area or, for large installations, land requirements. This parameter is not of great concern for a vehicle or a flashlight, but it may become extremely important if the energy-storage system is to be part of an electric utility system and is to be installed in or near a metropolitan area where real estate costs are a serious consideration.

Thus several characteristics affect the usefulness and effectiveness of a storage system. The characteristics mentioned above and a few others, with their appropriate units indicated, are:

unit cost (\$/J),
efficiency (%),

energy density (J/m³),
power density (W/m³),
specific energy (J/kg),
specific power (W/kg),
charged shelf life (years or seconds depending on application),
discharge time(s),
land requirements (J/m²),
useful life (years or cycles),
reliability (mean time between failures, mean time to repair), and
safety (probability of an injury per year).

Each of the various types of energy may be stored once a suitable storage medium has been found.

Mechanical Energy Storage

Mechanical energy may be stored as the kinetic energy of linear or rotational motion, as the potential energy in an elevated object, as the compressional or strain energy of an elastic material, or as the compressional energy in a gas. It would be very difficult to store large quantities of energy in linear motions because one would have to chase after the storage medium continually. However, it is rather simple to store rotational kinetic energy. In fact, the potter's wheel, perhaps the first form of energy storage used by man, was developed several thousand years ago and is still being used.

The stored energy in a flywheel is given by

$$E = \frac{1}{2} I \omega^2, \tag{17}$$

where I is the moment of inertia of the rotating mass and ω is the angular velocity. A tensile stress or load is created in the structural material of the flywheel as it spins. This stress, which increases as the angular velocity increases, limits the amount of energy that can be stored in a given mass of material. Materials with very high tensile strengths can store more energy per unit volume than can materials with low tensile strengths. It can be shown that there is a relationship between the mass M, the stored energy E, the strength σ, and the density ρ of a material under load.

$$M = \kappa \cdot \frac{E \rho}{\sigma}, \tag{18}$$

where κ is a constant that is greater than or equal to 1 and is determined solely by the geometry of the system. This relationship is very fundamental and applies to all forms of energy storage in which the energy is contained by a structural element that is put in tension as the energy is stored.

For flywheels that are not tapered near the rim, κ is greater than two ($\kappa \geq 2$) because the loads are almost always in the tangential direction. This type of stress is called a hoop stress. From equation 18, it is straightforward to get a rough estimate of the cost of a unit from the cost factors ($/kg) used by industries that fabricate various devices. For example, a fabricated cost of $3/kg may be used for steel with the relatively high tensile strength of 30,000 pounds per square inch (psi). In SI units tensile strengths and pressures are given in pascals (Pa). One psi is equivalent to 6,900 pascals; thus 30,000 psi is 207,000,000 Pa, or about 2×10^8 Pa. If the flywheel container and the motor drive and other parts double the cost, then a rough estimate of the storage cost is 2.2×10^3 J/$ or 4.6×10^{-4} $/J, which is independent of the storage capacity of the device.

An estimate of the energy stored per unit volume, E/V (J/m³) can be determined by substituting the value of E/M from equation 18:

$$\frac{E}{V} = \frac{E\rho}{M} = \frac{\sigma}{K}. \qquad (18a)$$

The energy-storage density in a steel flywheel with $\kappa = 2$ and a 2×10^8 Pa (30,000 psi) working strength is 10^8 J/m³, and about 5×10^7 J/m³ for a complete system.

Potential energy storage is most often utilized in the form of pumped, hydroelectric storage. Energy generated in an electric power grid during periods of low energy demand is converted into potential energy by pumping water from a lower reservoir into a higher one. The storage density depends on the elevation difference of the two reservoirs and is about 10^4 J/m³ for each meter of head. For a 100 m height difference for the two reservoirs, the energy density is 10^6 J/m³ and the specific energy is 10^3 J/kg.

Although the energy density and specific energy of pumped hydro is relatively small, the cost is low, which makes it attractive to electric utilities. For example, the Ludington, Michigan, pumped-hydro facility, which was constructed between 1968 and 1974, cost $350 million for about 14,000 MWh of storage and 2,000 MW of power capacity. This facility occupies about 2.5 square miles (640 hectares) of land. The storage cost was 1.4×10^5 J/$ or $6.9 \times$

10^{-6} \$/J, about one-seventieth the cost of the flywheels described above.

The storage of energy directly as strain in a material such as a spring has been used for many years in specific applications such as watches and clocks. Because of the small size of watch springs, expensive materials with working stresses of 300,000 psi (2×10^9 Pa) are used. The energy density and specific energy are quite small, however. For example, at the same stress level, the energy density in a spring is smaller than the energy density in a flywheel by a factor equal to the strain , ϵ, which is usually 0.01 or less.

An advantage of storing "strain energy" is that it has a long shelf life. Springs have been used for years in mine shafts on protective devices. When a cable fails the springs release and snap a stopping mechanism into a position to limit the descent of the cable car.

Energy storage in the form of compressed gas has been used since the late 1800s for special applications such as mining. Because massive containment vessels are required for compressed gases, it is not practical to store large quantities of energy in this form unless the vessel is very cheap, such as an underground cavern. The structural material is then the rock surrounding the cavern. This type of energy storage appears to be practical for the electric utility diurnal storage requirement in some locations.

Thermal Energy Storage

Thermal energy may be stored by elevating the temperature of a substance (increasing its sensible heat), by changing the phase of a substance (increasing its latent heat), or by a combination of the two. Both forms will see extended applications as new energy technologies are developed. Energy stored in sensible heat appears very promising for high-temperature storage of large quantities of energy at fossil-fired power plants. Oil will likely be used as the medium for this type of storage.

Many new homes are being equipped with solar heating and cooling systems that require diurnal energy storage. Materials being considered for this application include water, two- to four-inch diameter rocks, and phase-change materials such as Glauber's salt ($Na_2SO_4 \cdot 10H_2O$). These materials are all readily available, and water and rocks are obviously quite cheap, though the containment system may not be.

For a typical house in the high mountains and plateaus of the southwestern United States, where there is adequate sunshine in winter but the nighttime temperatures are quite low, the energy storage requirements are about 15 Btu/°F/ft$_h^2$ (in SI units 2.9 × 10^{5}J/°C/m$_h^2$) where ft$_h^2$ (m$_h^2$) is the heated floor space and the °F (°C) is the average temperature difference inside versus outside. For a house with an area of 2000 ft$_h$ (~200 m$_h^2$) the volume of water or rocks required to provide most of the heating for one day, with the outside temperature at 0°F will be 13 m^3 and 42 m^3, respectively. The cost of the water would be trivial, and the required 70 tons of rocks would cost only $700 to $3,000 depending on availability, giving an energy storage cost for the rocks alone of about 1.2 × 10^6 J/$ or 8.0 × 10^{-7} $/J. Unfortunately the cost of storage tanks, heat exchangers, and the space inside the building is considerable. These must be weighed against the ever-increasing cost of other heating methods such as natural gas, fuel oil, and electricity.

The cost of the oil for thermal storage at central power plants is based on the present cost of fuel storage vessels and heat exchangers. The storage density, which depends on the temperature swing, nominally between 530°F and 250°F (230°C and 121°C), is 4 × 10^8 J/m^3. At present fuel oil costs, this amounts to 2 × 10^{-6} $/J. Other costs, however, raise this estimate considerably. It has been calculated that additional equipment and oil treatment will bring this cost up to 1.2 × 10^5 J/$ or 8.4 × 10^{-6} $/J for a 2.6-GWh thermal storage plant on a nuclear boiling water reactor.

Thermal storage has also been considered for vehicle propulsion, but the cost and weight of the storage medium required for travel over distances greater than about 100 km make it an unlikely candidate compared to batteries and hydrogen.

Electrical or Electromagnetic Energy Storage

Energy may be stored in the magnetic field of a coil or in the electric field of a capacitor. The energy storage density in a capacitor is given by

$$E/\mathcal{V} = \frac{\epsilon (\Delta \mathcal{V}/\Delta x)^2}{2}$$

and is about 60 J/m^3 for air. Many materials have higher dielectric constants and can also withstand higher electric fields than air. Some of these, such as mica, glass, and oil-impregnated paper, have been used for capacitors for fifty to seventy years; others, such as Mylar, Teflon, titanium dioxide, and various titanates have been used only recently. The energy storage densities currently possible in small capacitors using barium titanate are about 5×10^5 J/m^3, whereas the theoretical limit is 10^7 J/m^3.

The most recent studies of large capacitive energy-storage systems have shown the oil-impregnated paper is the cheapest, most reliable dielectric. Costs of 10^{-2} \$/J have been achieved on large, 10 to 20 kV, 0 to 10 MJ systems. The single advantage of capacitive storage has been its rapid discharge capacity or equivalently high-power density. Most capacitors can be almost completely discharged in 1 to 10 microseconds. The energy density of capacitors is too low and the cost too high for most energy-storage applications.

Energy storage in magnetic fields was not economical until the recent, practical demonstrations of superconductivity. The development of materials having no measurable resistance to the flow of electric currents has increased the charged shelf life of practical magnets from a few seconds, which was the maximum possible ten to fifteen years ago, to days or weeks. Some small superconducting coils have been charged and have remained in the charged state for years without a detectable change in current or field.

Superconducting magnets capable of storing up to 9×10^8 J have been constructed for special nuclear physics applications. Some small magnets have been used for pulsed energy storage, but to date no large magnets have been constructed specifically to store energy for long periods.

The energy-storage density in a magnetic field of five tesla, which is easily achieved with many commercially available superconductors, is 10^7 J/m^3. In a very large magnet, such as those proposed for the diurnal energy-storage application, the superconducting coil occupies only a few percent of the field volume. As a result, the energy density relative to the superconductor volume may be as high as 10^9 J/m^3.

The cost of large superconducting magnetic energy-storage (SMES) units proposed for diurnal load leveling will not be proportional to size. The unit cost (\$/J) of larger coils will be less than the unit cost of smaller coils. It appears that with present technology SMES units storing more than 10^3 MWh will be competitive with other storage devices on electric utility networks. A SMES unit having the capacity of the Ludington pumped-hydro plant would require less than one square mile of land, one-third as much as Ludington; would have about the same initial cost; would have an efficiency of 95 percent rather than 70 percent and could be more easily sited. Smaller storage units, for example, for automotive propulsion, will not be economical with present technology. The advantages of this type of storage are efficiency, cost, fast response time, and, depending on environmental restrictions, decreased land requirements.

Most magnets, large inductors, and transformers use iron. Iron, however, is a very poor storage medium for magnetic energy. Its advantage is that a moderately high (up to about 2 T) magnetic field can be obtained with a relatively small driving force, usually the induction produced by a small current. Unfortunately this effect, which may provide a field multiplication of 1000 or more, saturates as the internal field approaches 2T. This enhancement then decreases as the external field increases, and, at 2T, the external and internal fields are approximately equal. At a field of 5 T iron would increase the stored energy in a given volume by 10 percent at most, or about 1 MJ/m^3. Storage of energy in the magnetic field of iron costing \$3/kg would be about 2×10^{-2} \$/J, 10^4 times more expensive than pumped hydro and other economical forms of energy storage.

The third form of electromagnetic energy, radiant energy, which includes light, radiowaves, and X rays is very difficult to store. It is possible, however, to store microwave energy for a short time in a cavity with superconducting walls. At present the energy density is a fraction of a percent of that possible with other techniques, the storage efficiency is only a few percent, and the equipment required to provide the appropriate environment is costly and complicated.

Chemical Energy Storage

Energy may be stored in systems composed of one or more chemical compounds that release or absorb energy when they react to form other compounds. The most familiar chemical energy-storage device is the battery. Energy stored in batteries is frequently referred to as electrochemical energy because chemical reactions in the battery are caused by electrical energy and subsequently produce electrical energy.

It is useful to speak of storage efficiency only for secondary or rechargeable batteries. Primary batteries, such as the dry-cell flashlight battery, are fabricated in a charged state and are discharged only once. The energy-storage efficiency of batteries depends on the battery design and type. The batteries for vehicle propulsion and utility energy storage are expected to have an efficiency of 70 or 80 percent for deep discharge.

Another important form of chemical energy storage is possible by using hydrogen. Hydrogen does not exist as a free element on the earth. It is, however, extremely abundant as a constituent of water, which may be decomposed by electrical or thermal processes to yield hydrogen and oxygen. In the past electrolysis has been used, with a relatively low efficiency, to produce hydrogen for special applications. If hydrogen is to be used for energy storage, then the total efficiency for the combined processes of hydrogen production, storage, and use must be increased to 60 or 70 percent. Assuming this is possible, there are several methods of storing hydrogen: as a liquid at a temperature of 20 K, as a compressed gas in cylinders or underground, or as a compound that is easily formed and decomposed, such as a metal hydride. Each of these has advantages and disadvantages. If hydrogen is to be used as a fuel for vehicles, then the specific energy should be as great as possible, and liquid hydrogen in a dewar is probably the best solution. A dewar is a device that uses a vacuum to insulate thermally cryogenic systems at temperatures below 77 K. The same principle is used in the vacuum or thermos bottle. The energy density is about 8.6×10^9 J/m^3 as compared to 3.6×10^{10} J/m^3 for gasoline. But the specific energy is 1.21×10^8 J/kg, which, even when the weight of a dewar is included, is greater than that of gasoline, 4.8×10^7 J/kg. If maximum efficiency is desired and weight is not a consideration, then a metal hydride storage system may be used. This type of storage has been proposed for electric utility diurnal storage. The energy-storage density of hydrogen in the hydride is approximately equal to the storage density of the hydrogen as a liquid. The storage of hydrogen as a compressed gas is not promising because the affinity of hydrogen for metals, which makes metallic hydride storage so effective, leads to hydrogen embrittlement in steel tanks containing high-pressure hydrogen.

An advantage of hydrogen over some other energy-storage methods is that it is easily transportable and can be used as a fuel some distance away from its production site. For example, part of the hydrogen produced by the excess capacity available at night could be stored in metal hydrides and used in fuel cells for power generation during the peak-load periods, part could be liquefied and used for automotive or airplane propulsion, and part could be used to augment natural gas for residential or commercial use.

Another form of chemical energy storage is ethanol, which

Table 1   A Comparison of Energy-Storage Technologies

Energy Storage Technology	Cost[a] ($/J)	Efficiency (%)	Energy Storage Density (J/m^3)	Max Power[b] Density (W/m^3)	Specific Energy (J/kg)	Max Specific Power[b] (W/kg)	Charged Shelf Life	Typical[c] Discharge Time (s)	Useful[d] Life (cycles or years)	Applications[e]
Mechanical energy										
Flywheels	$4 \cdot 10^{-4}$	70–90	$5 \cdot 10^7$	$5 \cdot 10^6$	$6 \cdot 10^3$	$6 \cdot 10^2$	min→days	$1 - 10^3$	20 yrs	RB/SA
Strain energy	$4 \cdot 10^{-2}$	50–90	10^6	10^5	$2 \cdot 10^2$	$2 \cdot 10^1$	years	$10 - 10^6$	20 yrs	SA
Compressed gas[f]	10^{-5}	60–70	10^7	10^3	—	—	weeks	$10^4 - 10^5$	20 yrs	EU
Pumped hydro storage[g]	$7 \cdot 10^{-6}$	70	10^6	10^2	10^3	10^{-1}	weeks	$10^4 - 10^5$	20 yrs	EU
Thermal energy										
Sensible heat										
High temp oil	10^{-5}	70–80	10^8	10^4	10^5	10	days	$10^4 - 10^5$	20 yrs	EU
Low temp-rocks	$8 \cdot 10^{-6}$	90	$3 \cdot 10^7$	$3 \cdot 10^3$	10^4	1	days	$10^4 - 10^5$	20 yrs	RH
Latent heat										
Various salts	10^{-5}	80	$3 \cdot 10^8$	$3 \cdot 10^4$	10^5	10	weeks	$10^4 - 10^5$	$10^2 - 10^3$ cycles	RH
Electromagnetic Energy										
Capacitors[h]	10^{-2}	90	10^5	10^{11}	$3 \cdot 10^2$	$3 \cdot 10^8$	days	$10^{-6} - 10^{-3}$	10^4 cycles–	SA
Magnets/coils[i]	10^{-5}	90	10^8	10^4	10^6	10^2	days	$10^{-3} - 10^5$	20 yrs	EU/SA
Chemical Energy										
Batteries[j]										
Pb acid	$2 \cdot 10^{-5}$	60–80	$5 \cdot 10^7$	$5 \cdot 10^3$	10^5	10	weeks	$10 - 10^4$	10^3 cycles	ES/SLI/AP
Other[k]	10^{-5}	65–80	10^4	10^4	$5 \cdot 10^5$		weeks	$10 - 10^5$	$10^3 - 10^5$ cycles	EU/SA/AP
Hydrogen[l]	$2 \cdot 10^{-5}$	30–70	$9 \cdot 10^9$	—	10^8	—	days/mos	$10^4 - 10^5$	20 yrs	EU/SA/AP
Ethanol		60–80	$2 \cdot 10^{10}$	—	$3 \cdot 10^7$	—	days/mos	$10^4 - 10^5$	20 yrs	EU/SA/AP

Table notes:

[a] The cost of energy storage is given for the entire system, including equipment required to convert the energy from and/or into a useable form. Actual costs may vary by a factor of two or three from the value given, which is a mean. The cost in \$/kWh is obtained by multiplying by 3600.

[b] Specific power and power density are maxima based on minimum possible discharge times, except for those technologies where the device is for the electric utility diurnal storage application.

[c] Minimum discharge times are usually for partial discharges. For specific technologies, smaller units can be discharged faster than larger units.

[d] For economic considerations, a lifetime of twenty years is equivalent to an infinite life. There is no indication that the various devices given a twenty year life expectancy in the table will wear out in twenty years.

[e] Abbreviations are given in the table for various applications:
EU, electric utility diurnal storage;
ES, emergency service—for electric power system failure, and so on;
SLI, automotive–starting, lighting, and ignition;
RB, regenerative braking;
RH, residential heating;
SA, special applications; and
AP, automotive propulsion.

[f] Data for compressed gas are based on the plant installed at Huntorf, West Germany.

[g] Data for pumped-hydro storage are based on the facility at Ludington, Michigan.

[h] Data are for oil-impregnated-paper capacitive energy-storage systems.

[i] Most data are for a 10^4 MWh superconducting coil constructed underground and used for diurnal storage on an electric power system. Energy density is based on excavated volume.

[j] Most data are based on diurnal energy-storage application. Automotive propulsion batteries will have higher power rating.

[k] Several advanced battery systems have been proposed for automotive propulsion and electric utility applications.

[l] Storage efficiency depends on type of hydrogen storage, and storage cost depends on the medium used. Liquid hydrogen storage is less efficient but cheaper than metal hydride storage. Power and storage capacities are almost unrelated for hydrogen storage.

may be produced through biological processes or from coal or other hydrocarbons. The energy density and specific energy for ethanol are 2.2×10^{10} J/m³ and 2.7×10^7 J/kg. Ethanol and a related alcohol, methanol, may be substituted for gasoline in internal combustion engines. Only partial substitution may be possible for existing engines, and complete substitution may require engine modifications or new engines. Methane can be produced by similar chemical processes. It can be transported with ease, can be added to natural gas supplies, and can be used in a chemical heat pipe, which makes use of the reversible reaction, $CH_4 + H_2O \leftrightarrow CO + 3H_2$. This reaction provides a convenient method of transporting low-grade heat but is not an effective means of storing energy.

Any reversible chemical reaction in which all of the reactants are solids or liquids may be considered for storing energy. In general the driving force for the reaction will be heat or electrical energy and, when the reaction is reversed, the output will again be heat or electrical energy.

Nuclear or Mass Energy Storage

Because it is difficult to produce specific nuclear species or isotopes in a controlled way and then to decompose them and produce other forms of energy, it is not practical to use nuclear energy as a form of energy storage.

A Comparison of Energy-Storage Technologies

The various types of energy storage are compared in Table 1 on the basis of the characteristics that affect the usefulness and effectiveness of a storage system. An attempt has been made to make this table as complete as possible at this time. Two characteristics, reliability and safety, have been left out because they are not true technical characteristics of the storage technologies but instead depend on manufacturing technique and quality control. All of the data in the table are taken from the articles herein and in *Electrochemical, Electrical, and Magnetic Storage of Energy* or from the articles listed in the bibliographies at the end of each of the Editor's Comments.

BIBLIOGRAPHY

Brown, H. L., Director. 1972. *Effective Energy Utilization Symposium.* Drexel University, Philadelphia.

Chalmers, B. 1963. *Energy.* Academic Press, New York.

Ford, K. W., G. I. Rocklin, R. H. Socolow, D. L. Hartley, D. H. Hardesty, M. Lapp, J. Dooker, F. Bryer, S. M. Berman, and S. D. Silverstein, eds. 1975. *Efficient Use of Energy.* American Institute of Physics, New York.

Garvey, G. 1972. *Energy, Ecology, Economy.* W. W. Norton and Co., New York.

Guyol, N. B. 1969. *The World Electric Power Industry.* University of California Press, Berkeley, Calif.

Hutchinson, F. W. 1957. *Thermodynamics of Heat-Power Systems.* Addison-Wesley Publ. Co., Reading, Mass.

Krenz, J. H. 1976. *Energy Conservation and Utilization.* Allyn and Bacon, Boston.

McMullan, J. T., R. Morgan, and R. B. Murray. 1976. *Energy Resources and Supply.* Wiley, London.

U.S. Dept. Defense and Calif. Inst. Tech. 1958. *Seminar on Advanced Energy Sources and Conversion Techniques.* Electro-Optical Systems, Pasadena, Calif.

Walsh, E. M. 1967. *Energy Conservation: Electrochemical, Direct, Nuclear.* The Ronald Press. New York.

Part I

PUMPED HYDROELECTRIC ENERGY STORAGE

Editor's Comments
on Papers 1 Through 5

Pumped hydroelectric energy storage, or pumped hydro, is the only type of large-scale energy storage that the electric utilities have used on a regular basis. Because the units are large and the operating period required for cost recovery is long, however, pumped hydro has been used only on the electric utilities. Although other storage technologies are being explored by the utilities, there does not appear to be any other proposed application for pumped hydro.

The three major requirements for a conventional pumped hydro site are sufficient water, terrain and rock structure adequate for both upper and lower reservoirs, and proximity to either the load or the generating plant. There are many locations where one or two of these criteria are met but not all three. In the midwestern United States, for example, the load exists almost everywhere and there is generally adequate water, but the land is almost uniformly flat. In the Rocky Mountains, there is great relief in the terrain, but the population density is low and little water is available. Thus neither of these areas is generally attractive for pumped-hydro installations.

The development of pumped hydro is a direct offshoot of the research by the electric utilities that led to the hydroelectric power plants early in the twentieth century. The concept of using pumped hydro as a means of leveling the diurnal load variations experienced by the utilities was first proposed in Germany around 1910. By 1930 there were several pumped-hydro units in Europe and a single 25 MW unit in the United States, in Connecticut. With time, the widespread use of thermal power plants (which achieve their highest efficiency and longest life when operated at a constant power level) and a degenerating load factor (the ratio of the minimum power delivered to the maximum power delivered during a fixed period) have led to extensive use of pumped hydro.

When pumped hydro became more important and the investment increased, it became apparent that less expensive alternatives should be explored. As a result present-day pumped hydro began to deviate from conventional hydroelectric power plants in the early 1960s. Whereas in earlier installations the turbine/generators were completely separate from the pump/motors, most recent units have a motor, a generator, and a reversible turbine on the same shaft.

Because the cost and efficiency of pumped-hydro installations generally improve as the hydraulic head is increased, the utilities search for sites with the maximum possible vertical separation between the reservoir and the turbine. An additional pressure that has caused the utilities to use ever-increasing heads is from environmental groups that have criticized the utilities for the large and sometimes ungainly reservoirs and the poor efficiency associated with low-head installations. Because most of the technically attractive sites for pumped hydro in the United States have already been used, the future concern of environmentalists, which mainly affects new installations, may have local effects but will not have a significant national effect. The Pacific Northwest may be an exception in that a great number of potential sites exist, but few have been used because of the extensive hydroelectric capability in this region.

A recent development, underground pumped hydro, has opened up the possibility of numerous pumped-hydro sites in previously unattractive locations. This new technology consists of an upper reservoir at ground level and a lower reservoir underground, usually at great depth. The major requirements are the same as for conventional pumped hydro but with a suitable underground cavern as the lower reservoir. Because a very large head can be used, the upper reservoir is generally small, and the site can be selected to accommodate environmental concerns as the exact location of the lower reservoir is not critical.

Paper 1, on pumped hydro, is a survey of the pumped-hydro installations in operation or under construction in 1968. The technology has not changed greatly since that time when a conventional, but rather unusual because of the use of two separate upper reservoirs, pumped-hydro plant was being constructed at Coo-Trois Ponts in Belgium. This plant and the considerations in the selection of the power generation machinery is the focus of Papers 2 and 3. Paper 4 relates the planning that went into the construction of one of the largest pumped-hydro installations in the United States. Subsequent papers in the annual proceedings of the American Power Conference, which is held each spring in Chicago, describe the progress and eventual operation of this facility. During the initial filling of the reservoir, the rock structure of the mountain was found to be very permeable, and extensive sealing was required before the system was made operational. The technology modifications required for an underground variation of the basic pumped-hydro system and the major factors that must be considered in selecting a site are described in Paper 5.

BIBLIOGRAPHY

Anon. 1972. *Power Generation Alternatives.* Engineering Division, Seattle City Light.

Ferreira, A. 1973. Initial Operating and Testing Experience—Northfield Mountains Pumped-Storage Project. *Am. Power Conf., 35th, Proc.,* pp. 955–965.

Fukasu, S., and Y. Ishii. 1966. Characteristic Features of Pumped-Storage Power Stations in Japan. *1966 World Power Conf., Proc.,* p. 1149.

Pfafflin, G. E., 1974. Future Trends in Hydro-pumped Storage Equipment. *Am. Power Conf., 36th, Proc.,* pp. 390–402.

Symposium on Pumped Storage. 1958. *Am. Power Conf. 20, Proc.*

Schimmelbusch, J. S. 1971. *Pumped Storage, a Bibliography, 1961–1970.* Bonneyville Power Admn. Rept., Portland, Ore., June.

Unsworth, G. N. 1975. A Review of Pumped Energy Storage Schemes. *Atomic Energy of Canada Ltd Rept. AECL 4926,* Pinaua, Manitoba.

1

Reprinted with permission from *Power Eng.* 72:58–63 (1968)

Worldwide pumped-storage projects

Experience with reversible pump/turbine units
has confirmed their flexibility. Unit sizes and
operating heads will continue to increase, subject
to commercial rather than technical considerations

By G. DUGAN JOHNSON, Chief Hydraulic Engineer, Allis-Chalmers, York, Pa.

Low cost peaking capacity becomes increasingly important as peak loads grow, sizes of base-loaded units increase, and reliability demands become greater. Interest in pumped storage has been growing at an increasing rate, and plants of up to 2000 Mw capacity are being planned.

Installed cost of pumped storage per kilowatt is mainly a function of site conditions, and tends to decrease with increases in head and unit capacity. This initial cost of construction has been the deciding factor for or against its installation in any case under consideration. Obviously, under these conditions, the use of reversible pump/turbine - generator/motor units is inevitable. The choice follows not only from lower rotating equipment costs as compared to *tandem* units (consisting of separate turbines, generator/motors, couplings and pumps on the same shafts), but also because of considerable cost savings in valving and civil construction, especially in the water passages to and from the units.

There has been a tendency to refer to reversible pump/turbine - generator/motor units as *pump/turbines*. Generator/motors are very little different in appearance from conventional hydrogenerators, but the pump/turbine is quite different from a conventional hydraulic turbine or pump. There are a number of detail differences in the generator/motor design, especially to allow for both directions of rotation,

but the only major differences are concerned with startup as a motor in the pumping mode. By comparison, the pump-turbine is quite different from a conventional pump, both because of the very large sizes and because of the adjustable guide vanes for turbine operation. It is different from a conventional turbine primarily because of the large impeller diameter required for pumping. In this paper, pump/turbine refers only to the hydraulic portion of the reversible unit.

Reversible units in operation or on order are listed in Tables 1 and 2. Table 1 lists Francis-type reversible single-stage pump/turbine - generator/motor units, and Table 2 lists adjustable-blade reversible single-stage pump/turbine - generator/motor units. Published (or *official*) ratings are given for all units, though it is recognized that they are not consistent in their relationships with the applicable model performance curves.

Perhaps the most surprising feature of these two tables is the large number of installations, which indicates that reversible units for pumped storage no longer can be considered as novelties. Rather, they appear to be well on the way toward replacing tandem units all over the world, with the possible exception of Germany. Historically, it is interesting to note that a German manufacturer pioneered both types of reversible pump/turbines in the decade

between 1930 and 1940, but that Germany has only one installation, Ronkhausen, of any consequence. The same manufacturer is generally credited with the original conception (about 1910) of pure pumped storage, utilizing separate pumping and generating units, as well as its combination with downstream peaking use in other power stations. It will also be interesting to see whether future pump-storage projects in Germany will continue to be equipped with tandem units or whether reversible units will be accepted in their stead.

In the decade between 1950 and 1960, the successful pioneering for Francis-type reversible pump/turbines was performed by a manufacturer in the U.S. (Allis-Chalmers), whereas the pioneering for the adjustable-blade diagonal-flow type was performed by a British manufacturer. Since 1960, pump/turbines have also been manufactured in Belgium, Canada, France, Germany, Italy, Japan, Sweden, Switzerland and the USSR.

As regards number of installations, Japanese pump/turbine manufacturers are a strong second to the U.S. in the Francis-type and are first in the adjustable-blade, diagonal-flow type. However, the British still lead in the latter category, when numbers and capacity of units are considered. The Swiss have shown the most interest in reversible machines on the continent of Europe, but many of them are small and with-

31

TABLE 1
FRANCIS-TYPE REVERSIBLE SINGLE-STAGE PUMP/TURBINE—GENERATOR/MOTOR UNITS IN OPERATION OR ON ORDER

No.	Name	Locations (a)	Type (b)	No. of Units	P/T (c)	G/M (d)	Pump M³/S	Pump H	Turbine Mw	Turbine Hₙ	Generator Mva	Generator Pf	Motor Mw	RPM N₀	RPM Nᵣ	Guide Vanes Yes	Guide Vanes No	Pump Start (e)	Loc. of Thrust Brg. (f)	Total No. of Guide Brgs.	Year in Operation
(1)	(2)	(3)	(4)	(5)	(6)	(7)	(8)	(9)	(10)	(11)	(12)	(13)	(14)	(15)	(16)	(17)	(18)	(19)	(20)	(21)	(22)
1	Pedreira "A"	BR	A	1	V	W	19.5	15.0	5.25		6.25	0.70	6.00	212	381	X		FV	A	3	1939
2	Pedreira "B"	BR	A	1	SMS[g]	W	42.5	28.0	12.0	28.0	16.5	0.85	13.5	138	250	X		FV	B	2	1947
3	Vigario	BR	A	4	CAC[h]	CGE	40.0	30.0	10.8	29.0			13.2	150	262		X	FV	A	3	1952
4	Santa Cecilia	BR	A	4	CAC[h]	CGE	40.0	14.7	5.20	13.7			6.50	167	286		X	FV	A	3	1952
5	Pedreira "C"	BR	A	3	CAC[h]	W	51.0	23.8	14.2	27.1	17.5	0.85	14.3	150	275	X		FV	B	2	1953
6	Flatiron	USA	A	1	AC	AC	10.2	73.2	9.00	88.4	8.50	0.85	9.70	300/257	555		X	FV	A	1	1954
7	Hiwassee	USA	A	1	AC	AC	111	62.5	62.0	58.0	70.0	0.85	76.5	106	165	X		RV	B	2	1956
8	Omorigawa	J		1	H	H	5.30	127.8	12.2	118			14.4	400		X		RV			1959
9	Lewiston	USA	D	12	AC	AC	96.4	25.9	20.9	22.9	25.0	0.80	28.0	113	203	X		FV	B	2	1961
10	Stafel	S	P	3	SB		3.32	212	8.15	232			7.70	1500			X[o]		H		1961
11	Provvidenza	I	P	1	AC[i]	ASG	17.0	262	52.2	259	65.0	0.85	52.0	375	575	X		SYN	A	3	1962
12	Hatanagi	J	A	1	AC	FD	19.3	103	51.8	102	58.8	0.85	41.8	200/167	330	X		RV	B	3	1962
13	Hatanagi	J	A	2	H	H		103	45.4	102			33.8	200/167		X					1962
14	Ferrera	S	P	2	SB	BB	4.00	41.9	1.50	44.0	3.30	1.00	1.84	600			X[p]	FV	A	2	1962
15	Vianden	L	P	2	V	BB	0.66	260	1.50	260	2.50		1.80	3000	4200	X		SYN	A	3	1962
16	Taum Sauk	USA	P	2	AC	GE	75.0	233	220	241	204	1.00	179	200	302	X		PM	B	3	1963
17	Mio	J		1	H	H		143	36.0	137			36.8	277		X					1963
18	Ikehara I	J	D	2	H	H	29.5	132	72.8	117	78.0	0.90	80.0	180		X		RV			1964
19	Smith Mountain	USA	A	2	AC	AC	111	62.5	65.0	55.0	69.5	0.95	75.7	106	155	X		RV	B	2	1965
20	Shiroyama	J	P	2	Tj	T	19.5	187	65.0	153	70.0	0.90	71.7	273	450	X		RV	A	3	1965
21	Shiroyama	J	P	2	H	H		186	65.0	181	70.0	0.90	72.8	300		X		RV	A	3	1965
22	Z'Mutt	S	P	2	SB	SECH	3.22	365	17.6	490	18.9		17.0	1500			X[o,p]	RV	A	3	1965
23	Yards Creek	USA	P	3	BLH	GE	41.0	223	110	200	125	0.90	113	240	383	X		RV	A	3	1965
24	Yagisawa	J	A	2	AC	T	80.0	85.0	87.5	97.0	85.0			150	250	X		RV	B	3	1965
25	Yagisawa	J	A	1	H	H		113	87.5	111			87.5	150		X					1966
26	Muna II	COL	D	1	Ei	T	8.70	26.8	2.43	31.0	2.87	1.00	2.87	400	720		X	FV	A	3	1966
27	Torrejon	SP	A	4	N	EE	63.0	24.5	23.3	32.3	32.0	1.00		107	245	X		RV			1966
28	Ikehara II	J	D	2	H	H	40.5	132	103	117	110	0.90	110	150		X		RV			1966
29	Cruachan	GB	P	2	B[k]	AEI	28.6	358	100	343	111	0.90	118	500	685	X		PM	A	4	1966
30	Cruachan	GB	P	2	EE[l]	EE	27.8	350	106	350	111	0.90	108	600	840	X		PM	A	3	1967
31	Cabin Creek	USA	P	2	AC	GE	39.0	323	166	363	167	0.90	135	360	540	X		PM	A	3	1967
32	Muddy Run	USA	P	8	BLH	W	74.0	130	113	108	111	0.90		180	285	X		RV	B	2	1967
33	Santiago	SP		2	KMW		9.00	241	23.1	216				500		X					1967
34	San Luis	USA	A	8	H	GE	39.0	89.0	24.0	60.0	39.0	1.00	47.0	150/120	225		X	RV	A	3	1967
35	Kiev	USSR		3					30.0	72.5											1967
36	Robiei	S	P	4	SB	SECH	9.75	340	40.8	390	45.0	0.80	42.7	1000	1380	X		RV	A	2	1967
37	Robiei	S	P	1	C[m]		2.30	285	10.0	390			8.10	1500			X		H		1967
38	Salina	USA	A	3	AC	W	65.7	74.7	46.0	71.6	48.0	0.90	52.6	171	300	X		RV	B	2	1968
39	Thermalito	USA	A	3	AC	AC	85.0	30.2	28.3	25.9	30.6	0.90	32.8	113	208	X		FV	B	2	1968
40	Oroville	USA	A	3	AC	W	53.0	181	89.7	152	115	0.85	130	190	284	X		SYN	A	3	1968
41	Shin-Narihagawa	J	D	3	H	H	106	57.2	78.0	94.7	79.0	0.95	72.8	144		X		RV			1968
42	Nagano	J	D	2	H	H	136	71.0	113	92.0	120	0.90	120	150		X		RV			1968
43	Ronkhausen	G	P	1	EW	S	21.9	270	60.0	266	80.0	0.83	70.0	500	676	X		PM	A	2	1968
44	Ronkhausen	G	P	1	SB	S	19.6	256	67.7	252	80.0	0.83	70.0	500	676	X		PM	A	2	1968
45	Ova Spin	S	P	2	SB	O	13.1	160	25.0	200	27.0	0.90	26.0	500/375	715	X		RV	A	2	1968
46	Seneca	USA	P	2	NN	W	90.8	214	162	197	220	0.90	195	225	359	X		SYN	B	3	1969
47	Villarino	SP	P	2	KMW	BB	29.0	404	135	382				600		X					1969
48	Villarino	SP	P	2	B[k]	BB	29.0	404	135	382				600		X					1969
49	Midono	J	D	2	H	H	90.3	59.8	63.0	79.2	65.0	0.95	62.5	150		X		SYN	B		1969
50	De Gray	USA	A	1	NN	AC	67.1	45.8	33.2	52.1	29.5	0.95	32.3	129	192	X		SYN	B	3	1969
51	Azumi	J	A	4	Tj	T	96.0	79.4	107	134	109	0.95	109	188	310	X		SYN	B	2	1969
52	Kisenyama	J	P	1	Tj	T	110	197	240	220	245	0.95	250	225	340	X		PM	B	3	1969
53	Coo-Trois Ponts	B	P	3	CAC[h,n]	ACEC	46.0	259	146	273	150	0.95	148	300	447	X		PM	B	3	1969
54	Cornwall	USA	P	8	AC	GE	72.3	311	257	320	235	0.90	220	257	390	X		PM	B	3	1970
55	Longwood Valley	USA	P	3	BLH	W	44.0	107	45.0	111	45.0	0.90	39.5	277	455	X		RV	A	3	1970
56	Northfield	USA	P	4	BLH	GE	93.0	226	257	227	235	0.90	217	257	390	X		PM	B	3	1970

out adjustable guide vanes; they are more like pumps than pump/turbines. Notable exceptions are Ronkhausen (1-43,44) and Robiei (1-36) which are good sized.

Table 2 includes only 13 adjustable-blade installations, with a total of six axial-flow and 24 diagonal-flow units. The maximum head shown for the axial-flow units is less than 14 meters, but diagonal-flow units have been built for heads up to 140 meters, which overlaps the head range of 30 to 400 meters for Francis-type units (Table 1). Most pure pumped-storage plants utilize heads above 180 meters and Francis-type units, with a maximum head indicated of about 400 meters for single-stage impellers.

Plants utilizing heads below 180 meters are usually associated with downstream peaking use or augmentation of natural stream flow for reuse in the same plant. Plants utilizing diagonal-flow adjustable-blade impellers for heads above 30 or 40 meters presumably have justified the added cost on the basis of better efficiency over a wide range of head and load. It is not possible to generalize regarding those plants involving downstream peaking use and/or augmentation of natural stream flow, except to state that each must be integrated into its particular system.

One of the main reasons that pure pumped-storage plants utilize the higher heads is that adequate upper reservoir storage can be achieved economically with a drawdown of about 10% (or less) of maximum head. A small percentage drawdown is especially desirable for single-speed reversible pump/turbines. The point of best turbine efficiency tends to occur at a higher head (or a lower speed at the same head) than does the best efficiency point pumping.

Actually, inlet and discharge system friction losses increase the total dynamic head (H) above the pool-to-pool head when pumping and reduce the net effective head (H_n) below the pool-to-pool head when generating. For this reason, minimum practical values of friction loss, which add to the effective operating head range, are also desirable.

In any particular case, a reasonably well-balanced pump/turbine design might have optimum turbine efficiency at maximum head and best efficiency pumping at minimum head. This is the basic reason for the inconsistencies in the published ratings of various machines. Some pumps are rated at minimum head (best efficiency and maximum discharge), some are rated at maximum head, while others may be rated anywhere between maximum and minimum, for any commercial reason. Similarly, turbines are rated at maximum head (best efficiency and maximum capacity), or at minimum head (for minimum dependable capacity, frequently the sizing criterion for the pump/turbine), or at any intermediate head.

The first large machine (55 Mw) in this category is Unit No. 3 in the Provvidenza (1-11) pumped-storage plant in Italy; this unit went into operation in 1962. Units Nos. 1 and 2 are horizontal tandem units (50 Mw each), including double-entry two-stage pumps for the maximum head of 290 meters. Consequently, this plant serves as a good comparison between the reversible and the tandem-type units. Unit No. 3 is the first large single-stage machine to pump successfully against a head so high.

Highest-powered (220 Mw each) reversible units now in service anywhere in the world are the two Taum Sauk (1-16) machines, pumping against a maximum head of 270 meters. When they went into successful operation in 1963, their generating capacity (255 Mw each) at maximum head (255 meters) was greater than that of any single hydro unit then in operation anywhere in the world. They are still two of the highest-capacity hydro units in operation outside of the USSR.

Next large installation in this category is Shiroyama (1-20, 21) in Japan (1965) where the four units (65 Mw each) were built by two manufacturers. In addition to building the two pump/turbines, each manufacturer also built the two generator/motors, i.e. two complete units. Apparently, this complete unit responsibility is standard practice in the Japanese hydroelectric industry, as it is elsewhere in the case of steam turbo-generators and in the

case of atomic power plants. This is one of three Japanese plants listed where the equipment contracts were divided between two manufacturers, a practice rarely found in the U.S., although not unusual in Europe. Also with initial operation in 1965 are the three 110-Mw units at the Yards Creek (1-23) plant.

In 1966-67, at the Cruachan (1-29, 30) plant in Great Britain, four 100-Mw units began operation against a maximum head of 370 meters, a high-head record at the time for large single-stage pumps. The total plant capacity of 400 Mw also made this the largest installation in Europe with reversible machines. Here, too, the contract was divided evenly between two pump/turbine manufacturers, one of whom also built two generator/motors, i.e. two complete units.

Early in 1967, the two 165-Mw units at the Cabin Creek (1-31) plant took over the high-head single-stage record with a maximum head of 380 meters. Also in 1967, the first of eight 110-Mw units at the Muddy Run (1-32) plant went into service. This is the largest installation in operation with reversible units; presumably it will continue to be until four of the eight units at Cornwall (1-54), or the four units at Northfield (1-56), are installed and in service, whichever occurs first.

Late in 1967, the first of four 40-Mw units at Robiei (1-36) in Switzerland took over the high-head single-stage record with a maximum head of 395 meters. As a matter of interest, the Robiei plant also contains a single isogyre (non-reversing) pump/turbine (1-37) of 10-Mw capacity under the same head as the four larger reversible units. In 1968, The Ronkhausen (1-43, 44) plant is scheduled to go into service in Germany. Although one was manufactured in Germany, each of the two 65-Mw pump/turbines was designed by a different Swiss manufacturer; both generator/motor units were designed and built by a single German manufacturer.

The Villarino (1-47, 48) plant in Spain, with four 135-Mw units scheduled for initial operation in 1969, will be the largest installation in Europe with reversible machines. They will operate against a maximum head of 410

TABLE 1 CONTINUED
FRANCIS-TYPE REVERSIBLE SINGLE-STAGE PUMP/TURBINE—GENERATOR/MOTOR UNITS IN OPERATION OR ON ORDER

No.	Name	Locations (a)	Type (b)	No. of Units	P/T (c)	G/M (d)	Pump M^3/S	Pump H	Turbine Mw	Turbine H_n	Generator Mva	Generator Pf	Motor Mw	RPM N_o	RPM N_r	Vanes Yes	Vanes No	Pump Start (e)	Loc. of Thrust Brg. (f)	Total No. of Guide Brgs.	Year in Operation
(1)	(2)	(3)	(4)	(5)	(6)	(7)	(8)	(9)	(10)	(11)	(12)	(13)	(14)	(15)	(16)	(17)	(18)	(19)	(20)	(21)	(22)
57	Salina	USA	A	3	AC	W	65.7	74.7	46.0	71.6	48.0	0.90	52.6	171	300	X		RV	B	2	1970
58	Grand Coulee	USA	A	2	NO		48.1	89.0	47.4	81.0				200		X			U	2	1970
59	Mormon Flat	USA	A	1	AC	AC	102	43.6	41.8	39.3	47	0.90	45.5	138.5	238	X		SYN	U	2	1971
60	Jocassee	USA	A	4	AC		176	89.6	154	89.6	186		159	120	194	X					1972
61	Clarence Cannon	USA	A	1	AC		128	22.9	32.0	22.9				75	139	X					1972
62	Carters Dam	USA	A	2	AC		126	106	129	105				150		X					1972

TABLE 2
ADUSTABLE-BLADE REVERSIBLE SINGLE-STAGE PUMP/TURBINE—GENERATOR/MOTOR UNITS IN OPERATION OR ON ORDER

No.	Name	Locations (a)	Type (b)	No. of Units	P/T (c)	G/M (d)	Pump M^3/S	Pump H	Turbine Mw	Turbine H_n	Generator Mva	Generator Pf	Motor Mw	RPM N_o	RPM N_r	Vanes Yes	Vanes No	Pump Start (e)	Loc. of Thrust Brg. (f)	Total No. of Guide Brgs.	Year in Operation
(1)	(2)	(3)	(4)	(5)	(6)	(7)	(8)	(9)	(10)	(11)	(12)	(13)	(14)	(15)	(16)	(17)	(18)	(19)	(20)	(21)	(22)
1	Baldeney	G		1	V		8.00	8.90	1.33	8.5	1.10	1.00	1.33	327/257			X[r]				1933
2	Traicao	BR	A	3	SMS[g]	W	50.0	7.00	2.57	7.00	5.50	0.90	4.80	150	300		X[r]	FV	A	5	1940
3	Sir Adam Beck	C	D	8	JI[q]	CW	130	22.8	34.0	25.3	31.0	0.95	33.3	92.3	208		X[s]	FV	B	2	1957
4	Sesquile	COL	P	1	SB	O	8.00	32.0	3.50	37.0	4.70	0.89	4.05	450	845		X[s]		A	2	1964
5	Ananaigawa	J		1	H	H		75.0	13.5	69.5				360		s	s				1964
6	Brazeau	C	A	2	D	CW	28.3	13.7	12.1	13.7				150			X[r]		A	4	1964
7	Valdecanas	SP	A	3	EE	EE	104	72.5	84.5	74.1	83.3	0.90	84.5	150	300	X[s]		SYN	B	2	1964
8	Kuromatagawa	J	D	1	F	F	23.9	75.0	18.0	73.0	19.0	0.90	20.5	333/300	710	X[s]		RV	A	3	1964
9	Z'Mutt	S	P	1	SB	SECH	2.30	110	2.13	135	3.30		3.00	1500		s	s	FV	H		1966
10	Forebay	USA	A	6	FM	GE	19.8	17.1			4.20	1.00	4.47	200	450		X[s]	FV	A	3	1966
11	Kagedaira	J		1	H	H			47.7	89.7						s	s				1968
12	Rifa	A	P	2	SB	WSW	14.0	29.0	4.75	35.0	5.80	0.86	4.78	300	700		X[s]	FV	B	2	1968
13	Takane	J	D	2	H	H	75.0	80.8	88.0	136	100	0.85	98.0	277		X[s]		RV			1969

NOTE (a):
A = Austria
B = Belgium
BR = Brazil
C = Canada
COL = Colombia
G = Germany
GB = Great Britain
I = Italy
J = Japan
L = Luxembourg
S = Switzerland
SP = Spain

NOTE (b):
P = Pure pumped storage
D = P + Downstream peaking use
A = Augmentation of natural stream flow

NOTE (f):
A = Above rotor of generator/motor
B = Below rotor of generator/motor
H = Horizontal shaft
U = Umbrella

NOTE (e):
FV = Full-voltage across-the-line induction motor start
RV = Reduced-voltage across-the-line (or reactor) induction motor start
PM = Pony motor, wound rotor type
SYN = Synchronous (or back-to-back) start, including "reduced-voltage, reduced-frequency"

NOTE (p): Two-stage units
NOTE (q): Basic design by EE (GB)
NOTE (r): Axial-flow impeller
NOTE (s): Diagonal-flow impeller
NOTE (c): Pump/Turbine Manufacturers
AC = Allis-Chalmers
B = Boving
BLH = Baldwin-Lima-Hamilton
C = Charmilles
CAC = Canadian Allis-Chalmers
D = Dominion
E = Ebara
EE = English Electric
EW = Escher-Wyss
F = Fuji Electric
FM = Fairbanks-Morse
H = Hitachi
II = John Inglis
KMW = Karlstads Mekaniska Werkstad
N = Neyrpic
NO = Nohab
NN = Newport News
SB = Sulzer Brothers
SMS = S. Morgan Smith
T = Toshiba
V = Voith

NOTE (d): Generator/Motor Manufacturers
AC = Allis-Chalmers
ACEC = Ateliers de Constructions Electriques Charleroi
AEI = Associated Electrical Industries
ASG = Ansaldo San Giorgio
BB = Brown Boveri
CGE = Canadian General Electric
CW = Canadian Westinghouse
EE = English Electric
FD = Fuji Denki
F = Fuji Electric
GE = General Electric
H = Hitachi
O = Oerlikon
S = Siemens
SECH = Secheron
T = Toshiba
W = Westinghouse
WSW = Wiener Starkstromwerke
NOTE (g): Now the hydraulic products div. of A-C
NOTE (h): Basic design by A-C (USA)
NOTE (i): Partial manufacture in Terni, Italy
NOTE (j): Licensee of A-C (USA)
NOTE (k): Basic design by KMW (Sweden)
NOTE (l): Joint design with SB(S)
NOTE (m): Isogyre = not reversible
NOTE (n): Partial manufacture in Belgium, thrust bearing in Germany
NOTE (o): Double-suction, pumping

meters, another new record for high-head single-stage pump/turbine units. All four 600-rpm generator/motors are being supplied by a single Swiss manufacturer, but two of the pump/turbines are being supplied by a Swedish manufacturer and two by a British manufacturer, all four of the same composite design. Also scheduled for operation in 1969 are two 200-Mw units at the Seneca (1-46) plant.

Although it would appear that a new record for unit capacity should be established when the 240-Mw Kisenyama (1-52) machine in Japan goes into operation in 1969, each Taum Sauk unit has about 5% more turbine capacity when the comparison is made under the maximum head at each plant. However, each Cornwall unit will have about 20% more turbine capacity than a Taum Sauk unit, both at maximum head. Each of the four 240-Mw units in the Northfield (1-56) plant, scheduled for operation in 1970, will also have more turbine capacity at maximum head than a Taum Sauk unit, i.e.

approximately the same as a Cornwall unit under similar conditions.

The Longwood Valley (1-55) plant with three 45-Mw units will represent a very low head (130 meters maximum) for pure pumped storage. Its economic feasibility was partially justified by its being combined with the necessary expansion of a large municipal water supply.

Three 145-Mw units at the Coo-Trois Ponts (1-53) plant in Belgium will be the highest-powered reversible machines in Europe when they go into service in 1969 or 1970. Also, the total plant capacity will be second in Europe only to Villarino, with Cruachan third. The pump/turbines were designed and are being partially manufactured in North America.

It is a very interesting fact that the purchaser's consulting engineers for Coo-Trois Ponts were also the purchaser's consultants for the Vianden (1962-63) pumped-storage plant in Luxembourg. The latter plant contains nine 100-Mw tandem units, each including a double-entry two-stage pump for the maximum head of 290 meters. These consultants on two of the most important pumped-storage plants in Europe changed their recommendation from conventional tandem units at Vianden to reversible single-stage units at Coo-Trois Ponts for almost exactly the same operating head range—certainly a definite indication of the trend away from separate pumps and turbines toward reversible pump/turbines.

Some auxiliary method of starting reversible units is necessary for pumping operation. Since the direction of rotation for pumping is opposite to that for generating, reversible units cannot be self-started hydraulically.

All pump/turbines are installed below the pump suction water level and are, therefore, fully primed just before a pump startup. Adjustable-blade pump /turbines can be accelerated from standstill to synchronous speed in a reasonable length of time with the impeller blades in the closed position, but Francis-type pump/turbine impellers are usually dewatered for rapid acceleration.

A full-voltage across-the-line induction motor start is the simplest and most economical method for unit capacities up to about 30 Mw in large

systems, and for smaller units in smaller systems. Acceleration from standstill to synchronous speed requires about 30 seconds.

For unit capacities up to about 100 Mw in large systems, the most economical method of reducing the system disturbance due to full-voltage start is to use a reduced-voltage (usually 50%) across-the-line start. Acceleration time is about two minutes.

For unit capacities above about 100 Mw, a separate pony motor is being used to accelerate the large machine to synchronous speed. Such a motor is small (5% to 10% nominal horsepower rating) compared with the main generator/motor; acceleration time will depend upon the moment of inertia of the rotating parts of the main units, but could be five to 10 minutes.

When full- or reduced-voltage starting is undesirable or not feasible, and it is desired to avoid the extra expense of a pony motor and associated equipment, synchronous starting may be arranged. In its simplest form, this method requires that a turbine (or another pump/turbine) be connected electrically at standstill to the unit being started. These two machines must be isolated electrically from the system during the accelerating period and the driving unit must have the torque-producing capability to accelerate the driven unit to synchronous speed.

Synchronous starting has the advantage that the disturbance to the system can be zero. Obviously, the accelerating time will depend upon the relative sizes of the two units, as well as upon dewatering (or not), but is usually in the range of two to five minutes.

The generator/motor is usually synchronized while the pump/turbine is dewatered. Therefore it is necessary to prime the impeller again before pumping can begin. This involves releasing the compressed air and allowing the water to refill the impeller while it is rotating at synchronous speed. Provisions should be made to throttle the compressed air release through a vent pipe in the pump/turbine head cover and to fill the impeller with water very slowly, from the periphery in to the *eye* diameter, by means of a high-pressure pipe connection to the penstock.

As soon as the pump is primed, it is desirable to start opening the guide vanes immediately to the most efficient

angle for pumping against the head existing at the time. This should be done at the fastest rate that can be tolerated by the hydraulic and electrical systems.

Similarly, when pumping is finished, the guide vanes should be closed at the fastest rate that can be tolerated. Provisions should be made for evacuating the water from *eye* of the impeller by means of compressed air through the pump/turbine head cover, after which the water in the impeller can be drained from the periphery at any desired rate.

Whether the thrust bearing should be located above or below the rotor of the generator/motor is not settled by these listings. Apparently, there is a technical preference for above, but the cost is greater than below because of the supporting structure; and most of the larger units have the thrust bearing below the rotor.

All of the units equipped with pony motors for starting in the pumping mode have a guide bearing above the rotor of the generator/motor and one design even has a guide bearing above the rotor of the pony motor.

Almost all of the larger units employ thrust bearings with forced-oil lubrication to overcome the problem of breakaway torque when starting as a pump, and all of the guide bearings, including those on the head covers of the pump/turbines, are oil lubricated.

The world's largest reversible machines, at Taum Sauk, have been in successful operation since 1963. The thrust bearings are below the rotors of the generator/motors; also, these were the first two units equipped for starting with pony motors in the pumping mode.

Startup problem and accidents are claimed in some quarters to be more common with reversible units than with other types. From this they deduce that there is some basic inherent weakness in reversible machines. This is not true. There are design problems, both basic and detail, that are different from conventional single-purpose machines, of course. In any particular case, if the designer fails to take one or more of these items into account, operating problems are the usual result. But such problems are not unusual in conventional single-purpose custom-built hydro equipment; they occur frequently in turbines and large pumps, and even in large valves and gates.

Again, consider the Taum Sauk units. They represented a fantastic extrapolation of experience at the time they were ordered in 1960. Not only were they reversible and large in capacity as well as in physical size, with a fair-sized pony motor on top of the generator/motor, but the plant was designed for complete automation and remote control. These latter features complicated and delayed the shakedown period which lasted several months. Part of the delay was also due to civil works problems beyond the scope of the manufacturers of the units.

Only two significant problems developed during startup. One was violent vibration due to a detail resonance condition that could not have been foreseen, and was easily eliminated by changing one of the two details involved. The other was a basic hydraulic problem. Upthrust increasing with overspeed would have lifted the dead weight of the rotating parts off of the thrust bearing at maximum possible overspeed due to a full load rejection during turbine operation. This was eliminated by installing valves and balancing piping on the pump/turbine head cover. One pair of valves is automatically opened and another pair closed by the mode selector switch, their positions being reversed for pumping as compared to generating.

The same upthrust problem appeared again at the startup of the Cabin Creek units, in spite of modifications intended to prevent it. This time it was eliminated by installing a single balancing pipe connecting the pump/turbine head cover and bottom ring; no automatic valving is required.

Specific speeds and heads do not show any definite trend. Actually, the highest specific speed permissible for any given installation depends upon the economically feasible submergence of the pump/turbines below the pump suction level, in conjunction with adequate structural design of the components of the machines. One of the outstanding differences between pump/turbines and conventional Francis turbines is the much lower runaway speeds of the former; this favors the mechanical design of the generator/motor rotors and allows for larger capacities at any given synchronous speed. It also encourages a higher specific speed versus head relationship for reversible units, as compared with conventional hydro-generators. The tendency to design for higher specific speed is furthered by the desire for minimum moment of inertia in the generator/motor rotor. This, in turn, minimizes the acceleration time during pump startup, which is particularly important in the case where a pony motor is utilized.

Each pump/turbine manufacturer will have at least one design suitable to recommend for any given installation. Each one of these designs will differ in such things as synchronous speed, depth of submergence required, and relation between turbine and pump output. However, the purchaser should not request proposals based upon each manufacturer's developed designs and attempt to evaluate them in regard to project requirements. Specifications in the invitation to bid should outline the important project requirements to be evaluated. In this way, the purchaser receives the benefit of each manufacturer's expertise being applied to get the best prformance for the particular project. Moreover, comparisons and evaluations of the various proposals received are then greatly simplified.

For an underground plant, additional depth of submergence is quite inexpensive, and pump/turbine manufacturers are encouraged to recommend even higher specific speeds. In this case, particularly, only well-recognized reliable manufacturers should be considered, and the purchaser must assure himself that adequate laboratory test data are available for turbine and pump performance, as well as cavitation characteristics and submergence requirements, for a model *essentially* homologous to the prototype units being proposed. He can then select the most responsive proposal with confidence, secure in the knowledge that all prototype performance guarantees must be confirmed by laboratory tests on a *completely* homologous model. Such tests are made before prototype manufacture has proceeded too far to permit minor modifications to be made.

Trend to higher capacities of units and plants (as well as higher heads) is well established, and it may be expected to persist in the continuing effort to minimize the installed cost per kilowatt. The ultimate limits of head and/or size for single-stage Francis-type machines will undoubtedly be determined by commercial, rather than by technical, considerations. Use of multistage Francis-type units for even higher heads is also being seriously considered.

Reversible units are remarkably flexible. Not only do they generate in one direction and pump in the other, but they can be available immediately as spinning reserve with the inlet valve open (or in one minute or less with the inlet valve closed); they can be started up from standstill and fully loaded in three minutes or less, subject to the limitation imposed by the water inertia and waterhammer characteristics of the supply and discharge conduits. They can be used as synchronous condensers in either direction of rotation for system power factor correction.

In effect, they have a double reserve capability when pumping: (1) as an interruptible load, releasing thermal or nuclear generating capacity; and (2) their own generating capacity, which can be available in several minutes. For fastest possible *turnaround,* a *pumping* unit can be tripped off by opening the circuit breaker and leaving the adjustable guide vanes open on a programed basis. Then, when the reverse runaway attains synchronous speed in the *generate* direction, the unit is put back on the line. When a purchaser specifies a need for rapid changeover from one mode of operation to the other, the various auxiliaries can be designed for the service specified, subject to the limitations imposed by the natural laws of physics.

This article is based on a paper presented at the World Power Conference in Moscow, August 1968.   **END**

COO-TROIS PONTS PROJECT— OFF-PEAK STORED ENERGY SUPPLIES PEAK POWER

Goetz E. Pfafflin*

*Daily, and even hourly, load swings
are offset by new
pumped-storage hydro project*

Where off-peak capacity is inexpensively available, pumped-storage has emerged as a convenient and economic facility for coping with the fluctuations of daily loads. The merits of pumped-storage were recognized some 50 years ago. There are 21 such plants in operation or under construction in the United States, another 21 in Germany and many more are operating in other parts of the world.

**Goetz E. Pfafflin, Hydraulic Products Division, York Plant.*

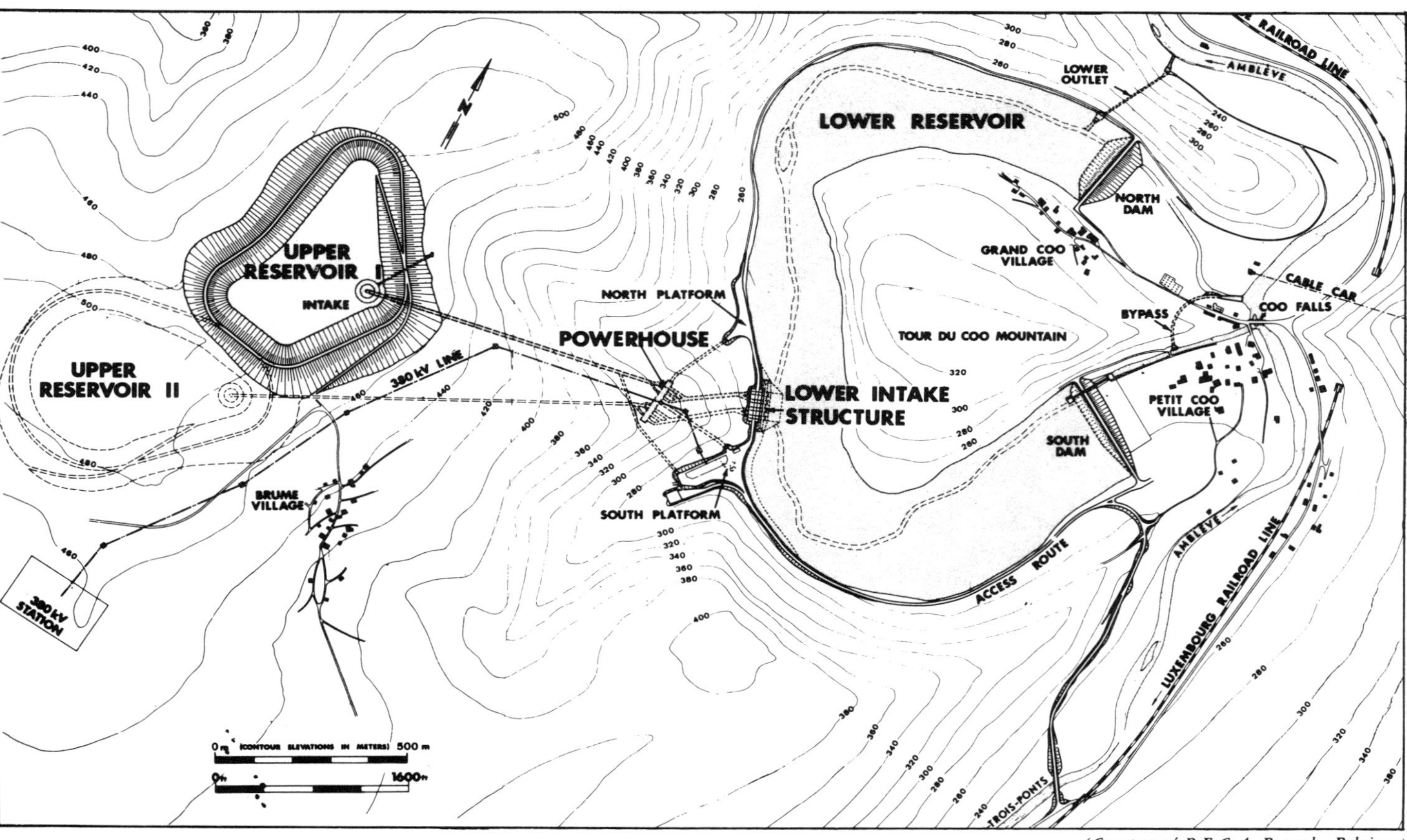

(Courtesy of B.E.C.A. Brussels, Belgium)

COO-TROIS PONTS pumped-storage hydro project stores energy in upper reservoir during off-peak power periods. This energy is used to generate power during peak load periods. **(FIGURE 1)**

The Coo-Trois Ponts pumped storage project represents a joint venture of six private Belgian utilities serving a market with a week-day power consumption in excess of 80,000,000 kwh.

The plant is being developed in two phases using one lower and two upper reservoirs (Fig. 1). Because of the obvious economic advantages, Coo-Trois Ponts has been designed around reversible machines, each phase consisting of three pump/turbine-generator/motor sets. One pressure shaft is provided per stage (Fig. 2).

The underground plant is located in the Ardennes in the Belgian province of Liege. It serves the inter-connected system at 380 kv. The general plant data is given in Table I. A typical system load cycle with pumping and generating characteristics superimposed on it is shown in Fig. 3.

Six modes of operation planned

The Coo machines have been designed for a number of modes of operation in addition to generating and pumping; while these additional modes do not influence significantly the design of the pump/turbines, they are of interest from an operating point of view. The following types of operation have been specified:

1. Generating
2. Pumping
3. Synchronous Condenser Operation, Generating Direction
4. Synchronous Condenser Operation, Pumping Direction
5. Spinning Reserve, Generating
6. Spinning Reserve, Pumping

The Coo pump/turbines are arranged for a quick generating start of 60 seconds from standstill to synchronized speed-no-load operation. Again, in the interest of maximizing the availability of the generating capacity to the system, the changeover time from pumping, absorbing approximately 400 Mw among the three units or 800 Mw among six units to turbining has been set for 90 seconds to synchronized speed-no-load, or 130 seconds to full load. The full generating load is approximately 400 Mw for three units or 800 Mw among six units, depending on head availability.

While the Coo-Trois Ponts pump/turbines have been arranged for a quick pump start of only 30 seconds from primed running to full power absorption, it must be noted that the availability of pumping power from the system

UNDERGROUND PLANT requires considerable tunnelling. Pressure shaft 1 supplies water for three pump/turbine-generator/motor sets. Pressure shaft 2 and its three units will be placed in service in the second phase of construction.   **(FIGURE 2)**

(Courtesy of B.E.C.A. Brussels, Belgium)

TABLE I
PLANT DATA

Installed Generating Capacity —
   Under maximum head . . . . . .830 Mw (6 units)
   Under minimum head . . . . . .615 Mw (6 units)
Pump Power Absorption —
   Under maximum head . . . . . .737 Mw (6 units)
   Under minimum head . . . . . .854 Mw (6 units)
Maximum Static Head . . . . . . .275.35M; 920 ft.
Minimum Static Head . . . . .235.65M; 772 ft.
Volume of Water Stored
   (2 Upper Reservoirs). .8x10⁶M³; 282x10⁶ cu. ft.
Normal Filling Period
   for Upper Pools . . . . .8.0 hours (6 units)
Full Load Generating Period . .6.5 hours (6 units)
Energy Stored in Upper Pools . . .4,800,000 kwh
Pump/Turbine Characteristics —
   Maximum turbine capacity . . . . . .194,500 hp
   Maximum net head . . . . .273 M, 895 ft.
   Synchronous speed . . . . .300 RPM
   Minimum submergence . . . . . . .18 M, 59 ft.
   Rated pump discharge. . . .46 CMS at 259 M;
                          1620 CFS at 850 ft.
Generator/Motor Characteristics —
   Max. generator capacity . .158,000 kva, PF 0.9
   Generating voltage . . . . .20,000 v.
   Maximum motor capacity . . . . . . .148,000 kw
Starting Motor Characteristics —
   Rated motor capacity. .11,600 hp (3 minutes)
   Rated speed . . . . .333 rpm
Transformer Characteristics —
   Number of transformers 6
   Transformer capacity . . . . .160 Mva
   Transformer ratio .20 kv/390 kv ± 15 percent

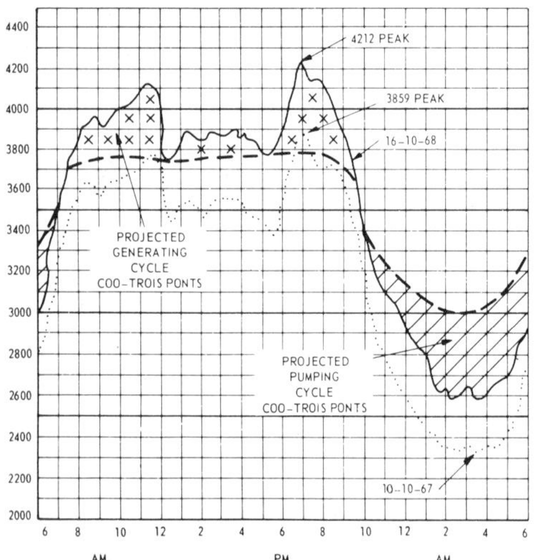

ACTUAL LOAD CURVE for October 16, 1968 shows projected pumping and generating cycles superimposed. Load curve for October 10, 1967 also shows typical load swings. **(FIGURE 3)**

is not always such as to permit this quick loading. The Cabin Creek design was based on a pump loading time of 420 seconds, after prime, based on the projected pumping load availability in the Colorado Public Service Company system (Fig. 4).

Several pumped-storage projects currently in the planning or design stage are to operate in relatively small grids, where such gradual loading is very important.

This flexibility, which is available to the designer of pump/turbines, represents a clear advantage over traditional tandem pump and turbine arrangements used in Europe very extensively for pumped-storage installations. Within limits the controlled rate of wicket gate opening can be adjusted, to provide a pump loading rate compatible with the system.

Francis type units chosen

The Francis type reversible pump/turbines are vertically arranged for top runner removal (Fig. 5). Among the interesting features of these reversible machines are the gate seals used to reduce leakage into the runner chamber during synchronous condenser operation and pump start. To eliminate the danger of galling between the gate ends and the head cover and bottom ring, often experienced in units relying on close clearances for the control of leakage, rubber-backed bronze seals are retained in the head cover and bottom ring to provide effective sealing.

Another feature of interest is the floating main shaft seal designed specifically for pump/turbines. While common soft packing provides the actual sealing contact with the rotating shaft, the seal assembly has a built-in controlled horizontal and vertical movement to accommodate the maximum shaft movement during transient conditions.

Gate restraining mechanisms are provided to prevent wicket gates from fluttering after shear pin failure. This energy absorbing design, provided in addition to gate stops, assures significantly more protection than can be derived from gate stops alone.

A forced-lubricated main bearing guides the one-piece main shaft.

The 87 in. spherical valves are designed with a longitudinal split. Upstream and downstream seals are provided for maintenance and normal operation respectively. The valves, based on a cast steel design, are rigidly coupled to the penstocks and flexibly to the spiral case inlets.

With three reversible pump/turbines connected to only one draft tube tunnel, draft tube gates became necessary. Flap type gates were selected in part because of their natural tendency to open in case of pressure build-up due to spherical valve and wicket gate failure while this gate is closed. An interesting design feature of this gate is in the music note seal provided around the periphery of the upper surface of the raised gate, allowing atmospheric pressure to act on that upper surface. The pressure differential thus acting on the gate provides a positive locking force in the raised position.

Normal pressure of the electro-mechanical governors is 600 psi. The sump tank, oil pumps and governing equipment are arranged in an actuator cabinet. Because Coo-Trois Ponts is to be automatically operated, provision for joint control has been made. The speed signal generator is mounted on top of the starting motor.

Pump start of these machines is achieved using an electric starting motor (or pony motor) which accelerates the pump/turbine-generator/motor through synchronous speed while the tailwater is depressed. Thus, no additional mechanical equipment had to be provided to facilitate operation types 3, 4, 5 and 6. The tailwater depression systems and runner seal cooling systems had to be provided, regardless of whether these units were to be run as synchronous condensers or not.

Contrary to general American practice in pump/turbine installations, Coo-Trois Ponts required quick start-up and changeover. The Belgian grid visualizes pumping operations, not only at night and weekends, but also during the mid-day lunch period when off-peak power is available. Such pumping operation is not uncommon in Europe because lunch breaks are simultaneous throughout a country.

Generator/Motor is modified umbrella type

As is common in Europe for generators and generator/motors of the rating of the Coo units, these modified umbrella machines (thrust and guide bearing below the rotor, single guide bearing above the rotor) are equipped with separate motor driven cooling fans. The single main shaft configuration eliminates the coupling below the thrust bearing bracket. The coupling between the main shaft and the rotor spider provides for 10 radial torque keys and 10 coupling bolts which are sized to support the main shaft and runner assembly when the generator/motor assembly is hydraulically jacked up without the main shaft having been uncoupled.

The 333 rpm starting motor was designed to provide three important functions:

1. The starting motor
2. The main excitation source
3. The emergency generator

Combining the functions of 1 and 2 provided for a reduction in overall height of the rotating assembly. Emergency power provided by the motor/exciter at 300 rpm is at 45 cycles per sec. The station auxiliaries have been sized accordingly.

Specifications for Coo-Trois Ponts required the thrust bearings to be furnished by the pump/turbine manufacturer. This requirement virtually eliminated a suspended arrangement and left the pump/turbine manufacturer to consider either of two arrangements. The first type uses a modified umbrella machine in which the thrust bearing would be supported on a lower bracket mounted immediately below the generator/motor. The second choice would be a head cover supported thrust bearing arrangement which would require the generator/motor manufacturer to furnish an upper and lower guide bearing.

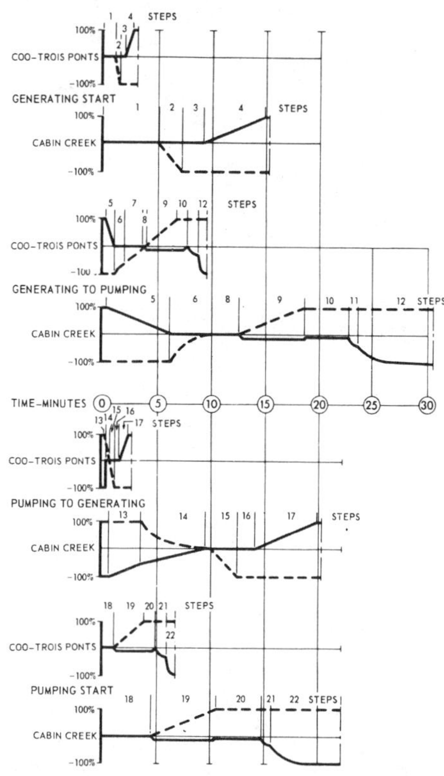

	CABIN CREEK	COO-TROIS PONTS
Rated Turbine Capacity (hp)	223,000	194,500
Rated Generator Capacity (kw)	150,000	142,500
Rated Motor Capacity (hp)	183,000	198,000
Rated Starting Motor Capacity (hp)	14,000	11,600
Synchronous Speed (rpm)	360	300
Rated Turbine Net Head (ft)	1,190	896
Rated Pumping Head-TDH (ft)	1,058	850
Rated Pump Discharge (cfs)	1,375	1,625

START-UP AND CHANGE-OVER TIMES for Cabin Creek and Coo-Trois Ponts reflect the difference in system requirements for these projects. **(FIGURE 4)**

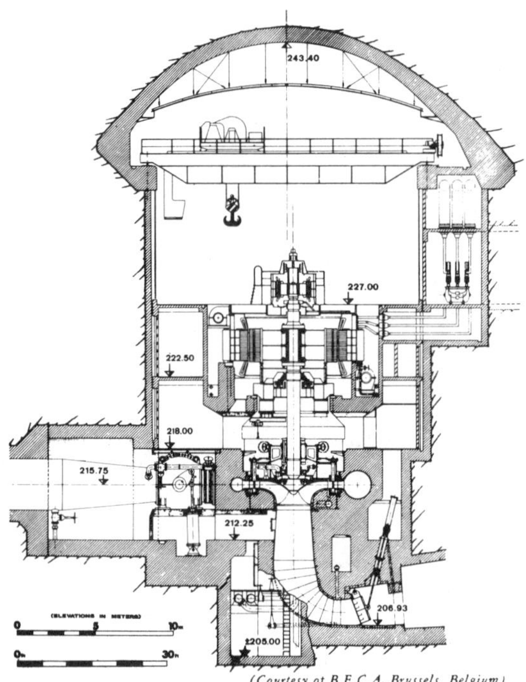

(*Courtesy of B.E.C.A. Brussels, Belgium*)

COO-TROIS PONTS generator/motors are of modified umbrella type and have thrust and guide bearings below the rotor and use one-piece main shaft. **(FIG. 5)**

The alternative of the head cover supported thrust bearing was ruled out in favor of a separate supporting structure which also allows significantly better service access to the main shaft seal, etc. (Fig. 6).

Thrust bearings supporting reversible pump/turbine-generator/motor assemblies have to be carefully considered, because they are subjected to service conditions significantly different from those supporting the conventional hydraulic turbines. Among the service conditions peculiar to pump/turbine service are the following:

1. Two directions of rotation.
2. Comparatively long times of rotation below synchronous speed when accelerating in a pump direction (Coo-Trois Ponts; three minutes compared to the Taum Sauk installation in the U.S. with ten minutes).
3. Frequent starts and stops, i.e., frequent transient operation under frequencies other than those easily identified as natural frequencies or forcing frequencies.
4. Transient loadings resulting from pump priming (in the case of starts with tailwater depression) and shut-off head pumping several times daily.

Each of these conditions has to be acknowledged in the design of the thrust bearings. Thus, for these reversible machines, a bearing that symmetrically supported the thrust shoes was selected. ▲

REFERENCES

1. "Pumped-storage: An Evaluation of the Progress and Experience to date with Large Reversible Units," G. D. Johnson, presented to the VIIth World Power Conference in Moscow in 1968.

2. "Das Pumpspeicherwerk Roenkhausen," B. V. Gersdorf, published in "Elektrizitatswirtchaft," Issue 24 in Germany, November 20, 1967.

3. "The Selection of Pumped-Storage Plant with Particular Reference to Tumut 3 Project," A. N. G. Bray and A. C. H. Frost, published in "The Journal of the Institution of Engineers, Australia, September, 1967.

4. "Experience in the Construction and Operation of Large Pumped-Storage Stations," E. Pfisterer and H. Press, presented to the VIIth World Power Conference in Moscow in 1968.

5. "Discussion on Turbine Specific Speeds," K. W. Beattie, presented to the Edison Electric Institute in May, 1967.

6. "Construction de la Centrale Hydroelectrique D'Accumulation D'Energie par Pompage de Coo-Trois Ponts," M. DuPont, G. deHouck and D. Greindl, published in "Electricite," Issue 137 Dec. 1968 and Issue 138 March 1969.

LOWERING DEVICE for thrust bearing facilitates bearing service without dismantling generator/motor. **(FIGURE 6)**

(*Courtesy of Siemens, Erlangen, Germany*)

3

MEETING PUMPED-STORAGE PROJECT REQUIREMENTS

Goetz E. Pfafflin*

The choice of pumped-storage equipment characteristics is based on careful engineering evaluation and precise model testing

Low load factors of pumped-storage plants emphasize the need for the optimization of all economic factors in the selection and arrangement of equipment.

The selection of the specific speed, and hence, synchronous speed of the pump/turbine, is thus subjected to considerable scrutiny. Speed not only directly affects the cost of the pump/turbine, but also affects other aspects of the hydro installation such as:

Goetz E. Pfafflin, Manager of Marketing Hydraulic Products Division, York Plant.

1. Pump/turbine size
2. Generator/motor size
3. Powerhouse size
4. Powerhouse elevation, which is particularly significant with surface powerhouses but can also be a serious consideration with cavern type plants because of geologic reasons and access shaft length for a given slope.

These are conflicting considerations because higher specific speeds (N_s) have a favorable effect on the relative cost of 1, 2, and 3, but an adverse effect on 4. The selection of N_s, however, cannot be based on cost comparisons alone. Practical operating experience with high specific speed reversible pump/turbines is limited, and the tendency is, therefore, to progress cautiously to higher speeds despite the economics that seem to favor them. This caution is particularly important when it is acknowledged that pumped-storage plants are considered as instantaneous reserves because at times of acute demand, pump/turbines can be made available as generating units practically instantaneously.

Figs. 1 and 2 illustrate experience in pump/turbine specific speeds through 1968 in the turbine cycle and pump cycle respectively.

RUNNER is for Francis type reversible pump/turbine for Coo-Trois Ponts project in the Ardennes in Belgian Province of Liege.

While the tender invitations for Coo-Trois Ponts suggested 333 rpm for these 145 Mw machines and while some European manufacturers offered equipment at 375 rpm, the consulting engineers selected 300 rpm as the synchronous speed. This speed is generally in line with past industry experience and is significantly more conservative than some other installations. However, at the time of award for Coo-Trois Ponts no high head reversible unit of more than 100/Mw was in operation at an N_s level higher than that selected for this pumped-storage plant. The arrangement for a complete Coo-Trois Ponts unit is shown in Fig. 3.

Model testing sets design

Ultimate acceptance of the pump/turbines at Coo-Trois Ponts will be based on prototype performance to be verified by field acceptance tests in accordance with the International Test Code. An extensive laboratory test program was performed with a model essentially homologous from the spiral case inlet to the draft tube outlet.

While model testing has always been an important step in the development of all major conventional hydro-electric projects, such model tests for reversible units of pumped-storage plants are even more important. Without accurate model test data, it would be impossible to optimize the station setting and the pressure shaft sizing, to determine the minimum main-motor capacity, and to evaluate the system design pressures. The model test program for Coo-Trois Ponts determined the following characteristics:

1. Turbine Performance — efficiency, power output and critical sigma.
2. Pump Performance — efficiency, discharge and critical sigma.
3. Four Quadrant Characteristics — torque vs. discharge and rpm vs. discharge characteristics for, turbining in turbine direction, pumping in turbine direction, pumping in pumping direction, and for turbining in pumping direction.
4. Miscellaneous — pressure fluctuations in draft tube and pump radial thrust, and wicket gate torque.

The need for accurately establishing the respective characteristics of items 1 and 2 are obvious. Items 3 and 4, however, are also important and draw attention to the complexity of the sigma considerations for pump/turbines in general and particularly for 2-pool pumped-storage plants. Fig. 4 illustrates the relationship between static, net and dynamic head, and tailwater elevation, as well as plant sigma.

Generally, it can be said for such plants as Coo-Trois Ponts that the higher the head the less is the machine submergence, and the lower is the plant sigma. Conversely, the lower the head the higher is the submergence. While the turbining sigma characteristics for low N_s machines such as are being built for Coo-Trois Ponts tend not to vary significantly over the range of heads incurred, the same cannot be

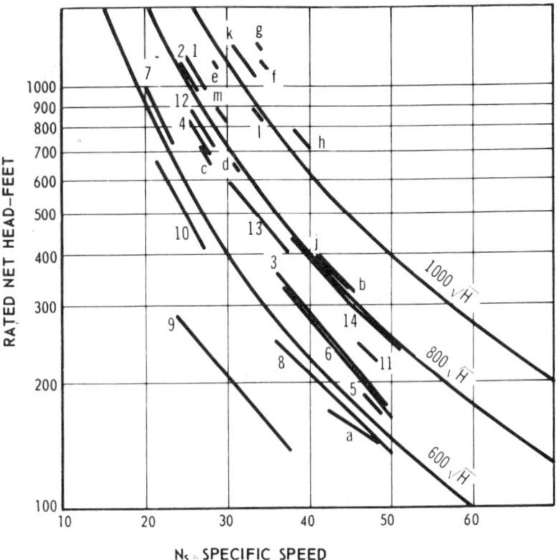

EXPERIENCE with Francis type reversible pump/turbine is given for turbining specific speed. **(FIGURE 1)**

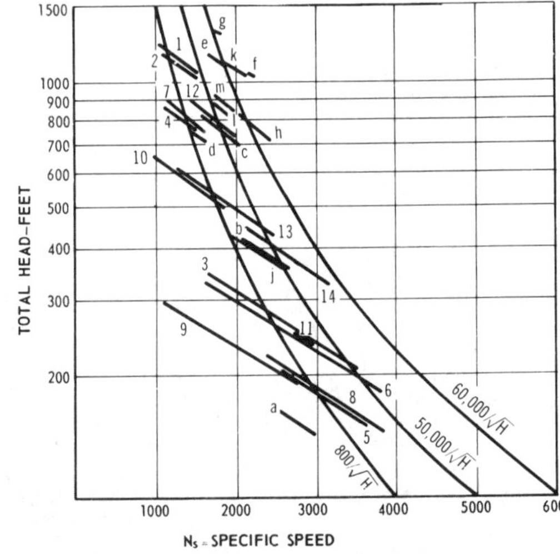

EXPERIENCE with Francis type reversible pump/turbine is given for pumping specific speed. **(FIGURE 2)**

1. Cabin Creek; **2.** Cornwall; **3.** Yagisawa; **4.** Taum Sauk; **5.** Smith Mtn.; **6.** Chubu; **7.** Terni; **8.** Hiwassee; **9.** Flatiron; **10.** Oroville; **11.** Salina; **12.** Coo-Trois Ponts; **13.** Shiroyama; **14.** Azumi. **a.** De Gray; **b.** Muddy Run; **c.** Kinzua; **d.** Yards Creek; **e.** Cruachan-500; **f.** Cruachan-600; **g.** Villarino; **h.** Northfield; **j.** Longwood Valley; **k.** Robiei; **l.** Rönkhausen (SB); **m.** Rönkhausen (EW).

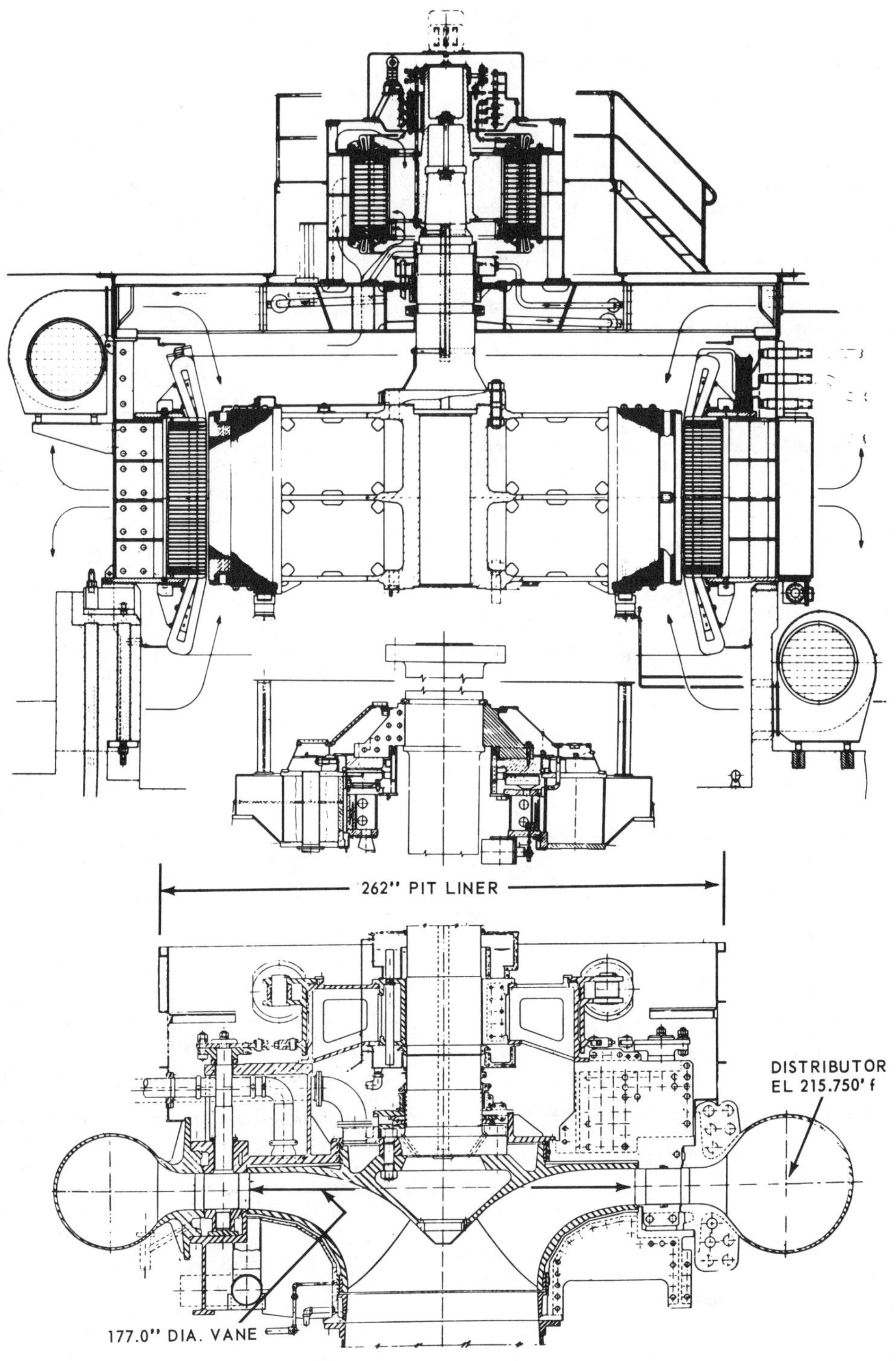

COO-TROIS PONTS PUMPED-STORAGE UNIT has starting motor/ exciter directly coupled to the generator.　**(FIGURE 3)**

said for pumping sigma characteristics. Because the unit speed of the hydraulic machine is selected as a compromise for the turbining and pumping cycles, there is significant change in pump critical sigma with a change in head.

Pumping against the minimum head and turbining under the maximum head tend to represent the two most severe operating conditions from a cavitation point of view and therefore need to be scrutinized carefully.

The four quadrant characteristics listed under 3 above are essential for the reliable establishment of transient behavior of the machine and for verification of the machine stability in the system (Fig. 5).

Pump/turbine transient behavior

Low load factors of pumped-storage plants tend to emphasize the importance of minimizing penstock diameters, generator/motor Wk^2, which may be desirable for acceleration of the machine in the pumping direction, and for pressure rise in the penstocks and pump/turbines. These aims are conflicting as illustrated by the pressure rise equation:

$$\frac{\Delta H}{H} = \Sigma \frac{L_0 V}{gH}\left(\frac{\alpha}{Wk^2} + \frac{\beta}{T_G}\right) \text{ where}$$

ΔH is the pressure or head change

H is the pressure or head of the installation

L_0 is the penstock length, generally a function of the topography and geology of the plant

V is the discharge velocity (for a given capacity and head), an inverse function of the cross sectional area of the penstocks

g, α, β, are constants

Wk^2 is the inertia of the rotating assembly

T_g is the governor time

Because pumped-storage plants generally are not required to contribute to system stability, T_g becomes a variable, the increase or decrease of which, in itself, does not affect significantly the cost of the installation. The same cannot be said, however, for the variables V, or $\dfrac{Q}{A}$ and Wk^2.

Q is the discharge rate in cfs

A is the net discharge area in sq ft

The question of optimizing pressure rise is further compounded by the throttling characteristics of low specific speed Francis type pump/turbine runners when going to over-speed.

Based on tentatively established penstock dimensions, transient characteristics can be established for a range of Wk^2 and T_g. Computer calculations of speed and pressure rise help to identify acceptable combinations of these variables.

For Coo-Trois Ponts, the consulting engineers compared overall costs for maximum pressure rise of 40 percent and 30 percent and finally selected the former based on the unit PD^2 of 4,700 TM^2 (Wk^2 equal to 28.9×10^6 lb ft^2) and T_g nominally 40 seconds. The

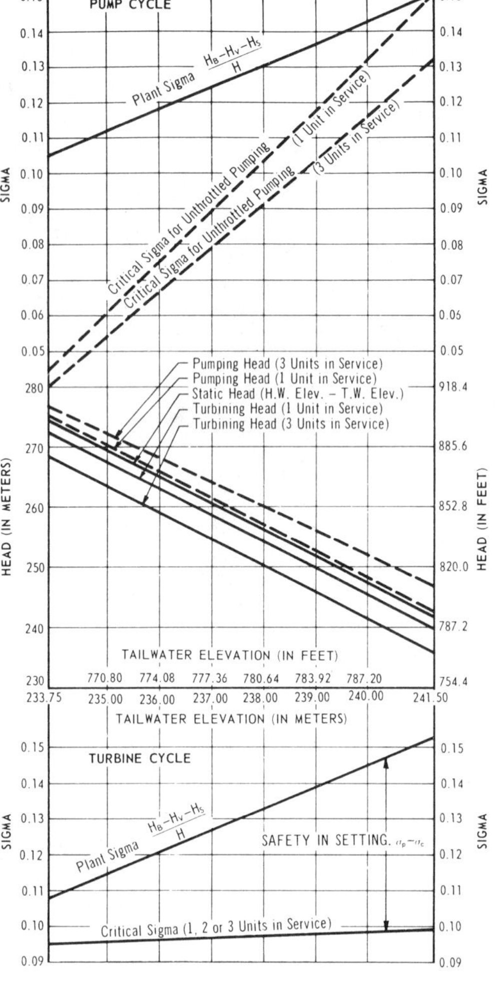

RELATIONSHIP of static, net and dynamic heads is given for tailwater elevation and critical sigma. **(FIGURE 4)**

SIGMA

Sigma (σ) is the ratio of the barometric head (H_b) minus the static suction head (H_s) at any point, usually taken for convenience as the elevation of the blade pivot axis for Kaplan runners and the elevation of the runner discharge for Francis wheels, to the total operating head (H) on the turbine.

Critical sigma is determined in the laboratory by testing the model with a constant wicket gate opening (and blade angle in the case of a Kaplan turbine) at a constant speed under a constant head while the elevations of headwater and tailwater are lowered step by step for successive runs until a definite "break," or change in performance, is observed.

MODEL of Coo-Trois Ponts pump/turbine was tested to determine characteristic of unit. **(FIGURE 5)**

throttling effect of the runner going to full dynamic over-speed resulted in water hammer exceeding 30 percent without the added throttling of the wicket gate closing. Even with the governor dead time delaying the gate closure and thus delaying the added throttling and consequent pressure rise, the selected Wk^2 and $\dfrac{\Sigma L_o V}{gH}$ would have resulted in excessive pressure rise. If 30 percent had been selected as the upper limit, Wk^2 and/or the penstock diameter would have had to be increased.

Fig. 6 typifies the various computer calculations used in the determination of maximum pressure and speed rise for Coo-Trois Ponts. With three machines served by the same penstock, maximum pressures and speeds resulted from simultaneous load rejection of all units. For that consideration, it was necessary to introduce a governor dead time of 9.8 sec so that the throttling effects of the accelerating runner and the closing gates act consecutively rather than simultaneously.

Pump power failure theoretically does not present a pressure rise or speed rise problem. However, Fig. 6 illustrates the discharge inversion through the machine resulting from power failure while rotating in the pump direction. The condition is a very unnatural and surely an undesirable condition.

Essentially the same calculations as performed for simultaneous turbine load rejection and normal un-

TYPICAL TRANSIENT BEHAVIOR of low Ns pump/turbine is given with actual governor dead time of 9.8 seconds. **(FIG. 6)**

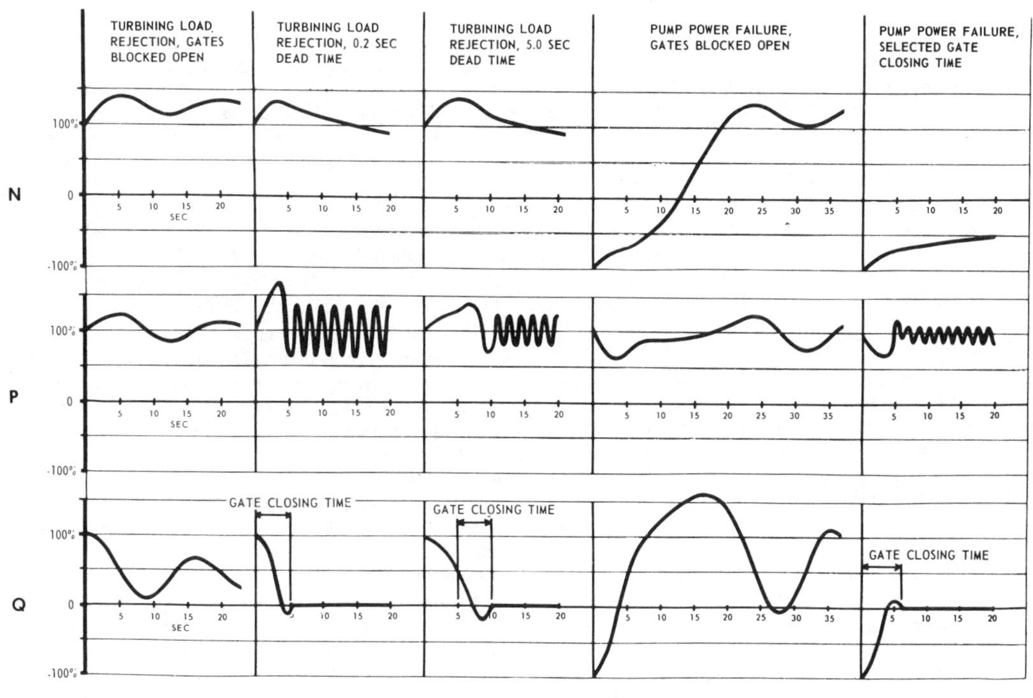

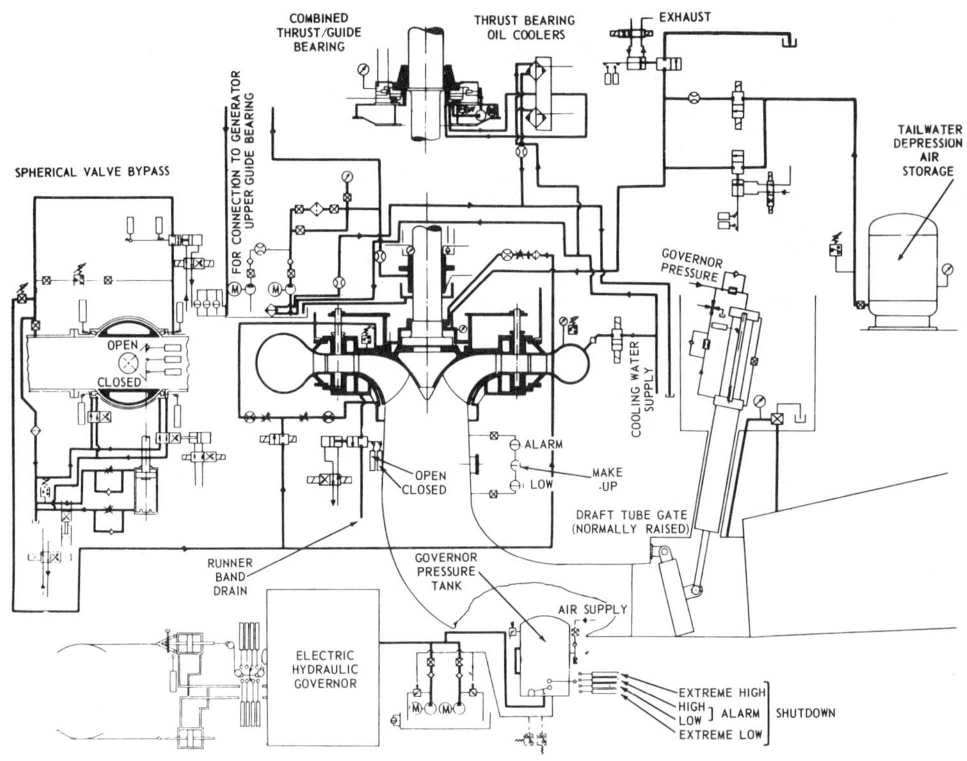

HYDRAULIC CONTROL SYSTEM for Coo-Trois Ponts hydro-mechanical equipment is shown schematically. **(FIGURE 7)**

loading used to determine maximum pressures on the upstream side of the turbine must be used for simultaneous pump power failure to determine maximum pressures on the downstream side of the turbine. When one unit is isolated with draft tube gates lowered for service and with atmospheric pressure in the draft tube elbow, simultaneous pump power failure of the other two units is critical and determines the design pressure for the draft tube gates. This problem obviously does not exist with surface power plants with independent draft tubes.

Control equipment requirements considered

Much consideration was given to the control equipment for the Coo-Trois Ponts machines to permit automatic and remotely controlled startups and change-over operations. In addition to the controls necessary for automatic operation, numerous safety devices and control-interlocks will be provided because extreme safety precautions are required by underground power plants (Fig. 7).

Pump-Start Sequence — Open cooling water valve. Start high pressure thrust bearing oil pump. Start guide bearing oil pump. Open runner seal valve. Set gate limit at "pump gate." Set speed-adjust out of range. Open spiral casing pressurizing valve. Cycle grease system. Prepare start motor circuit and check.

Open runner drain valve. Open depressing air valves and check. Energize start motor and check breakaway. Shut down high pressure thrust bearing oil pump. Initiate speed matching and check 95 percent synchronous speed. Synchronize main motor. Close spiral case pressurizing valve. Close depressing air valve. Close runner drain valves. Open runner air vent valves. De-energize start motor and check. Open inlet valve seal and check. Open inlet valve and check. Close inlet valve bypass valve. Close runner air vent valve. Energize shutdown solenoids. Open wicket gates to "pump gate."

Generate-Start Sequence — Open cooling water valve. Start high pressure thrust bearing oil pump. Start guide bearing oil pump. Cycle grease system. Set speed adjust to 50 cps. Set gate limit to 100 percent. Open inlet valve bypass valve. Open inlet valve seal and check. Open inlet valve. Energize shutdown solenoid. Close inlet valve bypass seal. Open gates to speed-no-load position and check breakaway. Shut down thrust bearing oil pump. Initiate speed matching, and check 95 percent synchronous speed and synchronous generator.

The pumping start dictates the bulk of the controls provided. Synchronous condenser operation and spinning reserve-starts follow the general pattern as for the pump-start without the subsequent priming through evacuation of the air from the runner chamber. ▲

4

NORTHFIELD MOUNTAIN PUMPED-STORAGE PROJECT PLANNING AND DESIGN

ANTONIO FERREIRA
Chief of Generation Civil Engineering
Northeast Utilities
Hartford, Connecticut

and

WALTER E. FISHER
Manager, Hydraulic Division
Stone & Webster Engineering Corporation
Denver

INTRODUCTION

The first of the four 250-MW units of the 1000-MW Northfield Mountain Pumped-Storage Project was declared in commercial operation on November 30, 1972. The remaining three units were declared commercial in 1973 and all units have been in successful operation since then. The Project is located in the towns of Erving and Northfield, Franklin County, Massachusetts. It is on the east bank of the Connecticut River about 30 miles north of Springfield and about 100 miles west of Boston (Fig. 1).

Northeast Utilities Service Company provided the engineering and construction-management liaison for the Project owners and licensees, The Connecticut Light and Power Company, The Hartford Electric Light Company, and Western Massachusetts Electric Company, which are the largest operating affiliates of Northeast Utilities. The Project was designed by Stone & Webster Engineering Corporation, which also provided the base construction management personnel. The civil works, including basically the tunnels, powerhouse excavation, and the upper reservoir dam and dikes, were constructed by the Morrison-Knudsen Northfield Associates. The electrical and mechanical work were subcontracted, with Stone & Webster acting as the construction and installation supervisors.

Generation planning studies performed by the Project owners in 1963 and 1964 had indicated that pumped-storage peaking capacity was the most economically feasible generation for meeting their short-hour, rapidly fluctuating, peaking power needs for the early 70s. The pumped-storage capacity was deemed even more readily adaptable to the needs of the larger New England load. The site was selected by Northeast Utilities engineers after studying the available sites within or adjacent to their service area using topographic and geologic maps. The object of the site studies was to find an upper reservoir site at a relatively high elevation close to an existing river, body of water, or valley that would be used or developed to form a lower reservoir.

The upper reservoir is on top of Northfield Mountain with the powerhouse and interconnecting flow lines excavated underground on the gently sloping west flank of this range which rises between the Connecticut River and its easterly tributary, the Millers River, a short distance upstream of their confluence. The elevation of the upper reservoir valley floor is about 750 feet above the level of the Connecticut River, approximately 8000 feet to the west, and about 550 feet above the bed of the Millers River, which is some 3000 feet to the east.

49

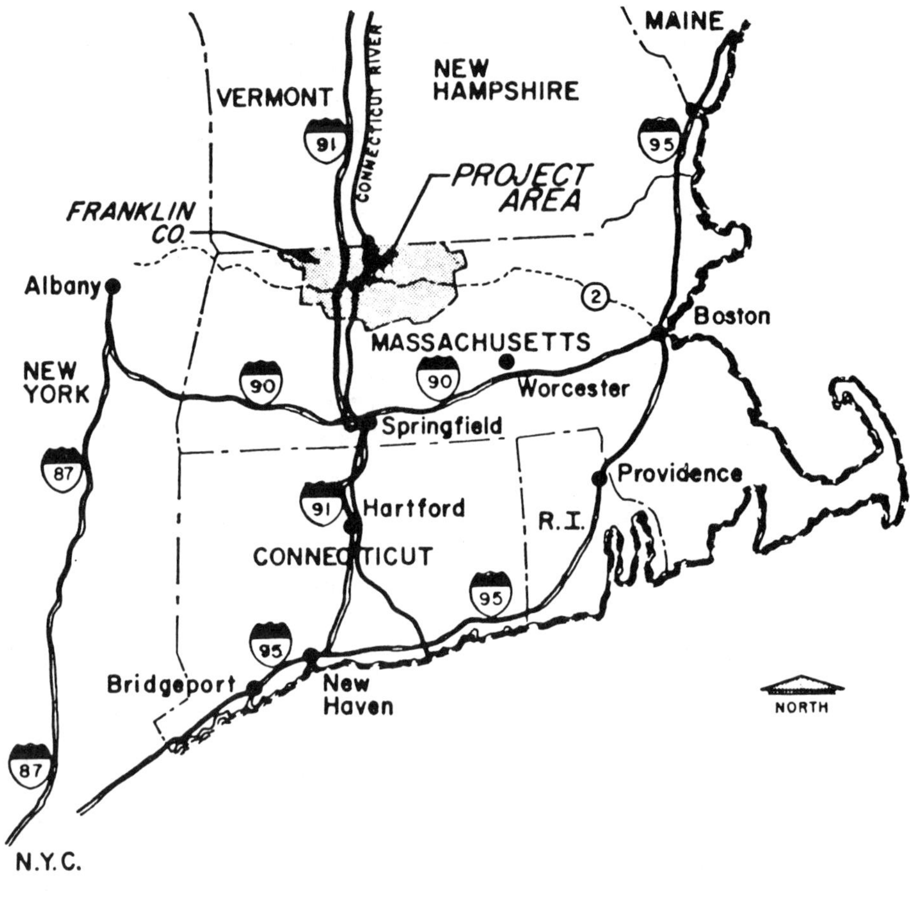

Fig. 1—Location map.

INITIAL INVESTIGATIONS, STUDIES, AND ESTIMATES

Comparative economic analyses, preliminary design studies, and field reconnaisance started in 1963. Literature and geologic maps of the area showed that Northfield Mountain, which is a part of the northwest limb of the Pelham dome consisted of granitic gneiss. Field inspections made by company engineers, Stone & Webster engineers, and geologic consultants indicated that the mountain top should be reasonably tight and stable. Where bedrock was not exposed, the soil cover generally appeared to average five to ten feet in thickness.

Construction of a rockfill dam across Briggs Brook, which drained the area, and of rockfill dikes at low saddles around the rim of the small drainage basin appeared to be sufficient to form an essentially watertight reservoir. The investigations also showed that it should be possible to excavate rock tunnels

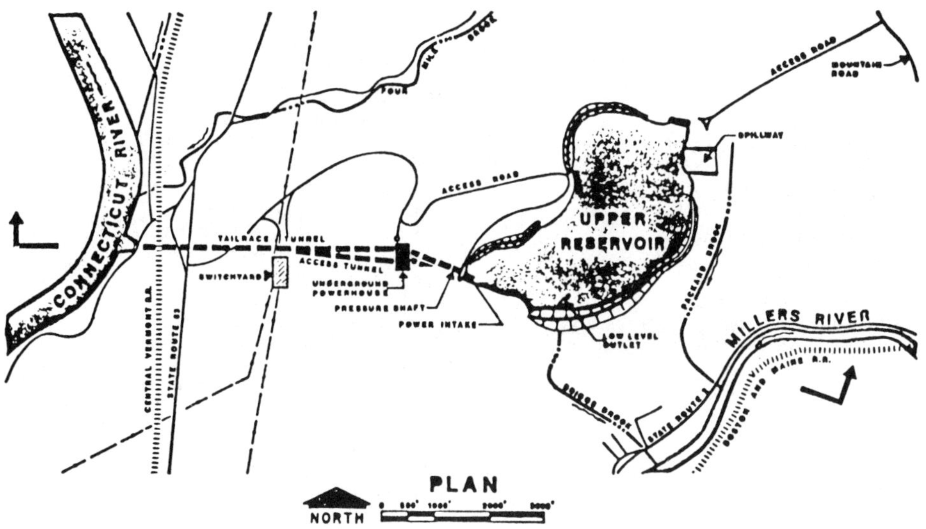

Fig. 2—Project layout.

connecting the reservoir to either the Millers River or to the Connecticut River (Fig. 2).

The design studies showed that although the distance from the upper reservoir to the Millers River was much shorter than to the Connecticut River, the reduced head and the high dam required to form a lower reservoir in the steeply sloping Millers River made it preferable to use the Connecticut River.

Adoption of the Connecticut River depended first on the feasibility of controlling surges and other transient flow conditions in the long two-mile hydraulic flow line and second, the ability to determine how the required storage in the 20-mile section of the river between the existing Turner Falls and Vernon hydroelectric plants could be obtained and managed.

The topography of the westerly flank of Northfield Mountain at the site was not suitable to the more conventional arrangement in which the powerhouse is located at the end of the flow line.

The intervening ground surface descends relatively rapidly from about elevation 1000 ft at the upper reservoir intake location to elevation 400 ft within a span of 3600 ft. In the remaining 2400 ft the surface descends by another 150 ft to elevation 250 ft at the tailrace canal. Except for about 2000 ft adjacent to the upper reservoir, rock cover over an unreinforced pressure tunnel located at a practicable depth would not be adequate to withstand the hydraulic loads. Additionally, a surge tank located within effective distance of a riversedge powerhouse would have an unsupported height in excess of 700 ft. Such a structure was not considered economically feasible and studies showed that, absent a surge tank, the desired quick-action operating criteria could not be met.

The first condition was resolved by locating the powerhouse underground close to the upper reservoir to reduce transient pressures in the penstocks and by the provision of underground surge tanks immediately downstream of the

TABLE I
FINAL DESIGN PARAMETERS

Capacity	1000 MW
Energy-Normal	6000 MWh (@ 1000 MW)
Reserve (Emergency Use)	2500 MWh (at or below 1000 MW)
TOTAL	8500 MWh
"Cold Start" to maximum capacity each unit	3 minutes maximum
Full Automatic Operation	Yes
Load/Unloading	Remote, from System Dispatcher
Start/Stop (normal)	Local and Remote
Emergency Generation (500 MW) (including unit startups if necessary)	Remote, from System Dispatcher
Pump Shutdown	Sequential automatic when upper reservoir is full
Upper Reservoir Emergency Spillway	Capacity equal to full pumping rate
Flood Flows in Connecticut River	Rate of discharge of flood flows at Turners Falls Dam not to be changed by Northfield operation at river flows exceed 65,000 ft³/sec (requirement of United States Army Corps of Engineers).

powerhouse to control surges in the long unreinforced tailrace tunnel.

Early design studies showed that lower reservoir storage could be obtained by raising the dam height and maximum pond elevation at the Turners Falls Dam, which diverts water to the Turners Falls Project hydroelectric stations. It was also determined that the existing timber flash boards would have to be replaced by spillway gates to provide control during higher river flow periods.

During the preliminary studies, the capacity of the proposed pumped-storage plant appeared to be in the range of 300 to 450 MW. Economic feasibility estimates prepared as the studies continued showed that the site could be developed for much higher capacities and that costs per kW reduced as the capacity increased. System production cost load simulation studies considering the energy input of the pumped-storage facility as a variable were performed to determine the economic size of the upper reservoir. The studies showed that system production costs continued to decrease with increased pumped-storage use up to a normal usage of approximately 6000 MWh. It was also judged that the plant should have an available emergency-use energy availability of 2500 MWh. The total usable energy content of the Project reservoirs was determined to be the sum of the normal and emergency requirements, some 8500 MWh.

FINAL DESIGN PARAMETERS

The final design parameters established for the Northfield Mountain Pumped-Storage Project are listed in *Table I.*

UPPER RESERVOIR

The dams and dikes forming the upper reservoir were initially designed as rockfill with concrete face and the application for FPC license was based on this design. A deposit of glacial till was discovered during construction of access and survey roads about two miles northeast of the reservoir. Economic studies showed that an impervious core rockfill

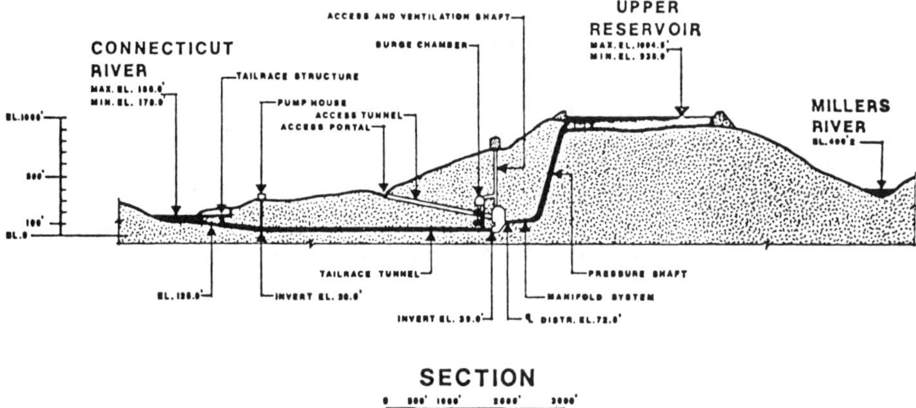

Fig. 3—Section through project.

construction would possibly cost slightly less and would otherwise be preferable. Bids were invited on either type. The successful bidder included the impervious core design. The maximum height is approximately 140 feet.

Rock for the dams and dikes was obtained from excavation of the intake canal. Excavation and fill were designed to balance.

The rock cut intake canal on the west side of the reservoir was provided to shorten the tunnel flow lines. The cross section was designed to limit the maximum velocity to two feet per second at maximum drawdown, with the object of limiting ice breakup during winter operation.

The emergency spillway was designed as an uncontrolled 465-foot long concrete weir constructed on a leveled rock excavated area on a saddle on the east side of the reservoir. It is capable of discharging 11,400 ft³/sec at a head of 3.15 feet, equal to the four-unit pump discharge at maximum gross head.

The reservoir water level would then be 2.3 feet below the dam and dike crest elevation. Because of the triple independent sets of automatic pump shut-off controls provided, plus two independent water level remote sensing/indicating equipment installation and a TV camera transmitting to the control room, significant discharges over this emergency spillway are only a remote possibility. The elevation of the center 20 feet of the weir is 1.53 feet lower, and is designed to discharge 200 ft³/sec to cope with the maximum probable rainfall runoff on the reservoir, should it be full and the plant be inoperable during the design rainfall period.

HYDRAULIC FLOW LINES

Lined and unlined rock tunnels were both investigated for the hydraulic flow lines (Fig. 3). Concrete lining was adopted with a water velocity of 25 feet per second, based on the undesirability of intermittent rock falls. With the rapid changes in flow rates and reversals of flow normal to a pumped-storage plant, it was believed that such falls might be of relatively frequent occurrence in an unlined tunnel and would reduce the reliability of the Project.

In design, all of the support loads and water pressure loads were assumed to be taken by the rock, none by the concrete.

Drain holes were provided through the tailrace concrete lining into the rock, and no waterstops were used at the construction joints in both the pressure tunnels and tailrace tunnel. The minimum thicknesses of concrete specified were based on providing a self-supporting circular pressure tunnel lining and an arched tailrace tunnel lining capable also of local rock support. No reinforcing was provided except at Y-branches in the pressure tunnels, at the bend between the inclined shaft and the approximately horizontal pressure tunnel and at the orifice T-branches.

Transitions were provided at both ends of the flow line to gradually reduce and accelerate velocities to regain most of the velocity head. Because of the relatively straight flow line at the tailrace exit, no hydraulic model was considered necessary, and its design was based on a 10 degree cone angle, similar to that used for draft tubes. Model tests at the Alden Research Laboratories of the Worcester Polytechnic Institute were made of the upper reservoir intake and transition to be sure that it would be free from vortex action at all reservoir levels.

The prototype has performed without vortices, as predicted by the model. Maximum velocity through the trash racks as determined from the model data was approximately six feet per second.

A 31-foot diameter inclined pressure shaft was used to shorten the flow line. An elbow at the lower end changes the direction of flow to a slope of 10 degrees and a Y-branch then divides the flow into two 22 foot diameter tunnels, which are in turn split by Y-branches into four 14 foot diameter tunnels. The design of the Y-branches was model tested for both directions of flow and with varying rates of flow in each branch

to make sure that no locally destructive velocities were present. Bent steel plate "armor" was cast in place on the vertical divider as protection against impact by rocks or other debris.

Downstream of the second set of Y-branches, the four 14-foot diameter penstocks or individual unit pressure tunnels are steel lined for 340 feet. Because they approach the powerhouse at an angle, they provide an effective 300 lineal feet of separation between the powerhouse and the rock around the concrete-lined pressure conduits, which are subject to full head water pressure, plus transient water hammer pressures, totalling approximately 605 lb/in.^{2} at the scroll case.

In view of the relatively good condition, tightness of joints, and freedom from erodible material in the rock, as shown by the diamond drill cores, this distance was judged sufficient to prevent any significant amount of seepage into the powerhouse cavern. The steel used for the lining was ASME A516, grade 70, chosen because it is relatively ductile and is easily welded. Yield stress is 38,000 lb/in.^{2} and tensile strength is 70,000 to 85,000 lb/in.^{2} Transfer of a substantial portion of the load to the rock was assumed through the concrete encasement, but the thickness of lining was designed for "free pipe" stresses based on 1.25 times the yield stress for the first 150 feet (47,000 lb/in.^{2}) and 32,200 lb/in.^{2} for the next 120 feet. At this point a 25-foot transition reduces the diameter from 14 feet to 9 feet 6 inches. Free pipe stress within the transition and the next 25 feet is 20,000 lb/in.^{2} and for the last 20 feet, 17,500 lb/in.^{2} The downstream 40 feet is coated with mastic material to prevent the transfer of longitudinal stresses into the rock adjacent to the powerhouse wall. The

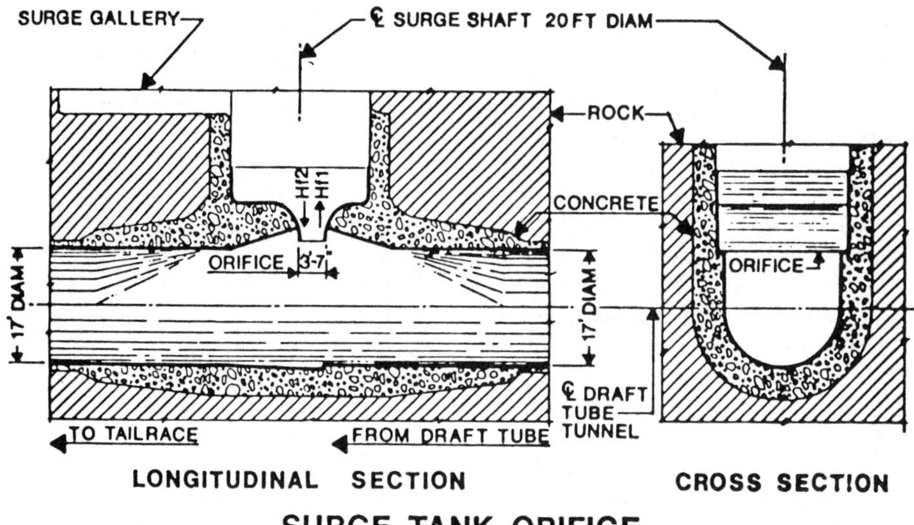

Fig. 4—Surge tank orifices.

end of the steel liner bolts to the sphere valve, which in turn is connected to the pump/turbine scroll case extension by a sliding expansion joint.

Based on the lowest seismic velocity of 12,000 ft/sec observed in test borings, the static Young's modulus for the undisturbed rock may be around 3×10^6 lb/in.^{2} Allowing for blasting cracks in the rock around the tunnel, an effective modulus of 1.5×10^6 lb/in.^{2} could be assumed. Using this modulus and assuming a gap between the lining and concrete encasement due to a 20 F temperature difference, the stress in the first 150 feet on 1-inch thick liner can be computed as being on the order of 18,000 lb/in.^{2} About 38 percent of the load is probably taken by the liner and 62 percent by the rock.

Surge shafts and galleries were required downstream of the pump/turbines to control hydraulic transients on changes in load and during startup and shutdown.

Four 20-foot diameter vertical shafts are provided, the T-connection to the draft tube being located 135 feet downstream of the unit center line. The shafts are interconnected by three galleries. The upper and lower galleries are approximately horizontal, while the center gallery is inclined. The upper and lower galleries limit the pressure increases and decreases in the unit draft tubes resulting from emergency four-unit startup, and following full-load pumping or generating loss of load or power input. The center gallery fulfills the Thoma requirements for stability during normal load changes and during independent plant operation. It is inclined to provide for coverage over the complete range of tailwater elevations. The vertical shafts were united with embedded steel mesh. The galleries were not lined, and inspections after startup testing showed that some previously blast-loosened rock was dislodged from the invert of the center gallery and had to be removed.

Hydraulic friction in the one-mile length of tailrace tunnel is relatively small (16 feet at full load generating) and, without damping, the three-minute surge cycle of the water in the system after a change in flow rate would have continued with little dimunition for a lengthy period.

Following large load swings, governor action to counteract the changes in head during these cycles could have resulted in further excitation, resulting in unstable operation and possible exhaustion of surge tank volumes. For this reason, steel orifices restricting flow into and out of the four surge shafts were installed at their lower ends above the draft tubes. It was desired that they be of the differential type, such that resistance to flow in the upward direction be greater than that in the downward direction. There was no great penalty in high draft tube pressures, but negative pressures much below atmospheric could result in separation of the water in the draft tube and the formation of void spaces. In the next half cycle these voids would collapse and result in destructive water hammer pressures on the bottom of the pump/turbine runners.

The orifice openings were required to be rectangular since they were also to serve as the access for the draft tube gates. Very little design data were available on the rectangular orifice desired, but it appeared possible to attain a differential ratio of at least two to one for opposite diretcions of flow. Model studies were therefore undertaken at the Alden Research Laboratories to determine the orifice characteristics for all combinations of flow at the T for different orifice shapes and widths. The result was an orifice with a ratio of 2.375, shaped as shown on Fig. 4. With the assistance of Dr. Henry Paynter of MIT,

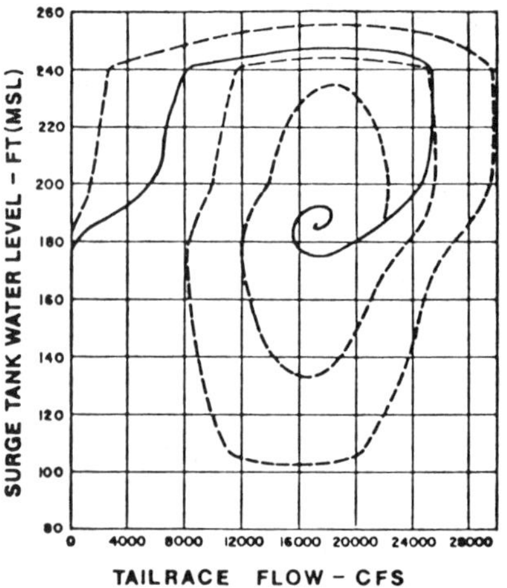

SURGE TANK WATER LEVEL vs TAILRACE FLOW
1000 MW LOAD ACCEPTANCE

large area water orifice ———
small area water orifice ———

Fig. 5—Surge tank water level vs. tailrace flow.

the results were described by a mathematical relationship which could readily be programmed into the computer program which had been developed, also with Dr. Paynter's assistance, to determine the overall response of the complete hydraulic system during transient conditions. From trials covering the most severe conditions, the minimum size orifice was determined, based on the condition that the minimum draft tube pressure should not fall significantly below atmospheric. It was then possible to establish the final volumes required for the upper and lower surge galleries. The effectiveness of the orifice in damping surge cycles is illustrated on Fig. 5.

Downstream of the surge shaft T-connections, the four 17-foot diameter circular tunnels transform at two Y-branches to two horseshoe tunnels equivalent in area to 22 feet diameter and through a third Y-branch to the horseshoe tailrace tunnel, with an area equiv-

TABLE II
GUARANTEED RATINGS OF PUMP/TURBINES

Speed

Synchronous	257 r/min	
Steady-State runaway at full gate	343 r/min	

Horsepower

Generating	Hp—Max Allowable	Eff.	Eff. at 75% Load
823-ft net head	367,000	89.5	90.6
770-ft net head	367,000	85.5	90.8
710-ft net head	320,000	85.2	89.6

Pumping Discharge		
Best Gate	cfs.	Eff.
835-ft dynamic head	2950	88.8
801-ft dynamic head	3230	89.3
767-ft dynamic head	3440	89.2
740-ft dynamic head	3600	88.9

Specified Speed

Generating	At Max Allowable Hp	At Best Gate
823-ft net head	35.3	32.5
710-ft net head	39.7	35.5
Pumping		At Best Gate
835 dynamic head		1904
740 dynamic head		2302

Recommended Net Positive Suction Head

Generating	74.0 feet
Pumping	83.0 feet

alent to a 31-foot diameter circular tunnel.

Stop logs are provided in the outlet transition structure to enable the tunnel to be unwatered. The end of the transition structure was also designed to enable the complete structure to be unwatered by means of bulkheads, supported by the concrete trash rack supports, under low tailwater conditions.

LOWER RESERVOIR

The requirement of the Army Corps of Engineers that flood discharges at Turners Falls not be significantly changed added to the complexity of the problem in that a workable lower reservoir storage and discharge management procedure had to be demonstrated to them. Backwater elevations in the 20-mile reach of river and at the upstream Vernon hydroelectric plant had to be determined for securing lands and rights.

Accomplishing these objectives required data collection which included detailed topographic aerial surveys of the river, river soundings, river level and flow measurements, and collection and analyses of existing river flow data.

Based on the data obtained, a 400-foot long hydraulic model of the river was constructed by the Alden Laboratory and, after calibration from the prototype data, it was used to determine the river and storage conditions with the Northfield Project in operation.

It was determined that to meet the Corps of Engineers requirements for flood discharges and at the same time maintain Project dependable capacity up to the 126,000 ft³/sec, 50-year flood flow, an increase in the discharge capacity at the Turners Falls Dam at lower pond elevations was required in addition to an increase in the maximum crest height of over 5 feet.

The Turners Falls Dam, which is in two spillway sections divided by a rock island in the center of the channel, was modified to meet the requirements by replacing the existing timber flash boards on the lefthand spillway with four 120-foot long by 13.3-foot high bascule gates and by replacing the existing spillway on the right-hand side with a new concrete gravity dam containing three 40 ft by 40 ft taintor gates.

It was also necessary to raise the floor level of the existing canal headgate house and to install new individual sluice gate hoists because of the increase in maximum pond height.

EQUIPMENT

Pump/Turbines

Because of the underground configuration of the Northfield Project, the cost of providing the deep submergence required by high specific speed units was relatively small. The reduced size of lower set units and the higher synchronous speeds would reduce both the powerhouse cavern dimensions and the size and cost of the generator/motors. An inspection of the Cruachan Pumped-Storage Project in Scotland, where four such units were installed, gave assurance that high specific speed units were practicable.

Discussions were held with manufacturers and when the specifications were prepared, no limitations on specific speed were included. Proposals received covered a variety of specific speeds, with synchronous speeds varying between 180 and 257 r/min. Evaluation of the proposals showed that use of the highest specific speed units with a synchronous speed of 257 r/min would be most economic. The machines purchased have guaranteed ratings as given in *Table II.*

Dr. Robert A. Sutherland had been retained as a consultant, and based on the results of his researches on cavitation on large pumps in service throughout the world, the manufacturer's recommended net positive suction head used for decision was increased from 83 feet to approximately 95 feet, based on four units pumping at minimum tailwater elevation. The centerline of scroll case was subsequently set at Elev. 72 feet, 111 feet below normal tailwater elevation (Fig. 6).

Reviews of operational problems experienced in large pump/turbines and turbines led to the inclusion of the following detail design features in the units:

1. Cushioned mechanical stops for wicket gate levers at end of travel.
2. Height of wicket gates to be larger than runner tip openings to prevent wicket gate entry into the runner.
3. Hydraulically-operated frictional clamps on wicket gate stems to be used to reduce or prevent vibration during pumping operation.
4. Steel-spring-operated frictional clamps on wicket gate operating mechanism so that in the event of sheer pin breakage, movement of a gate would continue to be under servomotor control, even though it was misaligned.
5. Air inlet piping into the draft tube in addition to that in the headcover.
6. Equalizing pipe connections between the space between headcover and runner and the runner and discharge ring in addition to that between the space between the headcover and runner and the draft tube.
7. Piezometer connections to enable pressures to be measured in various locations in the unit.

Sphere Valves

The 9-ft 6-in. diameter sphere valves are of normal design, with two exceptions. These are the requirements that

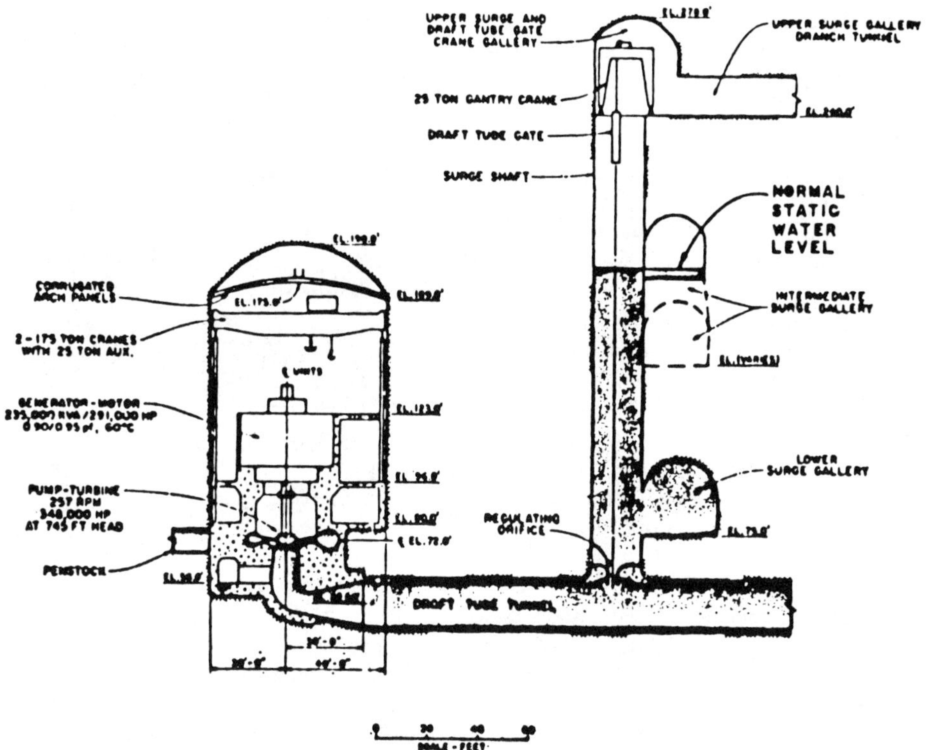

Fig. 6—Section through surge system and powerhouse.

the opening and closing time be 20 seconds or less, and that bypass valves be an integral part of the sphere valves. The valves provided are part of the downstream flanges and allow water from the valve casing to pass around the downstream seal ring into the scroll case. Closing the upstream sphere valve seal ring shuts off the supply of penstock water to the valve casing and thus to the bypass valves.

The rapid operating times were required to meet the three minute startup time requirement. The integral bypass valves eliminated high-pressure bypass lines within the powerhouse cavern, which commonly are subject to vibration problems.

The servomotor oil system operates at 1000 lb/in.^{2} and to prevent the possibility of diesel ignition, the space above the oil in the pressure tank is nitrogen filled.

Pump Starting

Design studies were made of all the then accepted pump start methods including across the line, reduced voltage, reduced frequency, hydraulic turbine, and synchronous from another unit. All of these were either found to be impracticable because of unfavorable effects on system voltage or because of high cost and complicated switching arrangements. The method finally adopted was the use of starting motors for each unit, capable of accelerating a unit to above synchronous speed within ten minutes. The horse-

[*Editor's Note*: Figure 6 is poor quality in the original.]

power required was 13,000 with the unit unwatered by use of a compressed air blowdown system.

Generator/Motors

Investigations were made to determine if generator voltages higher than the standard 13,800 volts would result in any significant savings, and it was found that they did not.

Each of the four generator/motors are 28-pole, 257-r/min, 13,800 volts. They are rated 235,000 kVA/291,000 hp, 0.90/0.95 p.f. at 60 C rise. At 80 C rise, they are able to operate continuously at 253,800 kW at 0.90 p.f. When pumping at unity power factor, each unit will develop 368,000 hp (Fig. 7).

Transformers and Cables

Many design studies were made as to the location of the 13,800/375,000-volt stepup transformers. Placing them on the ground surface at the top of the vent shaft or at the entrance to the access tunnel would have required lengthy and expensive generator leads. Placing the transformers in or adjacent to the powerhouse required 345-kV cables in the same locations, where they would be subjected to static oil heads of either 800 or 275 feet. Oil-filled 275-kV cables with a head of 1100 feet were in successful use in Scotland, but no cable manufacturer in the United States was prepared to guarantee oil-filled 375-kV cable for either of the heads required at Northfield. Because of the difficulty of supporting a vertical cable and pipe, the decision was made to use the access tunnel as the means of cable egress, and to install the transformers in reinforced-concrete enclosures within the powerhouse. The pipe-type cables are pressurized to 200 lb/in.² at the upper end, resulting in an oil pressure of about 300

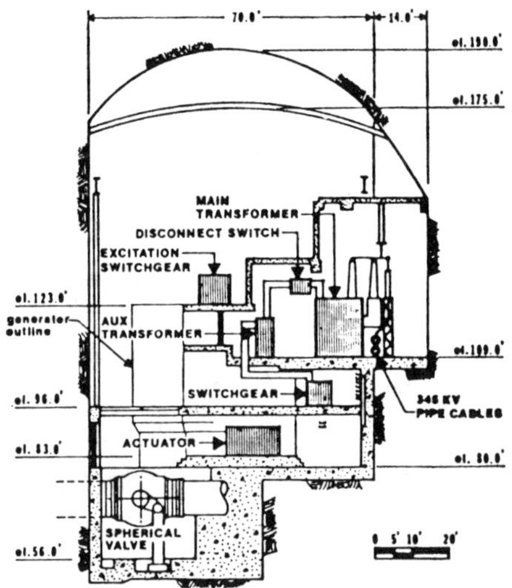

Fig. 7—Spherical valve and transformer vault.

lb/in.² at the lower end. The pipes are prestressed to avoid temperature expansion problems.

The two main power transformers are rated 500,000 kVA, 65 C, forced oil water cooled, 345,000 volt, grounded wye, with double low-voltage winding at 13,800 volts. Impulse voltage strength (BIL) is 900 kV and lightning arrestor rating is 276 kV.

Because of the long 345-kV cable connections between the transformers and the substation, a transient network analyzer study was used to determine the required impulse voltage strength (BIL) of the transformers, the lightning arrestor rating, and effect of sudden load rejection on sustained overvoltages on the system. The analyses showed that the higher overvoltages occurred with loss of system load with one unit generating at full load and failure of the controls to open the generator breaker. Before the voltage regulator could begin to act, the voltage at the machine terminals would increase due to loss of load and overspeed. The transformer would be over-

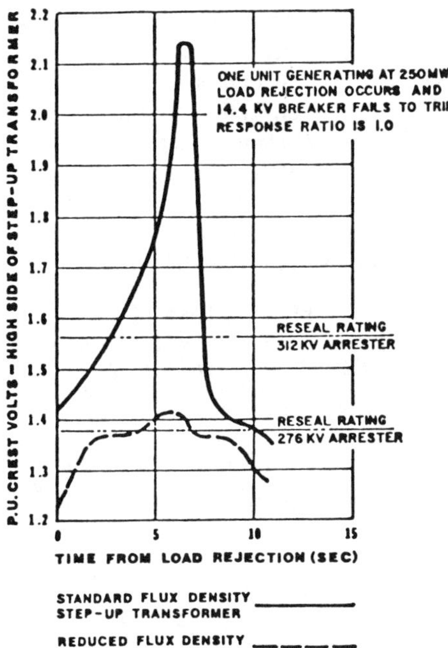

Fig. 8—Crest voltages on 345-kV side of stepup transformer.

excited and cause transformer saturation. This could lead to creation of voltage harmonics and the high capacitance of the 345-kV pipe-type cable could possibly resonate with the transformer reactance. It was found that transformers with additional iron to reduce flux density to about 15 percent below normal would greatly reduce the overvoltage, and hence insulation costs. This is illustrated on Fig. 8.

GOVERNORS

Electric hydraulic governors were chosen for Northfield. Power feedback is used to control the pump/turbine gate position. This means that if the governor load control is set at, say, 200 MW, the output will remain at that level regardless of changes in net head.

Joint load control is provided. All units on joint load control automatically maintain equal MW output to provide and maintain that selected on the MW master control on the main control panel. When a generating unit is started or stopped while on joint control, output of other units is automatically adjusted down or up so that total output remains constant.

Adjustable and variable automatic rates of loading/unloading are provided. The normal rate is adjustable between 40 and 200 MW per minute. Rates of 200 and 1000 MW may also be selected by pushbutton or by remote control when required. The rate normally used is 50 MW per minute, remotely actuated by the system dispatcher.

With power feedback, the speed regulation control setting is directly proportional to the change in load and system frequency. On a mechanical hydraulic governor, this is the droop control. If it is set at 10 percent, the unit output will change 10 percent for each 1 percent change in speed adjustment setting or in system frequency. Speed regulation is adjustable from 0 to 10 percent. If response to small changes in system frequency is not desired, the dead band control can be turned on. This will prevent response to frequency changes over the adjustable range of 0 to 0.25 Hz.

The governor and associated plant equipment also provide for the following automatic control features:

Pumping

1. Positioning of wicket gates to most efficient gate with changes in differential head.
2. Automatic pump startup, and loading to best gate.
3. Automatic pump shutdown.
4. Automatic sequential shutdown of units, based on:
   a. Signal from water level transmit-

ter—normal sequential automatic shutdowns in order preselected by plant operator. Warning signal is given, and plant operator can switch system off, if desired.

b. Signal from reservoir immersion switches, one for each unit, normal shutdowns at different elevations.

c. Opening of individual unit shutdown solenoid circuits by operation of float switches. Units trip out in sequence at different reservoir elevations.

Generating

1. Automatic startup and shutdown. After synchronization, units are automatically brought to minimum load (125 MW).
2. Limiting automatic unit loading to between 125 and 250 MW.
3. Change in loading rate to 200 MW per minute on receipt of signal from system dispatcher.
4. Automatic increase in plant output by 500 MW on receipt of signal from system dispatcher. This includes starting, synchronizing and loading all units not operating at time of signal. Rate of loading, 1000 MW per minute.
5. Automatic loading of plant to full output on receipt of second signal from system dispatcher.

BIBLIOGRAPHY

1. Samolis, R. P., "Electrical Features of the Northfield Pumped Storage Project," *IEEE Trans. Power Apparatus Syst.*, **PAS-88**, 1291-97 (1969) August.
2. Brennon, F. L. and St. Onge, G. A., "Unique Electrical Design Features of the Northfield Mountain Pumped Storage Project," IEEE Paper presented at the 1969 Winter Power Meeting, New York, January 26-31, 1969.
3. Gunwaldsen, R. W. and Ferreira, A., "Northfield Mountain Pumped Storage Project," *Civil Eng.*, **41**, 53-57 (1971) May.
4. Hardy, F. R., Jr., "Hydraulic Structures at the Northfield Mountain Pumped Storage Project," Paper presented at the 19th Annual Specialty Conference, Hydraulic Division of the American Society of Civil Engineers, August 1971.

5

Reprinted from pages 247–263 of *Energy Storage: User Needs and Technology Applications,* 1976 Eng. Found. Conf. Proc., Technical Information Center, ERDA, 1977, 424pp.

UNDERGROUND PUMPED HYDRO STORAGE

By

William S. Mitchell

Harza Engineering Company
Chicago, Illinois

Introduction

Pumped-hydro, as one form of energy storage, has proven itself over the years to be an economical and reliable means for providing utility station peaking power. Many systems are now in daily operation with more planned for future construction. But as workable as these type installations are, siting is becoming more problematic each year. A special topography is required where one of the surface reservoirs can be located on sufficiently high ground in the vicinity of a lower reservoir and a 'load center. Unfortunately, such coincidences of topography, favorable geology and load centers are not numerous. Where they are found to exist, there are also competing pressures of land use and environmental constraints. To circumvent this problem of siting, underground pumped-hydro storage is receiving special attention from potential users.

The concept of underground pumped-hydro storage evolved in Harza Engineering Company offices in 1960, and has been continuously studied since that time. The latest development relative to our work is license application to the Federal Power Commission for installation of a 1000 MW underground pumped-hydro plant at Mt. Hope, New Jersey. The application was filed by Jersey Central Power and Light Company, a component of General Public Utilities System.

Pumped-Hydro Systems

A simple comparison of the features of conventional pumped-hydro and underground pumped-hydro can be seen in Figure 1. In the conventional type system, two surface reservoirs are required. Flooded land area becomes significant unless one of the reservoirs happens to be a large, naturally occurring river or lake. The underground system requires only a single surface reservoir but has additional need of access shafts, a water shaft and an underground powerhouse and reservoir. Noted advantages of the latter system are:

- The possibility of using a naturally occurring, large body of water for the upper reservoir. Even if such is not available, the upper reservoir can be relatively small.

- Combining compressed air storage with underground
  pumped-hydro by using the air vented from the lower
  reservoir.

- Locating the lower reservoir at a depth suitable to
  the hydraulic machinery instead of the local topography.

The arrangement of equipment in the underground pumped-hydro
powerhouse is similar to a conventional pumped-hydro powerhouse except for the inclusion of the power transformers and
electrical buses, the latter for which SF_6 now is the most
economical type available. Comparison of hydraulic machinery
discussed in this paper can be made by reference to Figures 2,
3 and 4 which show elevations of the European developed tandem
pump/turbine machinery, and the single and multistage reversible
pump/turbine machinery, respectively.

Underground pumped-hydro can be arranged in either a single or
a multiple drop configuration, but economic studies favor two-
drops for the latter. The two-drop scheme is shown in Figure 5.
In this arrangement, there is a need for an additional powerhouse, intermediate level reservoir and water shaft manifold.

A more recent innovation is the coupling of the lower reservoir
vent shaft through a throttle valve to a combustion turbine. As
water is ducted from the water turbines into the lower reservoir,
the pressurized air in the lower reservoir air space is
displaced to the combustion turbine combustors, controlled by
throttle valve operation. Our studies have shown that this
type system is economically attractive and can provide large
amounts of peaking power for a relatively long period of time.
It can also be noted that the combined operation of water
pumping and air compression may well lead to faster emptying of the
lower reservoir, thus enabling longer periods of generation than
are now possible solely by pumping with the hydraulic machinery.

Costs
<u>Costs</u>

Costs for an underground pumped-hydro installation are influenced
by many factors: surface reservoirs, shafts, hydraulic
machinery, excavations, etc. The lower reservoir is the major
cost item for this type project. The next most important cost
item is the equipment.

The concept of pumped-storage has been known for over 50 years
(circa 1910), but the early stations were restricted in size
because of the limitations and relative complexity of equipment
that was available. The major break-through for large scale
pumped-storage came with the development of the reversible
pump/turbine, the first such unit of commercially significant
size being the 62 MW, 58 meter head unit developed by Allis-
Chalmers for TVA's Hiwassee Station and placed in service in
1956. Since then, pump/turbine ratings have multiplied six
times in capacity and nearly ten times in head. These gains
are the result of research and development by hydro equipment
manufacturers.

A reversible pump/turbine, as compared to the tandem arrangement of pump and turbine, replaces a separate pump and separate
turbine, connecting clutches, one penstock valve and the
controls for the valve and clutches. Therefore, it saves cost

by reducing size and complexity of the powerstation and
connecting waterways. For underground stations such items,
particularly the waterways, have relatively high unit costs.
Reduction in size and simplification to expedite construction
procedures have tremendous value. The reversible pump/turbine
thus provides savings beyond the cost of generating equipment
and auxiliaries and has to be discussed relative to those
savings. Thus, items of most importance, neglecting the upper
and lower reservoirs, can be considered as follows:

<u>Access</u> <u>Shaft</u> <u>and</u> <u>Water</u> <u>Conductors</u>

Shafts and conductors, three of which are needed, are for
personnel access and exit, water flow and venting of the
underground reservoir. Where conditions are suitable, personnel
shafts can also be used for ventilation, electrical bus and
control cables.

Geological formations to be traversed by the shaft will affect
costs regarding the extent to which they are water-bearing and
the extent to which the excavated surface is friable. Also of
importance are the depths of the various underground formations.

The methods chosen for shaft excavation will affect these
costs. For example, excavation can be accomplished by drilling.
Some situations may require the use of drilling mud, others may
not. Drilling and reaming, full-face shaft sinking, raised
drilling or pilot shaft drilling with benches blasted and
mucked from the cavern floor are other acceptable methods that
can be used to form the shaft. The first shaft obviously has
to be excavated downward with attendant relatively slow
progress; later shafts can be excavated more rapidly by
taking advantage of underground workings and a high-speed,
high capacity hoist in the shaft first excavated. Furthermore,
costs are affected by final shaft lining, whether it be done
by pouring mass concrete, placement of rings, ground freezing,
or using drilled-in concrete caissons. These latter methods
have been detailed in the paper, Developments in Underground
Pumped Storage, by A. E. Allen and W. E. Larson of Harza
Engineering Company, presented at the Edison Electric Institute
meeting of May 1974.

Shaft size is a parameter directly affecting costs; costs
tend to rise exponentially with both size and depth and
reflect the size of equipment used in the powerstation such
as the transformer, pump/turbine runner, pump/turbine head
cover, penstock valve and generator/motor stator sections and
rotor.

Shaft size is dependent essentially on the size of equipment
which must be lowered through the shaft. From our past studies
we have found that at high heads, the transformer is often
dimensionally the largest item to be handled. Where transformers
can be lowered in sections for assembly underground, shaft size
is set by rotating equipment size. To keep shaft size
relatively small, current planning is to assemble three, single-
phase transformers underground into a packaged unit. Each phase
transformer would be a separate, sealed tank lowered individually.
The physical size of a transformer is influenced more by
voltage capacity (dielectric distance requirement) than by MVA
rating. For a 300 MVA, 345 kV transformer, an 18-feet inside
diameter shaft is adequate. Such a shaft will pass the

components of a 250 MW pump/turbine unit designed for a 2,200 feet
hydraulic head, which match this transformer rating.

The pump/turbine stay ring can be lowered in sections, and while
so far it has not been found to govern shaft dimensions, it must
always be considered. There is also a minimum size of shaft to
accommodate ventilation air flow, elevators, hoists and
stairways.

Obviously, depth affects shaft excavation costs. It also
affects shaft equipment such as hoists, elevators and stairways,
electrical bus design, support and cooling systems. The above
costs are substantial since not less than two independent access
shafts are required for an underground powerstation to meet
ventilating and Federal safety (OSHA) requirements.

Figure 6 shows drilled shaft costs as a function of shaft
diameter for hard rock geological strata.

We have studied the relationship between shaft size, hydraulic
head, powerstation volume and generating capacity. For a given
operating head and number of units, shaft size and powerstation
volume increase little with increasing unit rating. Additionally,
assuming constant generating capacity, shaft size decreases
with increasing hydraulic head. The net effect is that for
various combinations of hydraulic head and generating capacity,
total shaft costs are relatively constant for an underground
pumped hydro project. Also, there is no significant relationship between shaft size and powerhouse volume.

We also find a general trend in costs favoring installation of
reversible pump/turbine machinery operating with a high head.
This is illustrated in Figure 7. It can be noted that there
is an appreciable decrease in total installation costs of a
large capacity power plant at 4400 feet compared to one at
3200 feet.

It is believed that the costs of drilling access and waterway
shafts can be reduced in two ways. First, by additional
research, for better drill steels and the use of grouting
chemicals for stabilizing poor quality formations above firm
rock in which the major structures must be located. Secondly,
through the use of improved techniques for detecting sub-
surface conditions and rock strata. Such techniques would
permit site selection before embarking on the slow, tedious
and expensive methods of boring, adit construction and
geophysical instrumentation whose use then could be confined
to design.

With regard to shafts, it should be noted that all waterways,
including machinery draft tubes, must be designed for full
head pressures, i.e. to the upper reservoir surface to protect
the system against misoperation.

The penstock shaft, lining and water conductor valve costs
are influenced in ways similar to those described previously.
However, there are some additional cost considerations.
Economic design velocity of water in the conduits is
related to system pressure rise and runaway speed, and this
will moderate shaft and valve size. There is also the question
of steel linings in the penstock. It is Harza's philosophy to
keep steel linings in shafts as short as will serve the purpose

of protecting the conduit and powerstation. Underground pumped-storage by its nature has minimum conduit length to minimize pressure rise.

<u>Powerstation</u>

Powerstation costs that should be considered are those for excavation, wall and roof treatment, concrete, lighting, plumbing and drainage, ventilation and geotechnical instrumentation. Of these features, excavation cost is the most important item. The second most important cost is wall and roof treatment. All these costs depend on the quality and stress condition of the rock material from which the cavern is formed.

For example, rock bolting at cost equivalent to $0.25/cu.yd. of excavation may increase excavation cost by an additional two to three percent. If the rock bolting requirement is found to be larger, the cost increase can be major.

Granite or granitic-type rock is most favorable for power-house excavation. If the rock is firm and little water is encountered, excavating is not unduly difficult and minimal grouting will suffice. Moreover, plumbing and drainage cost will be small.

Configuration and size of the powerhouse is selected to achieve an equitable balance between rock condition, minimum power-house volume and costs based on mass mining and quarrying excavation techniques.

As implied previously, powerstation depth affects costs, but the impact is small and can generally be neglected for the heads involved.

<u>Hydraulic Machinery</u>

We have found that the type of pump/turbine equipment selected for power generation has a significant effect on the final unit cost of the total plant.

Illustrated in Figure 8 are approximate costs for the hydraulic machinery, including the governors and valves needed to power the electrical generation equipment. The lowest cost units are noted to be single stage reversible pump/turbines. At the present time, 2100 feet head, single stage, reversible units are on order; 2400 feet head units are being planned. It is expected that 2500 feet operating heads will be developed in the near future.

Multi-stage pump/turbine machinery development is approximately 25 years behind the development of single-stage pump/turbines. In terms of current technology, the five-stage LaCoche units scheduled for operation in 1976 in France are pioneering efforts with ratings of 82.5 MW at 3100 feet head. These machines do not contain wicket gates. Other noteworthy machines under study in France are the two-stage, 328 MW, 3190 feet Grand Maison units with wicket gates and the 110 MW, six-stage, 3890 feet Super Bissorte units without wicket gates. The Edolo Project in Italy has been bid for 135 MW, 3990 feet units without wicket gates. As this type machinery is relatively new, capital costs are high. It can be expected

that with more widespread application of this machinery, costs
will approach those of Francis type reversible pump/turbines.

The tandem type units are normally considered for very high
head installations. But because of their high first cost,
they are being bypassed in power generation schemes in favor
of the more cost attractive, reversible pump/turbine used in
a two-drop scheme.

One of the significant parameters of pumped storage machinery
is the ratio of water pumping discharge rate to generating
discharge rate. The larger the ratio of pumping discharge
to generating discharge, the greater the dependable capacity
of the station computed on cycles of two to five days. We
have studied this pumping discharge relationship and found a
most favorable value of two - a pump/turbine having an electrical
pumping capacity twice that of its rated generating capacity.
This is based on computations which would provide 10 hours of
power generation per day with seven hours of pumping at an
overall efficiency of 70 percent.

With the very high heads desirable for underground pumped-
storage, pumping discharge using currently available machinery
is only about two-thirds to three-quarters of rated generating
discharge. The only way presently practical to improve the
ratio of pumping discharge to generating discharge significantly
is through purchase of oversized units. This affects not only
the investment cost for the unit but also the powerstation and,
where pump/turbine components govern shaft size, the access
shaft. If unit enlargement is limited, there may be need to
enlarge the system reservoirs to provide the necessary generated
energy.

Limitation of the water handling capabilities of a reversible
pump/turbine results from the characteristics of the water
itself in high velocity flows with high pressures and pressure
changes as it enters, passes through and subsequently
discharges from the pump/turbine runner. The problem can be
considered analogous, in part, to that of wind tunnels and
supersonic air foils. At one time the maximum velocity at
which a fluid could move was considered to be sonic velocity.

But analyses of shear factors, boundary geometry and other
related parameters led to breeching this barrier; and
supersonic velocities were attained in wind tunnels and in
aircraft.

A similar approach and in-depth type studies are considered
necessary for further development of very high-head reversible
pump/turbines. This will include study of the geometry of the
water passages with possible development of a new type
turbine runner. Consideration should also be given to
the control of water inflow to the machinery by means other
than wicket gates. Manufacturers of machinery are presently
performing studies and engaged in research oriented toward
increasing the usable head and size (power output) of pump/
turbines as well as the ratio of pumping discharge to
generating discharge. Manufacturers have the recognized
expertise for such work. However, they necessarily have
limited budgets. A cooperative research program involving
manufacturers, users and other interested parties in which
manufacturers would contribute their expertise could accomplish
much in increasing head, size and ratio of pumping discharge to
generating discharge for pump/turbine machinery.

Typical shaft, powerstation and equipment costs for various
types of pumped hydro installations are listed in Table I.
These reflect costs using a variety of hydraulic machinery.
Included are costs for conventional pumped storage which
requires surface reservoirs and a combined underground pumped
hydro-compressed air energy storage system.

The conventional 1200 feet pumped hydro system in the first
column uses an above ground powerhouse and three underground
penstocks between the upper reservoir and the powerhouse which
explains the high costs for Accounts 331 and 332. The two drop
single-stage system in the next column requires, among other
things, an intermediate reservoir and powerhouse. These costs
are included in Accounts 332 and 331, respectively. The one-
drop multi-stage system estimated in the third column has more
favorable cost for Accounts 331 and 332. However, this is
offset by the higher equipment costs in Account 333. The one-
drop tandem system in the fourth column shows increased costs
for the larger powerhouse and the pump/turbine equipment. The
last column is an estimate for a combined underground pumped-
hydro - compressed air storage system. The pumped-hydro has
two-drops with an intermediate reservoir and powerhouse and
two pump/turbines at each level. Above ground is a powerhouse
building housing six combustion turbines connected by a manifold
to the lower reservoir air vent shaft. This combined system
has high shaft, powerstation and equipment cost. But when all
costs of the total installation, reservoirs, land, etc. are
added, this system exhibits attractive power generating costs.

It should also be noted that Account 332 includes the costs of
the water conductor shaft, the upper intake structure, and the
draft tube outlets. Account 335 includes the cost of hoists,
elevators and other equipment that is ultimately installed in
the access shafts.

In comparison it can be seen that relative to the listed items,
costs for underground pumped storage are very attractive,
particularly in the case of single-stage reversible pump/turbines.
The high cost associated with installation of the tandem unit
attests to its limited popularity for pumped storage in America.

The costs shown in the table are representative of a 2000 MW
installation in hard rock. For a specific site, there may be
extenuating circumstances relative to topography, geology or
other features that can affect individual values. Each
installation must be considered within its own set of criteria.

Summarizing, those areas of research that can have impact on
future costs of underground pumped-hydro installations, we
note:

Excavation	-	New construction methods, improved drilling materials and new chemical grouting techniques for near surface use.
Equipment	-	Standardization of equipment relative to an economic head. The development of larger generating capacity, very high-head reversible pump/turbine machinery having a pumping to generating discharge ratio value between one and two; and developing a better understanding of the physical

characteristics of water in high pressure, high velocity flows.

Operation   -   Reliable, long term remote operation of underground plants.

The ease in siting of conventional pumped-hydro power plants where topography is favorable is becoming a thing of the past as competing land use and environmental protection pressures increase. Fortunately, we are seeing advances in equipment design and construction techniques that can contribute to the successful installation of underground pumped-hydro. In the not too distant future, we expect that there will be a general recognition and acceptance of underground pumped-hydro energy storage as an economic and reliable means for producing peaking power for the utility industry.

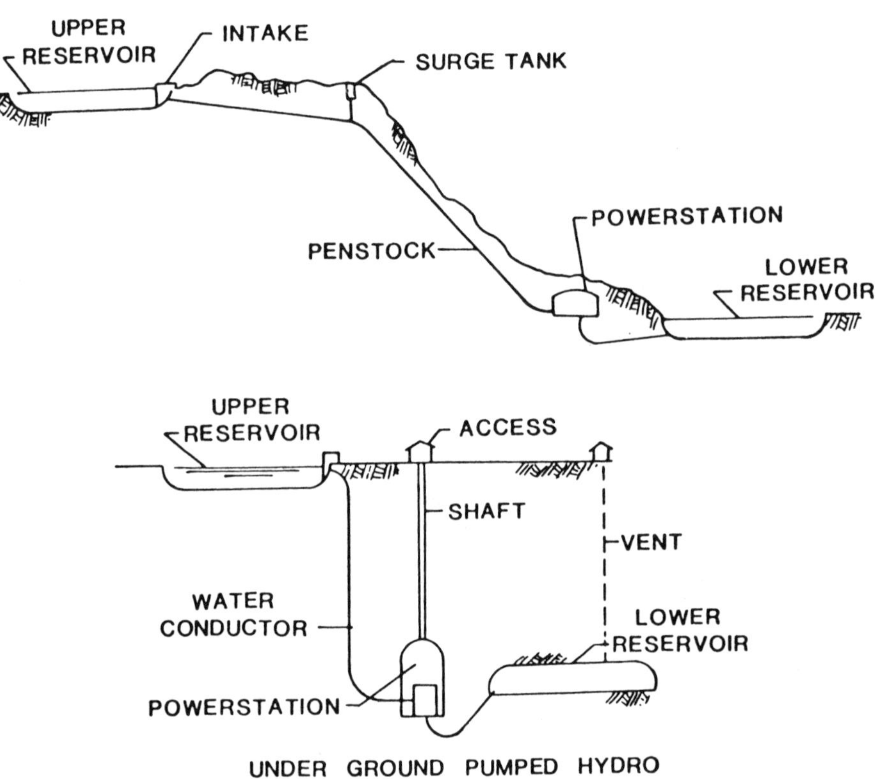

FIGURE 1. PUMPED STORAGE SCHEMES

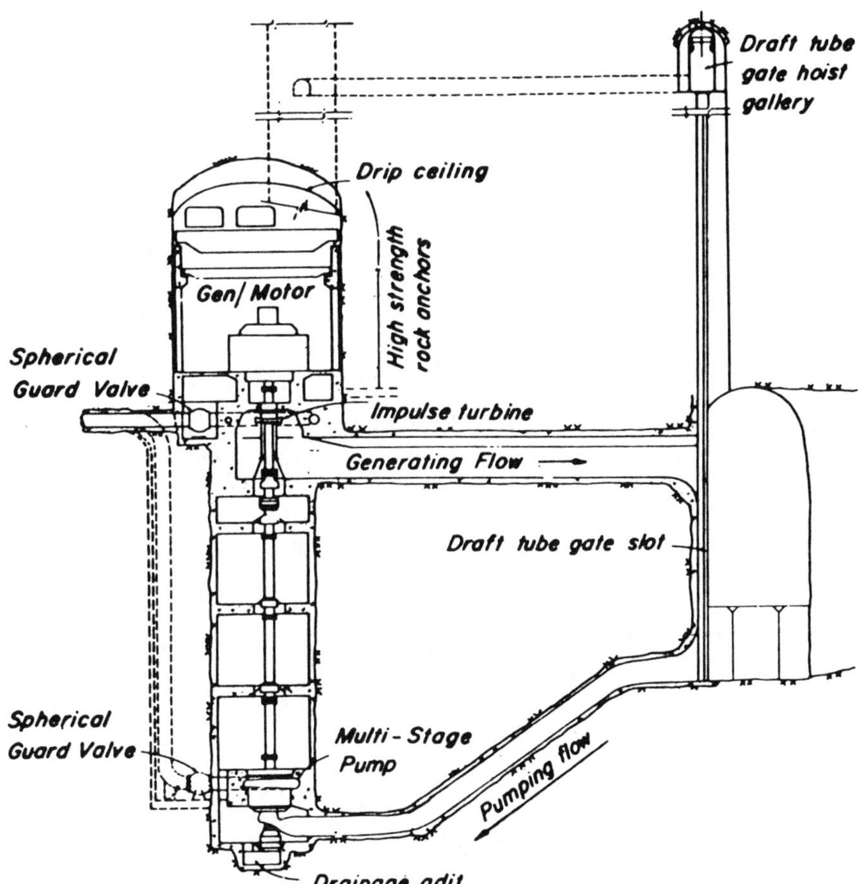

FIGURE 2. TANDEM PUMP-TURBINE INSTALLATION

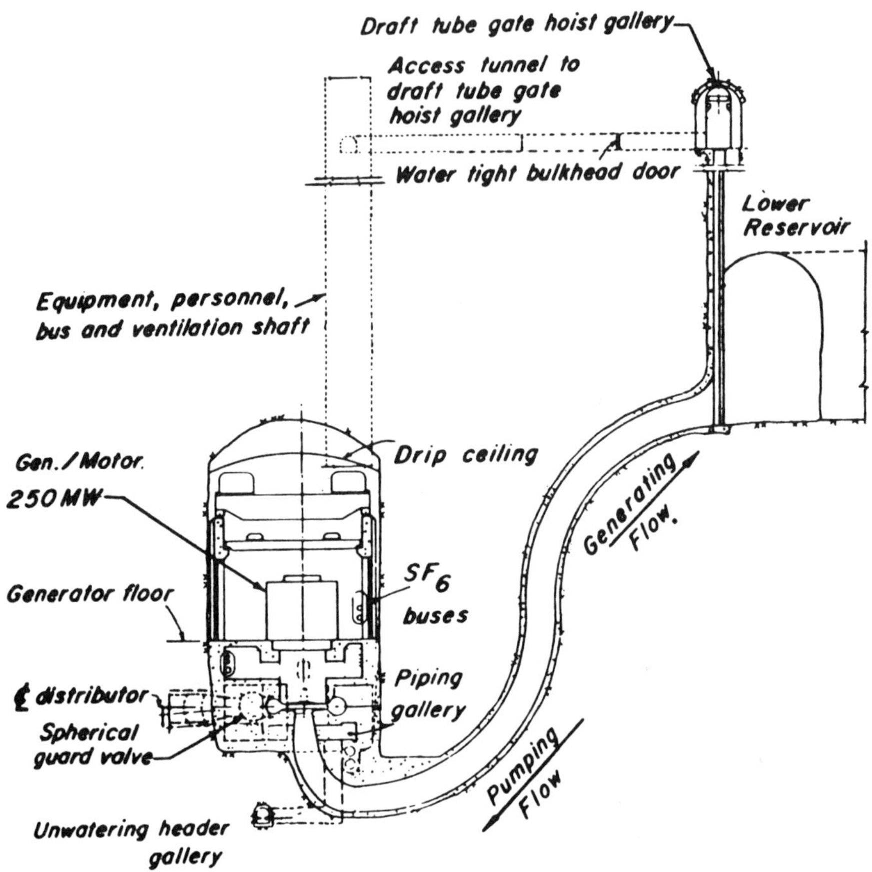

FIGURE 3. SINGLE STAGE REVERSIBLE PUMP TURBINE

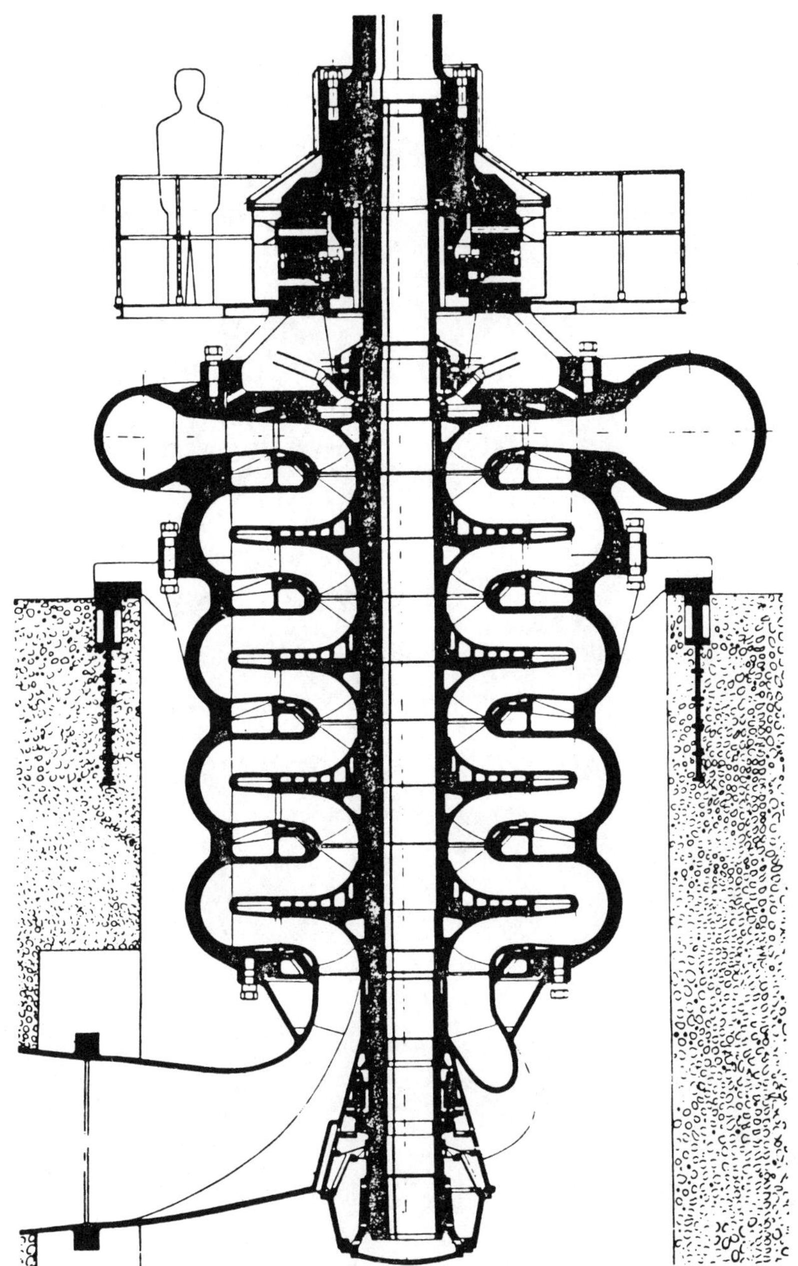

FIGURE 4. FIVE STAGE REVERSIBLE
PUMP/TURBINE (VEVEY DESIGN)

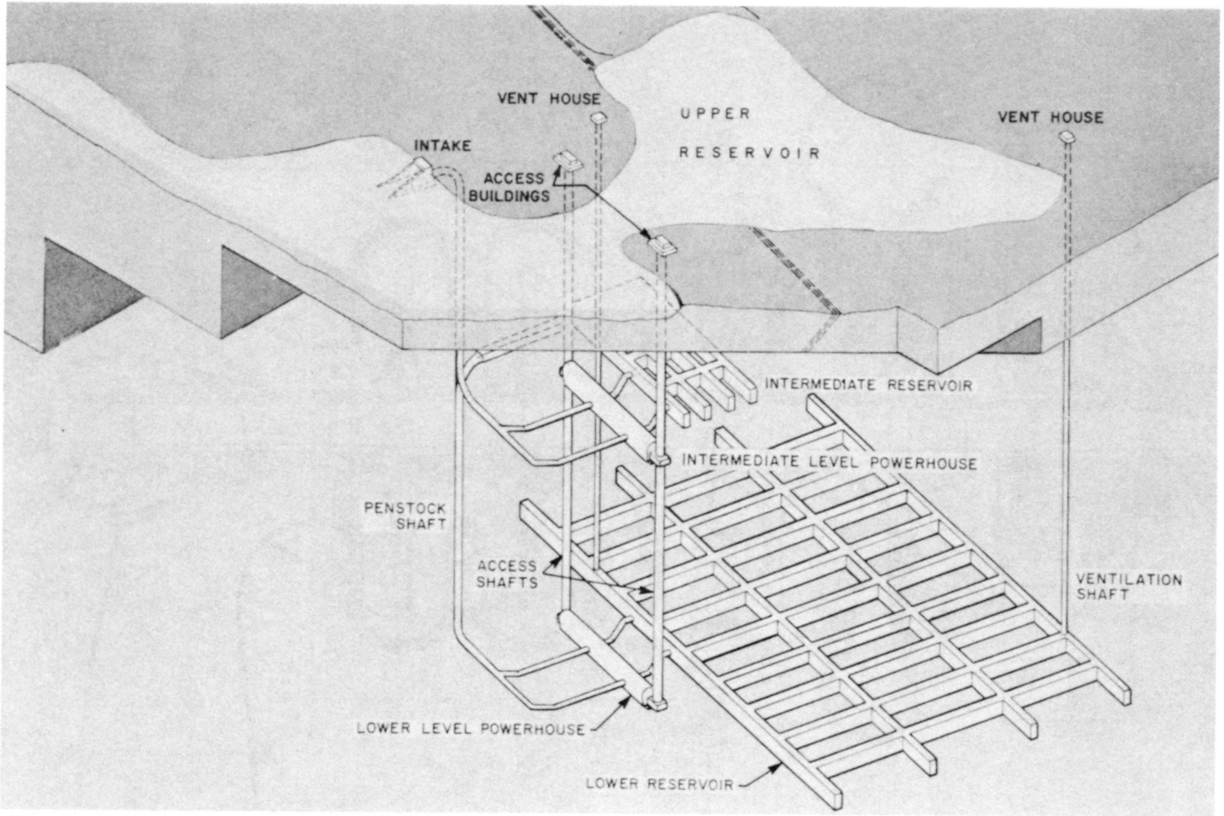

FIGURE 5. ISOMETRIC LAYOUT OF UNDERGROUND PUMPED-STORAGE PLANT

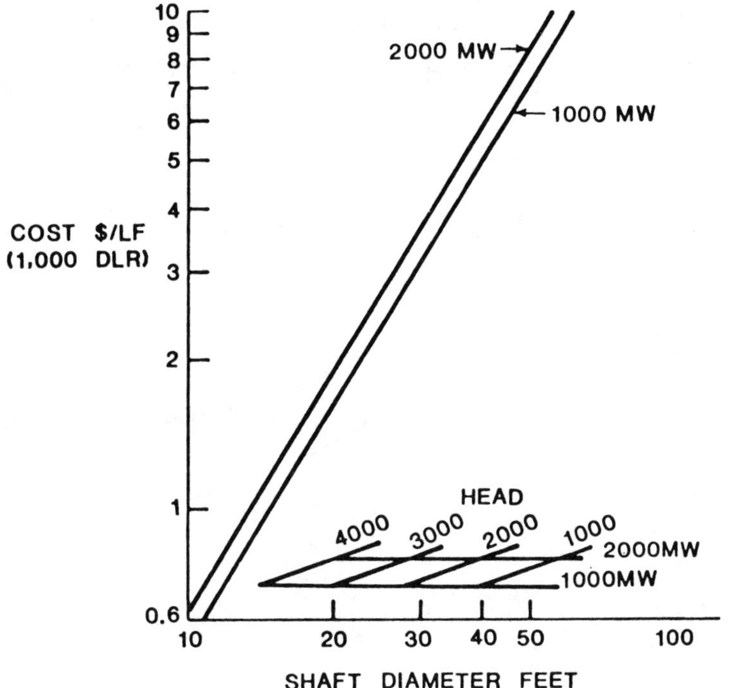

FIGURE 6. SHAFT COST PER FOOT FOR
VARIOUS SHAFT DIAMETERS
CONCRETE LINED IN HARD ROCK

(NOVEMBER, 1975)

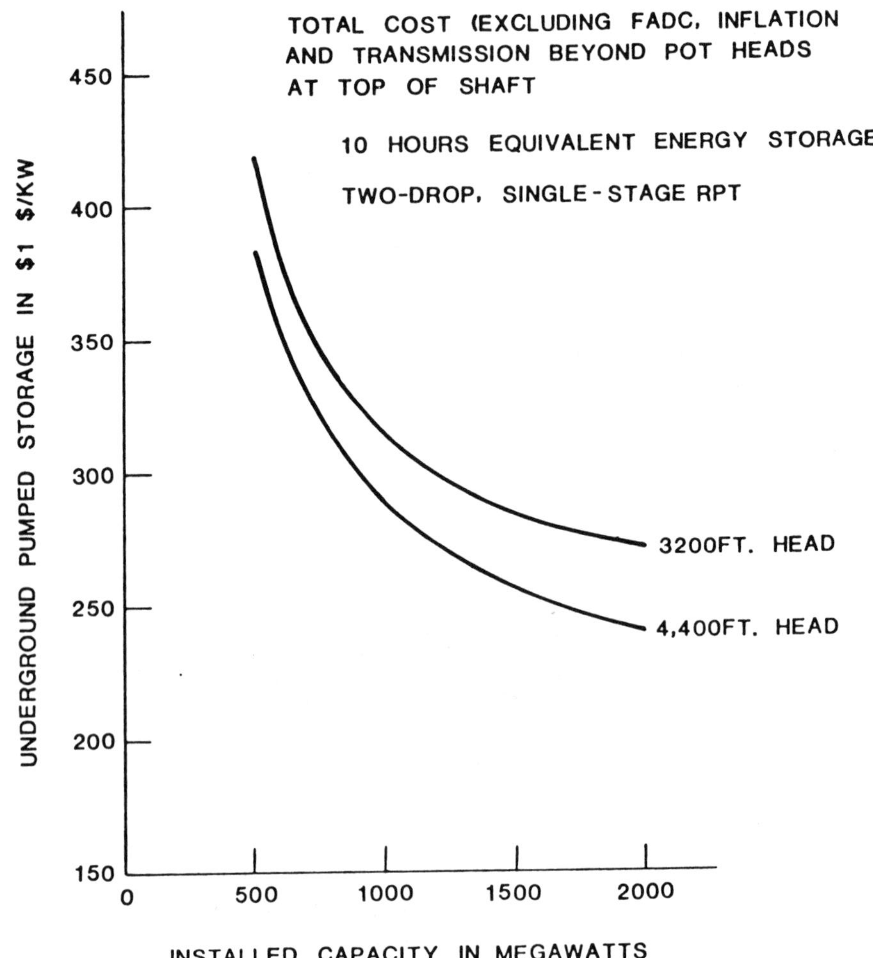

FIGURE 7. PLANT COST VS. HEAD

JANUARY, 1975

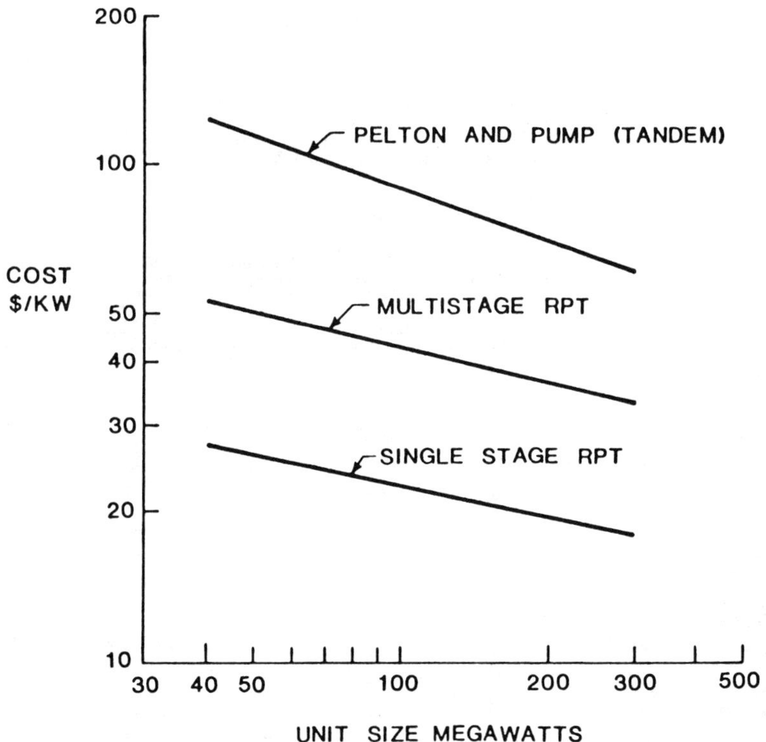

FIGURE 8. MACHINERY COST RELATIVE TO
UNIT SIZE FOR PUMP/TURBINES
(NOVEMBER, 1975

TABLE I

ESTIMATED COSTS FOR SHAFTS, POWERSTATION,
MACHINERY AND ELECTRICAL EQUIPMENT

2000 MW PUMPED STORAGE

DOLLARS PER kW - NOVEMBER, 1975

FPC ACCOUNT	ITEM	Conventional	Two-Drop Underground Single Stage	One-Drop Underground Multistage	One-Drop Underground Tandem	Two-Drop Underground Compressed Air Storage
		RPT	RPT	RPT	PT	
	Head	1200 ft	3200 ft	3200 ft	3200 ft	3400 ft
	Unit Size	6 - 350	8 - 250	16 - 125	8 - 250	4-230,6-194GT
331	Powerstation Structures and Improvements	15.92	13.00	12.30	15.28	15.62
331	Access Shaft	–	8.50	8.50	8.50	9.25
332	Water Conductor	56.30	15.96	13.60	15.40	14.50
333	Pump, Turbine Governor and Penstock Valve	17.14	17.95	28.30	41.60	10.10 P/T 28.20 GT
333	Generator, Motor	25.30	23.50	23.50	22.50	10.50 P/T 15.20 GT
334	Accessory Electrical Equipment	4.51	10.05	10.40	8.50	14.15
335	Misc. Powerplant Equipment	2.56	2.57	2.17	2.04	10.62
352/353	Substation and Equipment	4.82	4.84	4.86	4.75	4.82
	TOTALS	126.55	96.37	103.63	118.57	138.56
Operational & Maint. (dollars/kW)		1.50	2.16	2.16	2.16	1.87
Efficiency (percent)		70	70	70	70	57 (combined)
Construction Time (years)		6.3	7	7	7	6

Part II

COMPRESSED AIR ENERGY STORAGE

Editor's Comments
on Papers 6 Through 11

Gases have been compressed and stored for a variety of applications since the beginning of the nineteenth century. A major use of this technology has been for the storage and transportation of the gases themselves. Other uses include driving pneumatic systems, operation of some refrigerators, in particular those at cryogenic temperatures, and, more recently, underground compressed-air energy storage for the electric utilities. In the first of these applications, the compression is simply to make the storage more convenient, and the fact that the energy of compression is also stored is an unfortunate side effect. It is simply easier to store and transport many gases at pressures of hundreds of atmospheres in a small steel tank than at atmospheric pressure in a light but bulkier container. Consider, for example, the oxyacetylene torch. Compared to the chemical energy released in the flame, the compressional energy in the gases is insignificant and is completely thrown away in the expansion that occurs before combustion. The storage of gases as a matter of convenience is not a true form of energy storage and, with the exception of the underground storage of natural gas, which preceded and is relevant to the underground storage of compressed air, will not be considered here.

The first widespread use of energy storage in compressed gases was in steam engines. Steam has high compressional energy at temperatures well above 100°C (today up to 1,000°C) and at the resulting high pressures. Steam, however, is only a means of converting energy from a thermal source to the desired end product, which is usually either mechanical or electrical energy. Thus, unless there is a temporal separation between the generation of the steam and its use, little energy need be stored in the steam system. In fact, storing energy in steam for a long time is disadvantageous because the thermal conductivity to the surroundings cools the steam, thereby reducing the pressure and the stored energy. Of course, even a small system such as a steam locomotive may store enough energy to be a hazard if the boiler is not designed well. A boiler code developed by the American Society of Mechanical Engineers sets standards that apply to all new boilers constructed and to many other applications of pressurized vessels.

Compressed air was used in the late 1800s for operating tools and occasionally for storing energy. Some streetcar motors were driven by compressed air carried on board, frequently in conjunction with a source of heat to overcome the refrigeration effect that occurs when many compressed gases are allowed to expand. These compressed-air systems were in competition with steam until the beginning of the twentieth century but were slowly phased out because they were considered undesirable due to inefficiencies in the compression and recovery processes.

Mining provided for the earliest use of compressed gases for large-scale energy storage. In this application, which no longer exists, energy was delivered to a compressor at a continuous rate characteristic of the energy source, typically a remote electric generator, a steam engine, or a windmill. The compressed air was stored in tanks or in an underground cavern and then used for pneumatic mining equipment and hoists. The first machinery used in underground coal mines was driven by compressed air because this kind of machinery had no sparks or flame that could ignite methane and cause underground explosions.

Many industrial and commercial establishments have small compressed-air systems that drive suspension systems and pneumatic tools, including grinders, hoists, and jacks. These systems are commercially available from several manufacturers and are used not for their energy storage capability but rather for their cost-effectiveness and energy conversion capability, which results from the use of one intermittently run electric motor to power many mechanical tools. The construction industry relies heavily on pneumatic tools because a petroleum-fueled portable air compressor is a convenient source of power when it is difficult to connect to electric power lines.

The largest compressed-gas systems for energy storage in the future will be those associated with peak shaving for electric utility applications. Many utility companies use gas turbines, which are essentially jet engines, to produce electric power for a few hours each day. These engines consist of a compressor, a turbine, and an electric power generator. The air enters the compressor, where its pressure and temperature are increased, before it is mixed with natural gas or a low-viscosity petroleum distillate such as kerosene before being burned in a combustion chamber. As the hot, high-speed gases leave the combustion chamber, they cause the turbine to rotate and drive the compressors and the generator. About 29 percent of the combustion energy goes into electricity produced by the generator and 65 percent first goes into driving the compressors and then becomes waste heat.

By adding a motor to the system and separating the compressor and turbine, for example by a clutch, it is possible to compress air with off-peak electrical energy from coal, nuclear, or hydroelectric plants; store the high-pressure air underground; and then operate the gas turbine during the midday peak-load period with precompressed gas. In this operating mode, power output is increased but a larger generator is required. A major benefit of compressed-air energy storage (CAES) is that coal and nuclear fuels consumed in relatively efficient central-power stations are substituted for the oil or gas that is usually burned in the gas turbines.

Underground caverns, aquifers, and depleted gas fields are widely used to store natural gas. As a result, aquifers, mining and excavating techniques, and fluid flow in reservoirs have been studied extensively over the past four decades by the petroleum and natural gas industry. This technology base led to an immediate appreciation of the problems and economics of storing large quantities of compressed air underground to aid in peak-power generation.

The efficiency of CAES is only 50 to 60 percent. This relatively low value is due to the refrigeration effect in the gases and the pressure drop in the reservoir. As a result, the energy required for compression is nearly twice as great as that available in the compressed air when, after storage, it enters the turbine and is mixed with the fuel. Several underground compressed-air storage systems have been proposed for specific electric utility applications, but most either have not appeared economically competitive with other systems or their effectiveness was not proven. As a result the governments of the United States, Great Britain, and West Germany are active partners in the research into CAES. One large CAES plant has been constructed at Huntorf, West Germany, and is now operational.

Paper 6 describes in a flowery and vindictive manner the early use of compressed air and the difficulty its proponents had in introducing it into systems where steam was already well entrenched. The first large-scale use of compressed air for energy storage, which was for pneumatic mining and hoisting equipment, is the subject of Paper 7. The use of salt dome caverns for natural gas storage, described in Paper 8, was relevant to the decision to go ahead with the first compressed-air storage project for electric utilities, which is described in Paper 9. This article examines the design and initial stages of construction of the first CAES system. The air is stored in a solution mined salt dome. A CAES system designed around another storage medium, an aquifer, is the focus of Paper 10.

Paper 11 looks at the need for storage, compares CAES to pumped-hydro storage, and provides a general overview of this technology. It includes discussions of the heat lost in compressing air, the efficiency of the process, and system cost.

BIBLIOGRAPHY

Allen, K. 1972. Eminence Dome—Natural-Gas Storage in Salt Domes Comes of Age. *J. Pet. Tech.*, Nov. 1972, pp. 1299–1301.

Anon. 1972. *Power Generation Alernatives.* Engineering Division, Seattle City Light.

Anon. 1957. *Gas Storage at the Point of Use.* Stone and Webster Engineering Corp.

Bersticker, A. C., ed. 1963. *Symposium on Salt.* Northern Ohio Geological Society, Inc., Cleveland, Ohio.

Bush, J. B. 1976. Economic and Technical Feasibility Study of Compressed Air Storage. *Energy Res. Dev. Admn. Rept. ERDA 76–77.*

Coogen, A. H., ed. 1974. *Fourth Symposium on Salt.* Northern Ohio Geological Society, Inc., Cleveland, Ohio.

Gibbs, C. W., ed. 1968. *Compressed Air and Gas Data.* Ingersoll-Rand Co., New York.

Katz, D. L., and K. H. Coats. 1968. *Underground Storage of Fluids.* Ulrich's Books, Ann Arbor, Mich.

Katz, D. L., D. Cornell, R. Kolayaski, J. R. Elenbaas, F. H. Poettman, J. A. Vary, and C. E. Weinaug. 1959. *Handbook of Natural Gas Engineering.* McGraw-Hill, New York.

Katz, D. L., and E. R. Lady. 1976. *Compressed Air Storage.* Ulrich's Books, Ann Arbor, Mich.

Olsson, E. K. A. 1970. Air Storage Power Plant. *Mech. Eng.,* Nov. 1970, pp. 20–24.

Schul, L. F., and A. K. Hegeman. 1961. *Gas and Air Compression Machinery.* McGraw-Hill, New York.

Tirotsov, E. N. 1978. *Natural Gas,* 3d ed. Gulf Publishing Co., Houston.

Workshop on Compressed Air Energy Storage System, Proc. 1975. Energy Res. Dev. Admn. Rept. ERDA 76-124.

6

Reprinted from pages 685–690 and 696–699 of *Am. Soc. Mech. Eng. Trans.*
15:685–704 (1894)

A NOTE ON COMPRESSED AIR.

BY FRANK RICHARDS, NEW YORK CITY.
(Member of the Society.)

In offering the following paper for the consideration of The American Society of Mechanical Engineers, which embraces in its membership the ablest of investigators of mechanical efficiencies, it is not intended to challenge minute and exhaustive criticism, but rather to present the subject in such a way that it may attract as widely as possible the attention of the general mechanical public. I may be able to make it appear that the compressed air practice of the day is not sufficiently advanced to invite or to warrant the minute and careful analyses which belong, for instance, to the marine engine, to the pumping engine, or to the locomotive.

I think that compressed air has not received the attention which it deserves, and that where it has secured some attention it has not always had fair play. There are many men in these days, and many intelligent engineers among them, who will not even consider the claims of compressed air as a means of power transmission, because their minds have been so filled with the idea that its use entails enormous losses. That consideration settles it for them, and that is the libel from which compressed air suffers, so that it does not get a fair chance even to show what it can do. It is only another case of giving a dog a bad name ; and in this case it is a very good dog with a very bad name. It is the worst kind of a case to set right. It is but common justice to tell the straight truth in the matter and to spread it as widely as possible.

The position of compressed air before the mechanical public, and especially the American mechanical public, has been a peculiar one all the way through. It has had no disinterested,

---

* Presented at the Montreal Meeting (June, 1894) of the American Society of Mechanical Engineers, and forming part of Volume XV. of the *Transactions.*

85

all-around friends to look after its interests, nor interested ones either. There have been no men, and still less any company of men, who have made the application of compressed air their business and have looked after it. Where is the general compressed air company performing for compressed air such functions as more than one company is performing for electricity? and why is there not such a company? Of the builders of air compressors not one of them has been, not one of them is to this day, responsible for the economical application of compressed air *after the compression*, or apparently cares anything about it. Where compressed air has been used in this country, and where any thought has been given to economy in its use, the air compressor, it would seem, has been almost exclusively studied and talked about. In the progress of the steam-engine it has sometimes seemed that the steam-boiler has hardly received its proper share of attention. In the existing writings upon steam-engine economy it is probably safe to say that the engine engrosses ten times as much of the matter as the boiler does. In the case of compressed air the boiler, or compressor, gets ten times as much study and discussion as the engine or motor which uses the air. There are such vagaries of injustice in civilized communities.

And this impression which has got abroad, of the waste of power in the use of compressed air, has, curiously enough, been promoted and disseminated by the air compressor people more than by anyone else. We may say this with perfect safety, for it strikes so generally that it hits no one in particular. The compressed air literature accessible to the general public consists principally of air compressor catalogues. The argument of the average catalogue runs like this : "If you are going to use compressed air for any purpose, look out for the enormous losses of power to be encountered and which you are sure to experience *if you don't buy our compressor*." And the result has been that many have " caught on " to the terrible tale of the waste of power, and have helped to spread it far and wide. The argument is of course not maliciously meant, but it has done more work and somewhat different work from what was intended.

Fig. 201 tells the whole story of compressed air. The piston F is fitted to the cylinder E so that we may assume it to move freely and without leakage. The piston being at A, as shown, and the cylinder being filled with air at a pressure of one atmosphere, and at normal temperature, a sufficient weight is placed upon the piston to force it down into the cylinder and compress the air contained in it to a pressure of say six atmospheres. The volume being inversely as the pressure, the piston should go down to C. We find that it actually only goes down to B, and the reason is that while the air is being compressed the operation of compression also heats it, and the heat distends or expands the air, and its volume is consequently considerably greater than it should be upon the assumption that the volume is always inversely as the pressure. Supposing both the piston and the cylinder to be absolute non-conductors of heat, and that the air heated by the compression loses none of its heat of compression, then if the weight which forced the piston down be taken away the piston will be driven back to its original position at A, and the air contained in the cylinder will have resumed its normal pressure and temperature, and will have done as much work, or will have exerted as much force, by its return as was employed in the act of compression. If while the piston was at B, and with the weight upon it sufficient to balance the pressure of six atmospheres, the air by any means had been cooled to its original temperature, the piston would have fallen to C, and the law that the volume varies inversely as the pressure would have held good, as then the initial and the final temperatures would have been the same. The air being thus cooled to its original temperature, and the piston being at C, upon removing the weight from the piston it would return only to D instead of to A. When the piston arrived at D the pressure of the air in the cylinder would have fallen to the original pressure of one atmosphere, and the piston at D would be balanced between the equal pressures above and below it. As the air is heated during the operation of compression, so is it equally cooled during its expansion, and when the piston reaches D the air in the cylinder is then at atmospheric pressure, because it is then much colder than it was at the beginning; and it is

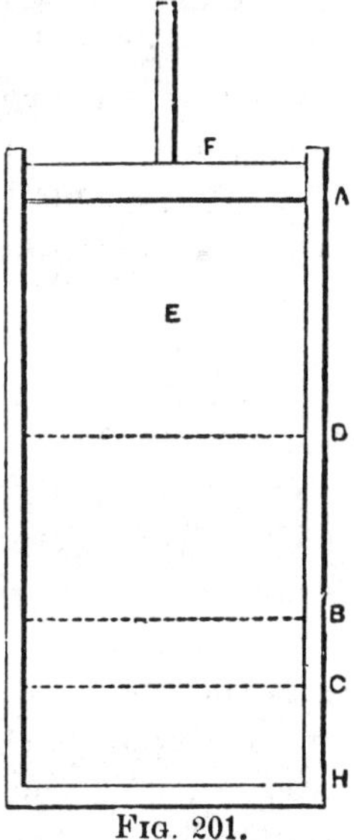

Fig. 201.

solely because of this loss of heat that the pressure falls so early, and that the piston does not return to A where it started from. If while the piston is at D the air can by any means recover all the heat which it has lost, the piston will return to A as before. The distance DA compared with CA represents the total possible theoretical loss of power in the compression and the re-expansion of air.

We may now consider the practical aspect of the case. And, beginning at the very beginning, with the practical compression of the air, we cannot but notice how readily the air lends itself to economy in its use. By the way in which it is possible to connect the air-compressing cylinder with the steam-engine in the best air compressors of the present day, no power is lost in operating the piston of the compressing cylinder; or, stated more accurately, the entire friction of the air-compressing apparatus is so compensated for by the saving of friction in the engine that it is not only reduced to zero but it becomes a negative quantity of considerable magnitude. This may be proved by incontrovertible evidence. The air compressor, it will be noted, comprises a complete steam engine, with all its component operating parts, steam piston, crosshead, connecting-rod, crank-pin, crank-shaft, complete valve motion, etc., and in addition to these there is the air piston to operate, and it is natural enough to suppose that the operating of the air piston puts just so much additional friction upon the engine, with a call for so much more power to overcome it. But the arrangement of the air piston in a straight line and upon the same rod with the steam piston saves enough from the friction of the engine proper to more than compensate for the friction of the air piston, and the operating of the air piston costs practically "less than nothing." In a regular stationary engine which transmits its whole power through the crank shaft, and by its driving pulley, gears, or otherwise, wherever it may be wanted, the whole power exerted is felt in friction upon the crosshead slides, connecting-rod boxes, crank-pin, main bearings, etc. In the air compressor most of the power of the steam-engine is transmitted directly from the steam piston to the air piston without the intervention of the engine friction. Not only is the effective steam pressure so immediately transmitted, but the force due to the inertia of the reciprocating parts, especially the heavy crosshead which is usually employed, is sent directly to the air piston through its piston-rod.

The losses due to the friction of the ordinary stationary steam-engine usually amount to 10 or 12% of the indicated horse power. In an Ingersoll-Sergeant air compressor exhibited at Chicago, with engines of the Corliss type, the mean friction, as determined by Professor Jacobus, was 5%, so that, crediting the saving of friction to the air cylinders, the friction cost for operating them was −5 to −7%.

The diagram, Fig. 202, scale 40, is intended to show the practical possibilities in the use of compressed air at 75 lbs. gauge, or six atmospheres. The line $a\,b$ is the adiabatic compression line, or the line of compression, assuming that no heat is taken away from or lost by the air during the compression. The initial temperature of the air being 60°, the final temperature would be

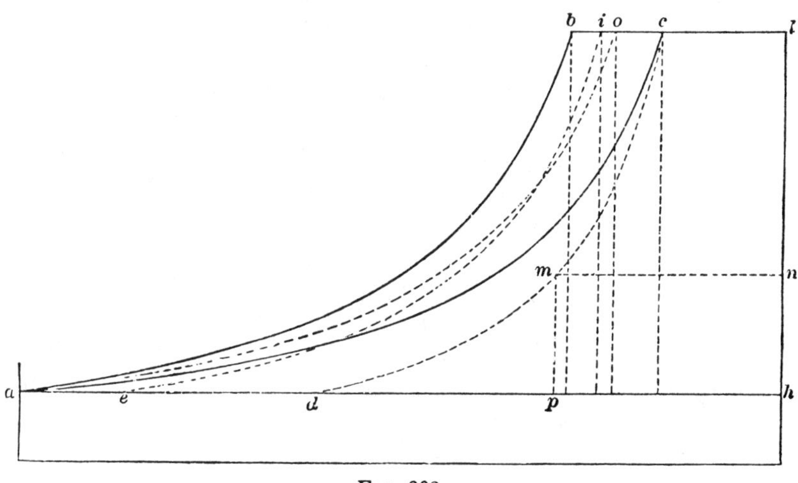

FIG. 202.

about 415°, and the final volume is .28 of the initial volume. The line $a\,c$ is the isothermal compression line, which assumes that all the heat of compression is got rid of just when it is produced, or that the air throughout the compression remains constantly at its initial temperature. The final volume in this case is .1666 of the initial volume. Remembering that these lines, $a\,b$ and $a\,c$, represent the compression of the same initial volume of air, it is evident that there is quite a difference in the power employed in the two cases, and herein lies the loss, or the possibility of loss, of power in the operation of compression. The mean effective pressure, or resistance, of the air for the stroke on the adiabatic line $a\,b\,l$ is 35.36, while the mean effective pressure for the isothermal line $a\,c\,l$ is but 27, or only 76% of the former. The comparison should, however, be made the other way. The adiabatic mean effective pressure is 130% of the isothermal mean effective pressure, and the 30% is, of course, the additional, or, as we might say, unnecessary, power employed. Neither of these lines, $a\,b$ or $a\,c$, is possible in practice. Air cannot be compressed without losing some of its heat during compression, so that the actual compression line must always fall within the line $a\,b$. On the other hand, it is equally impossible to abstract all the heat from the air coincidently with the production of that heat, so that the actual compression line must always fall outside the line $a\,c$. The best air compressor practice of to-day is very near the line $a\,o$, or the mean of the adiabatic and the isothermal lines. The actual line is generally outside of this, and seldom inside of it. It would be impossible to produce a line exactly coincident with this in practice. If we produced a line giving the same mean effective pressure as $a\,o$, it would probably run above $a\,o$ for the first half of the stroke and perhaps a little below it at the last. If the compression were "compound," or done in two cylinders, there would, of course, be a break in the continuity of the curve. The mean effective pressure for the line $a\,o\,l$ is about 31.5, or still nearly 17% in excess of the mean effective pressure for the line $a\,c\,l$. As this is for what must be considered exceptionally good practice, the loss of power for the average practice in air compression may be put at 20%. Lest some impatient ones should drop the subject here, or lest some rival of compressed air should pick up this and run away with it, we might insert a reminder here that all this loss is not necessarily final. It is proper to compare the actual compression line $a\,o\,l$ with the ideal isothermal line $a\,c\,l$, because the volume $c\,l$ is the volume available when the air is put to use. Though at the completion of the compression stroke there is always some of the heat of compression remaining in the air, and its volume is greater than $c\,l$, that heat is always lost in the transmission or the storage of the air, and the available volume is never practically above $c\,l$.

[*Editor's Note*: Material has been omitted at this point.]

DISCUSSION.

Prof. D. S. Jacobus.—The author of the paper just read calls attention to the fact that in the use of compressed air for motive power purposes the reheater adds greatly to the work developed.

A short time ago the question arose as to the effect of passing the air through a mass of hot water to perform such heating, as is done in the Mekarski and Hardie compressed air street-car motors. When this is done a portion of the water is evaporated, and the steam thus mingled with the compressed air does work in the motor cylinder. It was found that for the average conditions, shown by tests of the Hardie motor on the Second Avenue Railroad in New York City, where air was admitted to the motor at 120 lbs. pressure above the atmosphere, that the water evaporated into steam did over one-third of the work. In other words, if dry air were brought to the temperature of the hot water the work done would be less than two-thirds that for the moist air. In this case a trifle over one quarter of a pound of water was evaporated per pound of air.

If the maximum rate of evaporation of water shown in the tests of the Hardie motor was maintained, or about one-half of a pound per pound of air, the work done by the moist air would be twice that for dry air admitted to the motors at the same temperature. The effect of the presence of moisture is to increase the volume of the working fluid entering the cylinder, and to maintain a higher temperature during expansion than that which would occur with dry air. Both of these effects tend to increase the work developed per pound of air. Calculations have been made to show the temperatures of the exhaust for different temperatures of the hot-water bath, and for various amounts of water evaporated. The relative amounts of water and air storage on street-car motors for the best commercial results have also been estimated, together with the distance that motors will travel under given sets of conditions.

The entire subject will probably form a paper to be presented at a later date.

Mr. C. O. Heggem.—Compressed air as a power transmission has been in use at the works of Russell & Co., Massillon, Ohio, for the last six or seven years, beginning with a 600 lb. hoist and increasing from day to day until at present there are in operation 26 five-ton cranes, one cupola stock elevator, and a large number of small hoists, varying from 400 to 1,000 lbs. capacity. Shears and punches in the boiler shop, where, out of the way of the line shafting, are being operated by compressed air. The accompany-

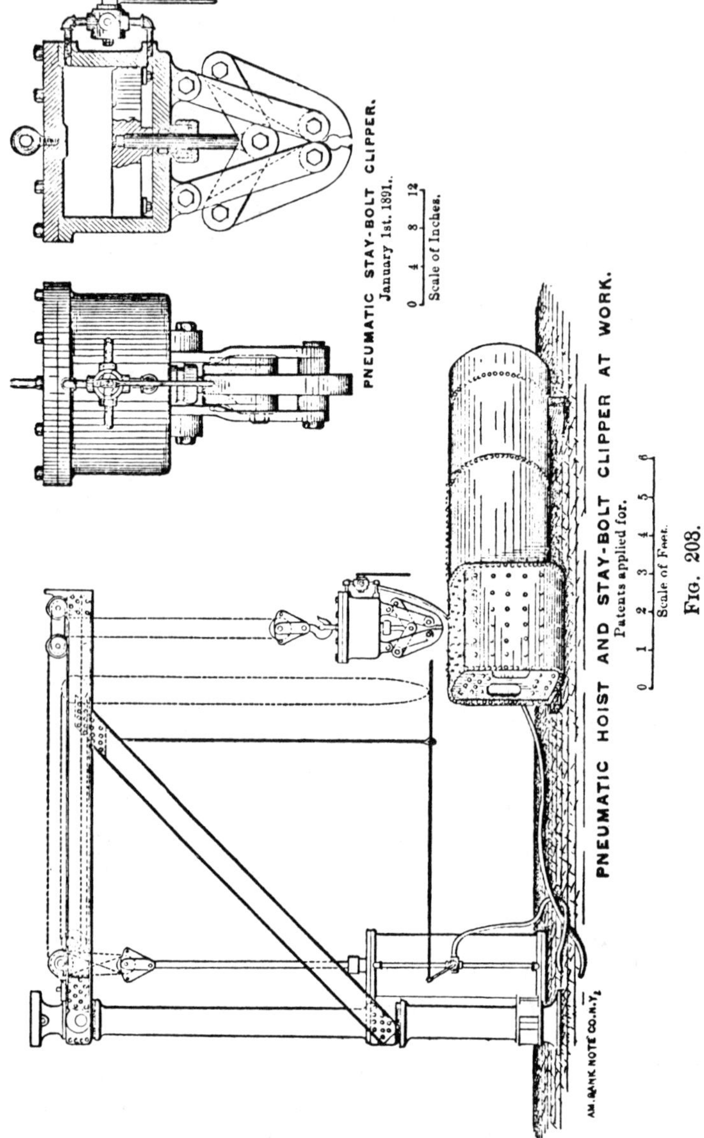

ing print (Fig. 203) shows the style of crane and one form of shear as applied, cutting off the stub ends of stay bolts in locomotive

boilers. Here, by the aid of compressed air, two boys may cut off 700 to 800 stay bolts a day, which is far in excess of what could be accomplished with hammer and chisel as formerly used.

It is in the foundry that compressed air is more especially of service, as its elastic action makes it very convenient in drawing patterns, closing flasks, etc.

The air is supplied by three compressors:

One 10 × 13 belt-driven Clayton.

One $7\frac{1}{2}$ × 7 duplex steam-driven Clayton.

And a smaller steam-driven Knowles.

The belted compressor is run constantly during the working hours, and supplies the most of the air required.

The steam-driven Clayton is set to start whenever the air pressure falls below 60 lbs., and, consequently, acts simply as an assistant to the other compressor.

The Knowles is kept as a reserve in case of breakdown.

The air is stored in three reservoirs of 250 cubic feet capacity each, and these reservoirs are placed in such localities of the works where the greatest amount of air pressure is required. Owing to the intermittent use of the air, these comparatively small compressors are able to furnish a sufficient amount and at a cost of something like $3 per day of ten hours.

The piping aggregates about 2,500 ft., and is $2\frac{1}{2}$ inch in the main line, with $1\frac{1}{2}$ and $1\frac{1}{4}$ inch branches.

Two of the reservoirs are about 1,000 ft. apart, but the pressure in each, as shown by the gauge, is the same.

In laying pipe for compressed air in the ground, care should be taken to go pretty deep in order to insure immunity from frost, as pipes have been known to freeze in this climate two feet below the surface.

Prof. D. S. Jacobus.—In Mr. Richards' interesting discussion of the possibilities of compressed air as a means of transmitting power to motors, there is one statement to the effect that there is no company in the United States which makes a business of applying compressed air, to which I wish to take exception. Those of us who are engaged in sewerage work are acquainted with just such a concern, the Hydro-Pneumatic Sewerage and Water Supply Company; and visitors to the Columbian Exposition last year had an opportunity of seeing twenty-six of its sewage ejectors or pumps working by compressed air, which handled daily a quantity of sewage equivalent to that of a city of from 100,000 to 400,000 population, according to the attendance. As few mechanical engineers follow sewerage work closely, it may not be out of place to state that these ejectors are automatic devices used in flat places for pumping sewage or water. In such localities it is often difficult to construct gravity sewers of any considerable length, owing to the fact that the trenches soon reach such an excessive depth that the work is difficult and it is hard to discharge the sewage conveniently. In such cases ejectors are employed. The sewage is collected in a number of convenient places from which it is forced in a series of pulsations into main sewers at a higher level. The compressed air for operating the ejectors is supplied through small iron pipes from a single compressor station, often located in the pumping station of the local water-works.

Such appliances are necessarily small ; at the Columbian Exposition the capacity of the ejectors was from 60 to 600 gallons a minute. High efficiences are not to be expected, therefore, and those obtained appear very creditable when it is recalled that the liquid pumped is generally unscreened sewage. At Lowestoft, England, some tests made by Prof. W. C. Unwin on a small plant showed an efficiency of nearly 49 per cent. from the steam end of the compressors to the discharge pipe of the ejectors. At Rangoon, under very unfavorable circumstances, a plant having twenty-five ejector stations and seven miles of air pipes showed an efficiency of 33 per cent., although the quantity of sewage handled was so light as to make the test of little value except as an indication of what such apparatus is good for under exceptionally poor conditions. Nevertheless, these figures compare pretty well with those of the pumps of 15 per cent. efficiency mentioned by Mr. Richards, and show that a careful study of all the conditions has enabled compressed air to be used with fair economy, as well as entire sanitary success, in pumping a liquid which is particularly difficult to handle.

Mr. Frank Cawley.—In view of the close study which the author of this paper has given to the subject, as well as the opportunities which he has had at command for observation, I had hoped that he would go more fully into the economic use of compressed air. That air may be used with economy for certain purposes, as compared with other power transmission mediums, I am positive, yet I have never had an opportunity of experimenting in this line.

[*Editor's Note*: Material has been omitted at this point.]

7

Reprinted from pages 826–827, 830–831, 833, 836, 840–854, and 887–888 of *Am. Inst. Min. Eng. Trans.* **46**:826–888 (1913)

The Compressed Air System of the Anaconda Copper Mining Co., Butte, Mont.

BY BRUNO V. NORDBERG, MILWAUKEE, WIS.

(Butte Meeting, August, 1913.)

THE high cost of coal in Butte and the development of large amounts of cheap electric power from the Missouri river caused the Anaconda Copper Mining Co. in 1908 to make an investigation as to the possibility of adapt-

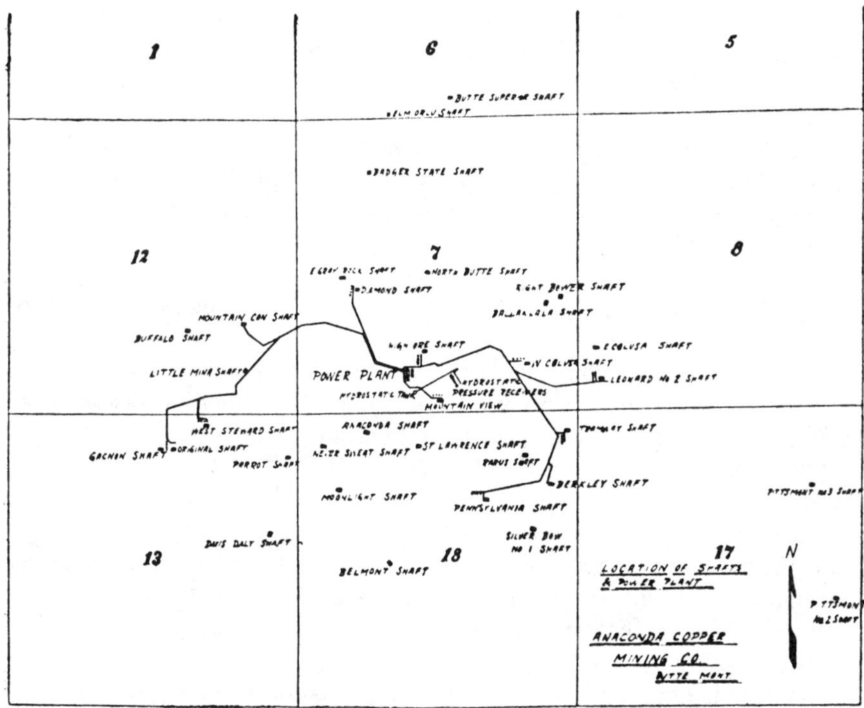

FIG. 1.—MAP OF BUTTE MINING DISTRICT.

ing this cheap electric power for the operation of their many large hoisting plants. At this time electric power had already been adopted for practically all other power purposes around the mines, such as pumping, compressing air, etc.

The map (Fig. 1) shows the location of the mines of the Butte mining district. These mines are located within a circle of approximately a 1-mile radius. The shafts are all vertical, the maximum depth being 2,000 to 2,800 ft. at present, but sinking to greater depth is in progress in several shafts which are expected, in the future, to reach an ultimate depth of at least 3,500 ft. The hoisting is done from a great many levels between 1,000 ft. and bottom of the mine.

Every mine has its own dry house, which requires a boiler plant.

For the purpose of making this investigation it was determined to make extensive tests of four representative steam hoists of the district, namely, at the Speculator shaft of the North Butte Mining Co., and the Diamond, High Ore and Mountain View shafts of the Anaconda company.

At the Speculator shaft the hoist is a Nordberg, having steam cylinders 32 and 32 by 72 in. with Corliss valve gear. The drums are smooth, 12 ft. in diameter, carrying 1.5-in. round rope in two layers, operated by means of clutches.

At the Diamond shaft the hoist was originally built by the Risdon Iron Works but had been twice rebuilt by the Anaconda Copper Mining Co. It had steam cylinders 30 and 30 by 72 in. with Corliss valve gear and link motion reverse. It was equipped with reels carrying flat rope 0.5 by 7 in.

At the High Ore shaft, the hoist was built by the Union Iron Works. The cylinders were 30 and 30 by 72 in. with poppet valves operated by wrist plate motion with link reverse. This hoist was also equipped with reels carrying flat rope 0.5 by 7 in.

The Mountain View hoist was a Webster, Camp & Lane with cylinders 28 and 28 by 72 in., having Corliss valves with link motion reverse. It was equipped with reels carrying flat rope 0.5 by 7 in.

In passing it should be noted that several of the hoisting engines at Butte have to be located very close to the shafts, which necessitates reels and flat ropes to eliminate an otherwise prohibitive fleeting angle.

The tests were made by H. W. Dow and B. V. Nordberg, Jr., representatives of the Nordberg Manufacturing Co. Those directing the tests deserve great credit for the persistency with which they worked to get a sufficient number of readings to give an accurate record of the hoistings, without which it would not have been possible to determine the different factors entering into the power problem of the hoisting system.

The main object of these tests was to determine the power required and the nature of the load. In order to do this it was necessary to determine the exact time and duration of every trip, the load hoisted and the depth from which it was hoisted in or out of balance, as well as the speed conditions during each trip.

[*Editor's Note*: Material has been omitted at this point.]

The magnitude of the dead loads in a hoist is an important factor in the mechanical efficiency of the hoisting operation, as it influences the friction and the dynamic energy that has to be supplied to the hoist in getting up to speed and released when retarding the hoist. Ordinarily this dynamic energy is lost in form of heat on the brake shoes.

Occasionally it is required to lift unbalanced loads from various depths. In some mines such unbalanced loads come most frequently from the bottom of the mine. At North Butte considerable unbalanced hoisting was done from 2,200 ft. depth. Hoisting the skip and cages weighing 11,000 lb., one-half of the rope, 4,000 lb., and the load of 12,000 lb., or a total of 27,000 lb. from 2,200 ft. gives us:

$$\frac{27000 \times 2200}{33000} = 1800 \text{ shaft horsepower}$$

the work being performed in one minute.

At the Mountain View some of these unbalanced loads represented about 2,000 h. p. at the shaft. The average work per day of these hoists ranges from 25 to 110 shaft horsepower, based on the productive work.

The chart (Fig. 2), in which the effective work performed at the shaft is plotted over the dial of a clock, gives a good idea of the continuity of hoisting at these mines. The charts show the work done at the Speculator Shaft Jan. 11 and 12, 1909, and is a good average of the operation at the mine. Only the productive work, that is, the work represented by the hoisting of ore, is plotted.

From the tachograph diagrams and the log, the weight hoisted and the depth were determined, also the time during which the work was done. Thus, for instance, there were hoisted from 12:00 to 12:45 a. m. 31 loads from 1,800 ft. depth, or 155 tons, representing an average work of:

$$\frac{155 \times 1800}{45} = 6200 \text{ foot tons per minute} = 376 \text{ shaft horsepower during that period.}$$

The variation in the load during the period under consideration as determined from continuous indicator diagrams (Fig. 3), time distance curves (Fig. 4) and tachograph cards (Fig. 5) is shown in Fig. 6.

Frequently unbalanced hoisting was done when one compartment was obstructed. Figs. 7, 8 and 9 show tachograph card, power curve and continuous indicator diagrams taken March 8, 1909. Following a period of hoisting ore in balance from the 1,800 ft. level the station tenders were lowered to the 2,200 ft. level and ore hoisted from there out of balance.

A glance at the diagrams showing the power when hoisting in balance from 1,800 ft. depth (Fig. 6) shows that the horsepower fluctuates from +2,300 to — 1,600 i. h. p., the average during this period being 630 i. h. p.

[*Editor's Note*: Material has been omitted at this point.]

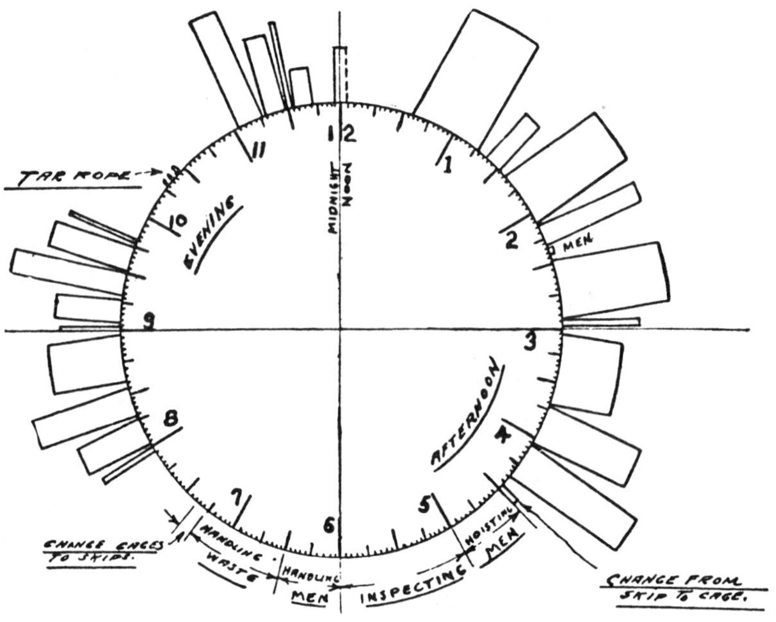

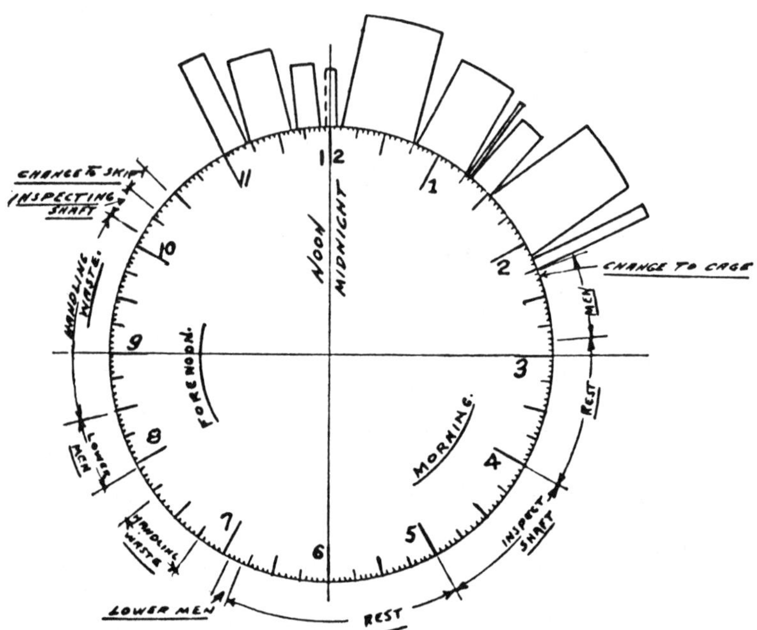

FIG. 2.—DIAGRAM SHOWING PRODUCTIVE WORK AT SPECULATOR SHAFT, NORTH BUTTE MINE.

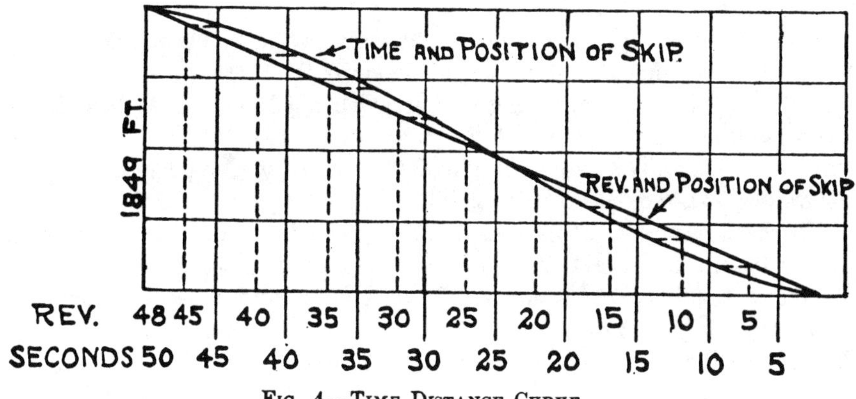

Fig. 4.—Time Distance Curve.

Fig. 5.—Section of Tachograph Card.

Fig. 6.—Balanced Hoisting from 18th Level, Speculator Shaft, North Butte
Mine, Jan. 12, 1909, 12 : 10 to 12 : 14 : 40 a. m.

Average indicated horsepower per trip = 880.
Maximum indicated horsepower =2,380.

This represents very nearly the average balanced hoisting condition at these mines.

From the diagrams of unbalanced hoisting from 2,200 ft. depth, Fig. 8 shows that during 34.5 min. the horsepower fluctuates from + 2,700 to − 900 h. p., the average during this period being 160 h. p. It has already been pointed out that most of the unproductive work is done out of balance. Occasionally a load averaging 2,000 i. h. p., with a peak 2,700 i. h. p. follows right after one averaging 900 i. h. p.

Much energy was also used in hoisting waste from lower to upper levels out of balance. In fact, it was found that although in all the large producers of the Butte mines an auxiliary hoist was installed to relieve the main hoist of non-productive work, still the non-productive and unbalanced hoisting performed with the main hoists covered fully as much, or more energy, than the productive work. That is largely due to the fact that most of the unproductive hoisting is done out of balance and that the "dead loads," i. e., ropes, skips and cages, are usually heavy. Thus, the skip and cage attached thereto at the Speculator shaft weighed about 11,000 lb. while the load in the skip was 10,000 lb.

The following table gives a good idea of the continuity of hoisting and the ratio of productive to non-productive work. The work done is expressed in effective horsepower: For 24-hr. run,

$$\text{Effective horsepower} = \frac{\text{Total W't hoisted (lb.) x distance in ft. hoisted}}{\text{Total actual time (min.) hoist is in motion x 33,000}}$$

Effective Horsepower.

	Productive or Hoisting Ore	Non-Productive			Ratio Prod. Work to Non-Prod. Work	Percentage of time during which hoist is doing positive work
		Men	Material	Total Non-Productive		
Speculator.....	565	433	758	1,191	1:2.12	24.0
Diamond......	706	449	630	1,079	1:1.52	19.6
High Ore......	456	713	813	1,526	1:3.34	12.5
Mountain View.	508	493	631	1,124	1:2.22	21.6

[Editor's Note: Material has been omitted at this point.]

Accurate records are available showing the work done and energy used by the steam hoists at the largest producing mines at Butte. It would take up too much space to exhibit these records in their complete form. The examples given above will, however, give a good idea of the magnitude and nature of the load to be handled by the Butte hoists.

When, in 1909, it was decided to operate these hoists by power, transmitted electrically to Butte, the question was discussed as to how much these existing hoisting conditions could be modified and improved. Any system of dispatching whereby the starting of two or more hoists simultaneously could be prevented was objectionable from a mining point of view. The periods of heavy hoisting, with corresponding periods of light or no hoisting, as illustrated by Fig. 2, could not be materially changed without seriously interfering with the underground operations.

It was then planned to regulate the hoisting so that these heavy periods would not coincide at all the mines as they did at that time. Up to the present time, however, this condition has not been changed materially.

Furthermore, the mining company desired the hoists not only to operate under the conditions already outlined for the steam hoists, but to possess considerable flexibility to allow for greater hoisting depths in the future and also for greater rates of production. After consideration of all these facts the Anaconda Copper Mining Company decided not to attempt to employ any method of operating the hoist by the application of electricity directly to them but to adopt the compressed air system of hoisting designed by the writer.

The power system for operating the hoists of the Anaconda Copper Mining Company mines comprises:

(a) An electrically operated air compressing plant in which air is compressed to 90-lb. pressure.

(b) The electric current is generated by water power located at a distance of 150 miles from Butte. In order that a maximum amount of energy can be transmitted through such long transmission line it is important that the work is perfectly equalized and all peaks in the load eliminated. This requires an air storage of great capacity and one in which the stored energy can be held without loss for extended periods.

(c) Air reheating plants at the different shafts.

(d) The hoisting engines were provided with new cylinders designed for economical use of the compressed air. The old and wasteful auxiliary hoists were replaced with new ones, specially designed for compressed air.

The air storage plant was connected with the rock drill system so that during the idle periods when little or no air was used by the hoists, the air could be turned into the rock drill system.

At present upward of 25 per cent. of the total capacity of the compressor station is sent into the mine. This not only helps out the mine system but provides a high load factor for the compressor station.

Description of the Plant.

*The 100,000-volt substation at Butte is located near the center of the district in which the power is distributed. The substation building is 150 ft. by 50 ft. in plan and 50 ft. high. It is a brick-walled, steel-framed structure with concrete floors and roof.

There are installed at present two banks of single-phase transformers, rated at 3,600 kw. per bank and two banks rated at 7,200 kw. per bank, connected in delta on both high and low tension sides. They step the voltage down from 102,000 to 2,500, at which voltage it is distributed to customers. The transformers are installed in fireproof compartments, entirely shut off from the rest of the building by brick walls and opening only out of doors. The transformers are mounted on wheels and can readily be run out on to a flatcar which stands on a track running parallel with the building in front of the row of transformer compartments. This arrangement furnished a convenient method of handling the transformers, both at the time of installation and afterwards, in case it is necessary to make repairs.

On the gallery above the transformer compartments are located the electrolytic lightning arresters. On the gallery opposite are the 100,000-volt line switches. Possibly the most unique feature of the electrical layout is the 100,000-volt bus construction. For flexibility in switching, duplicate busses are provided. The busses themselves are made of 1.5-in. iron pipe suspended by standard line insulators from the roof trusses of the building. The three conductors of each three-phase bus are suspended one above another, each being supported by the next one above. The connections to the lines are also of iron pipe, making the bus structure as a whole quite rigid and well adapted to the use of suspension insulators.

The switchboard is in two sections, one section operating all line and transformer switches, which are remote controlled; the other section taking care of the 2,500-volt feeders, which are controlled by hand-operated automatic switches.

All electrical apparatus in the substation was supplied by the General Electric Co.

The load supplied in Butte is confined entirely to the mines, the power being used chiefly for the operation of motor-driven air compressors and electrically driven pumps.

*Substation—reprinted from an article by Max Hebgen in the "General Electric Review."

The load is very nearly uniform for twenty-four hours each day throughout the year. The load factor is, in fact, close to 90 per cent.

Description of Compressor Station.

In the compressor house which is located close to the substation there are installed six Nordberg two-stage compressors, three of which are fitted with variable capacity valve gear for automatically varying the capacity, while the other three run at a fixed capacity. The compressors are directly connected to Westinghouse synchronous motors and each compressor runs at a speed of 76 rev. per min. Fig. 12 shows an interior view of the compressor house. The cylinder dimensions of all compressors are 30 and 50 by 48 in. The pistons of these compressors are supported outside the cylinders, on crossheads running in oiled guides, the piston rods being very large in diameter so as to insure minimum deflection. Only the packing rings touch the cylinder walls. The variable capacity compressors have Corliss inlet valves and automatic discharge valves, the capacity being regulated by the closure of the inlet valves at different points of the stroke by means of a releasing mechanism under control of a pressure regulator. The constant capacity compressors are fitted with positively operated Corliss valves for inlet and discharge. These compressors show a high efficiency. A set of indicator cards from these compressors is shown in Fig. 13. The capacity of the compressors is at 76 rev. per min.—7,650 cu. ft. of free air per minute. The atmospheric pressure at the power plant is 12 lb. absolute and the normal air pressure 90 lb. gauge, or the same as used on the rock drill system, but when cards (Fig. 13) were taken the discharge pressure was 98.2. The indicated work is 1,099 h. p., while the theoretical work of perfect two-stage compression with perfect intercooling and no pressure losses is 1,045 h. p. referred to the actual air compressed.

As the starting of these compressors, if done by the electric motor in the ordinary way, would throw peaks of considerable magnitude upon the electric system, a new method of starting was adopted whereby such peaks are entirely avoided. The valve motion of the compressors was fitted with a reversing gear whereby the valves, for the purpose of starting, are so adjusted as to turn the machine into an air motor. In the Butte power system there is always a large volume of air stored and maintained under 90 lb. pressure. The compressors are thus started and brought up to speed by the energy in the stored air. When up to speed and in synchronism with the generators 150 miles away and with other compressors running, the electric current is switched on and valve motion reversed. The machine immediately starts to compress air. Thus, when starting no electricity is used and when the air compression begins with the com-

Fig. 12.—Interior of Compressor Plant.

pressor up to speed and in synchronism, the power is only the normal power necessary for the compression. This starting gear works very satisfactorily. At the end of April and beginning of May, 1912, Butte was visited by several severe electric storms which caused interruption in

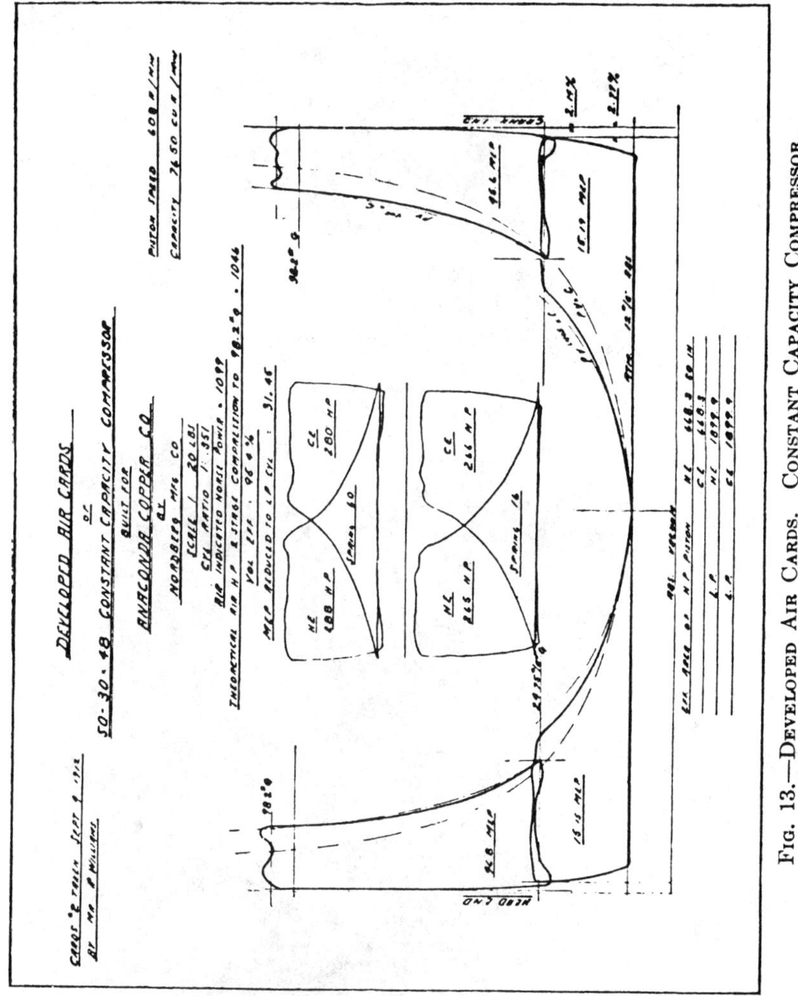

FIG. 13.—DEVELOPED AIR CARDS. CONSTANT CAPACITY COMPRESSOR.

the operations of the compressors several times. At that time there were four compressors installed, three running and one in reserve. On May 2, 1912, during a particularly severe storm, all three compressors were stopped by lightning. The shutdown was not of long enough duration to interrupt the hoisting, and the time required to start the three compressors was less than four minutes from the moment when the first

FIG. 14.—VIEW OF VARIABLE CAPACITY COMPRESSOR.

Fig. 15.—View of Constant Capacity Compressor.

compressor started until all three were in step, reversed and compressing air into the system.

The compressors are connected to a number of large air receivers and these to the distributing main in a manner as shown in Fig. 17, from which can be seen that any of the receivers can be shut off from the system for inspection or repairs, there being a valve on each side of the receiver for that purpose. Figs. 14 and 15 show respectively the variable and the constant capacity compressors.

Air Storage.

In order to equalize the load on the compressor station, it was necessary to provide for two kinds of peak loads on the hoists, first—the peak due to the acceleration of any hoist. These peaks are very large but last for a few seconds only. To take care of this condition, storage receivers of sufficient capacity to liberate sufficient air for the acceleration of the hoist without excessive pressure drop were installed at each hoist. The highest rate of air consumption at the large hoists is at the rate of about 60 cu. ft. of compressed air per second for 5 sec. This condition determines the air receiver capacity necessary, each requiring four receivers of 2,100 cu. ft. volume. It will be readily understood that as the pipe lines from the compressors to the different hoists on the system are of considerable length that the capacity of the pipe lines are of little value as storage reservoirs unless accompanied by a prohibitive pressure drop. The installation of the above receivers therefore allows the pipe lines to be designed for average flow.

The amount of energy that can be stored in the receivers or expansion tanks is, however, limited. Calling the volume of compressed air supplied to each receiver during any given length of time v, the volume drawn from the receiver by the hoist during the same time v_1, and the volume of the receiver V gives us

$$V = \frac{v_1 - v}{1 - \dfrac{1}{r}} \quad \text{and} \quad r = \frac{V}{V + (v_1 - v)}$$

if r is the ratio of the final to the initial pressure in the receiver during the same time.

During the acceleration period when the maximum rate of air consumption may reach 60 cu. ft. per second for 5 sec. $5 \times 60 = 300$ cu. ft. for 5 sec., we get, if air is supplied to the receiver at the rate of 15 ft. per second, under which condition there is no appreciable drop of pressure in the pipe line:

$$V = 4 \times 2100 = 8400 \text{ cu. ft.} \quad v = 5 \times 15 = 75 \text{ cu. ft.} \quad v_1 = 300 \text{ cu. ft.}$$

$$v_1 - v = 275.$$

$$r = 8400 \frac{8400}{8400 + 275} = \frac{8400}{8675} = 0.97$$

so that, if the initial pressure in receiver is $90 + 12 = 102$ lb. absolute, the final pressure will be $102 \times 0.97 = 99$ lb., or the pressure drop 3 lb. As the air in the pipe line is accelerated during the acceleration period of the hoist, and therefore attains a higher rate of speed than 15 ft. per second, the pressure drop during the period will actually be less than 3 lb. The pipe lines are made of such diameter that they will allow a flow of 2,700 cu. ft. of compressed air per minute or 45 cu. ft. per second with a pressure drop of 5 lb. The heaviest unbalanced loads call for an air consumption of 40 cu. ft. per second and as they are hoisted in one minute of time, 2,400 cu. ft. of compressed air will during that time flow into the hoist cylinders.

The pressure drop will, therefore, be less than 5 lb. as less air is taken out of the receiver than can be supplied to it from the pipe line with 5 lb. drop.

Each of the large hoists is fitted with four storage receivers aggregating 8,400 cu. ft. capacity and there are similar receivers installed at the compressor plant. As all these receivers are connected by the pipe lines they will act together in case of any heavy demand of air, no matter if this air is demanded at one hoist or at several hoists simultaneously. In order to show in how far remote these receivers can equalize the demand for power if heavy hoisting is going on simultaneously in several mines, we will assume that in three of the largest producers hoisting is in progress from 1,800 ft. depths and that in each 36 loads are raised in 36 min. This is a condition that actually exists and must be met, and each trip requires 1,012 cu. ft. compressed air. Three compressors are running, each supplying 880 cu. ft. compressed air per minute, or 2,640 cu. ft. per minute for the three.

There are at the different mines connected to the system 40 air receivers and on the compressors 12 receivers, or a total of 52 receivers, each of 2,100 cu. ft., so that the aggregate receiver volume will be $52 \times 2,100 = 109,200$ cu. ft. We have thus:

$$v = 2640 \times 36 = 95040$$

$$v_1 \; 1012 \times 3 \times 36 = 109296$$

$$V = 109200$$

$$r = \frac{V}{V + (v_1 - v)}$$

$$v_1 \; v - = 14256$$

$$r = \frac{109200}{123456} = 0.886$$

At the end of the 36-min. period there is thus a pressure on the system of $102\times0.886 = 90$ lb. absolute. The air pressure has then dropped 12 lb. during that time. From the above can be seen that air receivers placed in the pipe line are not very effective when it is required to store much energy. Such receivers can liberate only a limited volume of air $=$ $V\left(\dfrac{1}{r}-1\right)$ depending on the drop of pressure. In the foregoing example the receiver volume was 109,200 cu. ft., with a 12-lb. drop and 90 lb. initial gauge pressure the volume of air liberated is

$$= V\times\left(\frac{1}{0.886}-1\right) = V\times(1.113-1) = V\times0.113$$

or only a little over 11 per cent. of the receiver volume.

In this respect the action of such air receivers (expansion tanks) is identical with that of a flywheel in an Ilgner transformer. In both cases only part of the stored energy can be liberated. There is, however, no loss incurred when storing energy in the form of compressed air if the receiver is tight, while in case of a fast running flywheel considerable power has to be consumed to overcome friction and windage, which power is absolutely lost. The action of the flywheel in this respect is similar to that of a leaky air receiver from which there is a constant leakage of such magnitude that, if filled with air, the air would in a few minutes be lost. An illustrative example of this will be given further on.

The foregoing shows how the instantaneous peaks, due to acceleration of hoists, were taken care of, but in addition to these instantaneous peaks there were longer peaks due to rapid periods of hoisting as illustrated by Fig. 2.

In order to get an idea of what peaks would arise from such conditions the writer estimated that if all the mines of the Butte district were connected to the power system and operated with compressed air, and if the operation could be so regulated that the periods of high air consumption would not overlap each other at the different mines, the resultant air consumption would be as per diagram Fig 16. This diagram is based upon investigation at different mines in the year 1909 and includes 27 hoisting engines of various sizes and capacities. According to this chart the highest rate of air consumption would be 37,200 cu. ft. free air per minute, and the average, 27,660 cu. ft.[1]

---

[1] It can be seen from this chart that there are 10 peaks in the air consumption curve above the average and 10 depressions. The different peaks and depressions are in reality the average of a great number of peaks and depressions of very great magnitude, resulting from the simultaneous acceleration of several hoists and simultaneous periods of short duration during which the air consumption is very low. Some of these peaks figure as high as 100,000 cu. ft. of free air per minute during a few seconds of time. Such air consumptions, as have been stated before, are taken care of by the expansion tanks.

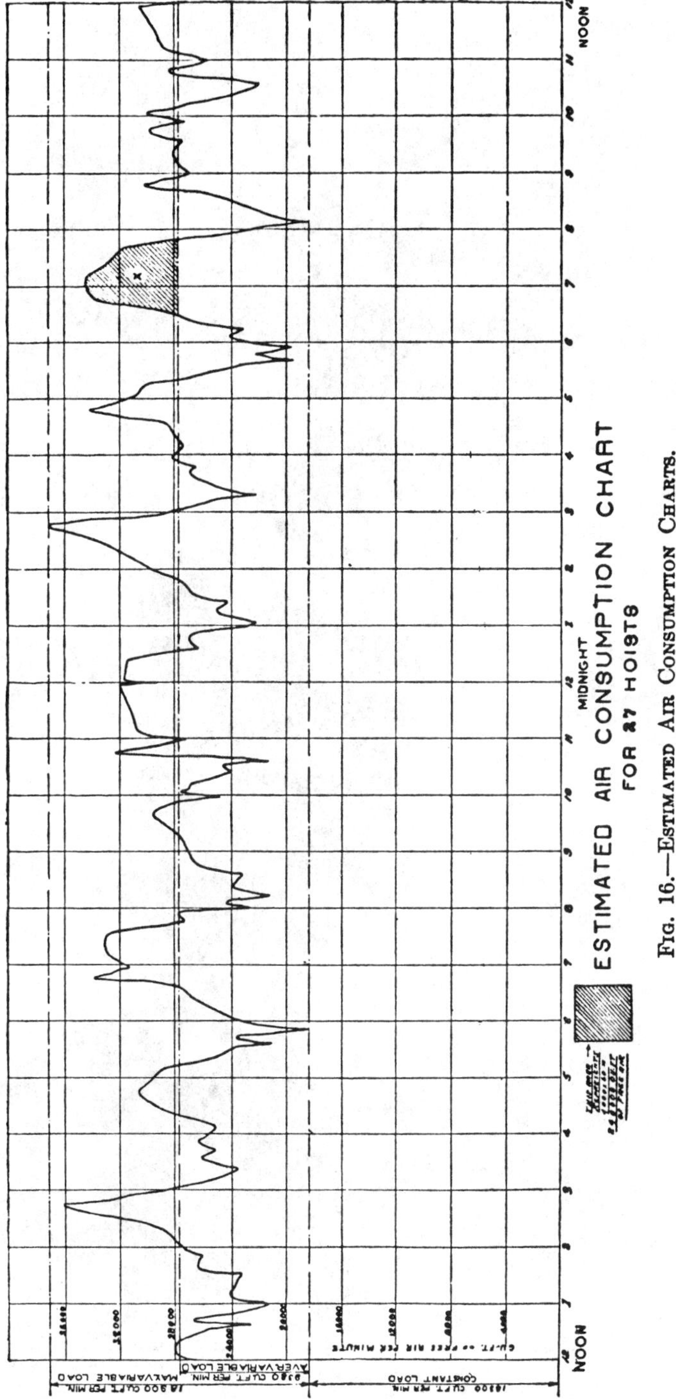

FIG. 16.—Estimated Air Consumption Charts.

Fig. 17.—Outside of Compressor Plant.

This chart can be considered to represent the most favorable distribution of the work with minimum peaks in the air consumption. The largest peak (X) in the diagram represents a volume of 44,000 cu. ft. of air

FIG. 18.—HYDROSTATIC TANK.

FIG. 19.—HYDROSTATIC AIR RECEIVERS.

at 90 lb. pressure and gives the measure of the minimum storage capacity required in order to equalize the load on the system. The energy represented by this peak is about 1,060,000,000 foot pounds.

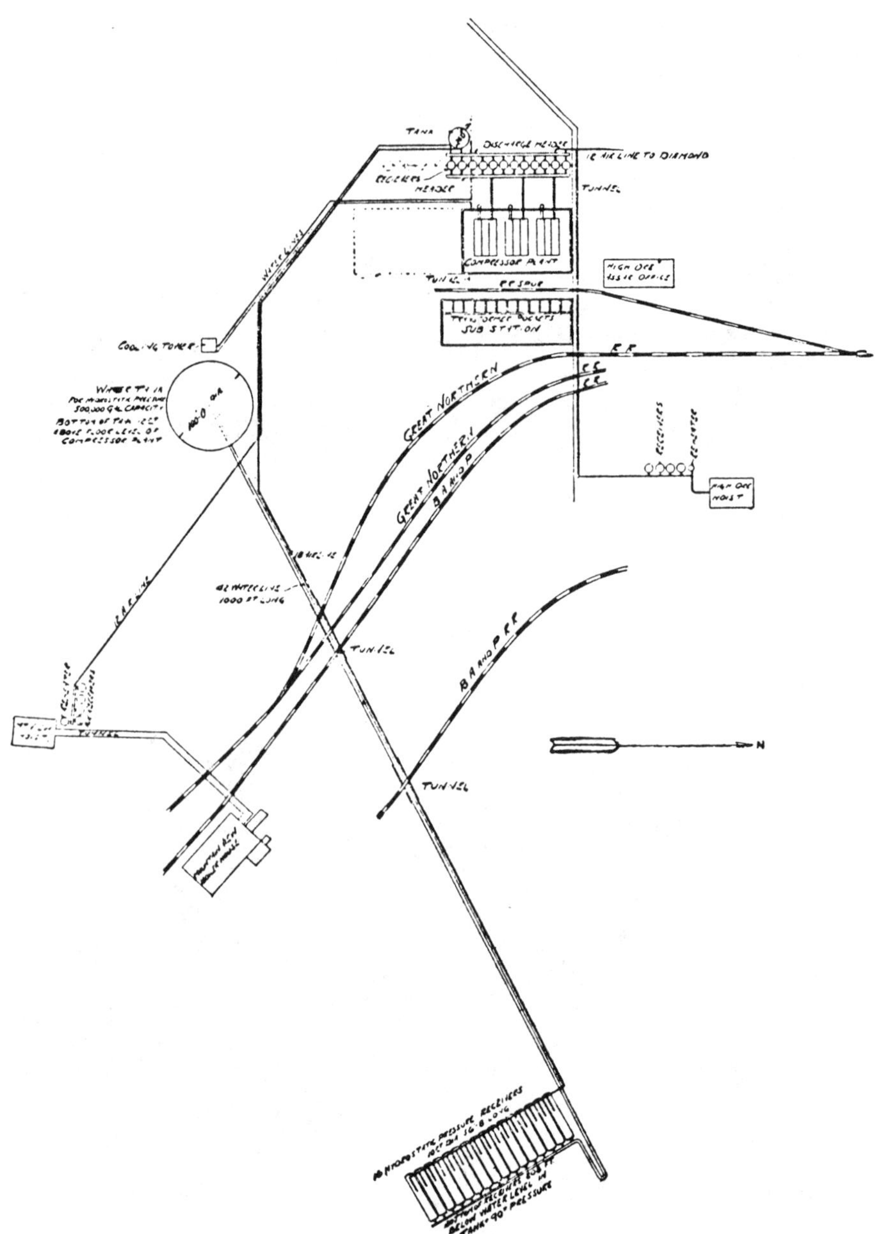

Fig. 20.—Butte Hoist Compressor Plant and Hydrostatic Pressure System, Butte, Mont.

To provide for this peak of 44,000 cu. ft. capacity with expansion tanks would mean that with a pressure drop of 10 per cent. ($v = 0.9$) we would then have $(v_1 - v) = 44{,}000 = V(\frac{1}{r} - 1) = 0.1V$ or $V = 440{,}000$. Such a storage receiver would be prohibitively large.

If, therefore, a uniform load on the electric motors operating the hoisting engines is aimed at, most of the energy used must necessarily be stored energy; hence, the storage device must not only have great capacity, but must also retain the stored energy without loss. A body of water located at a certain elevation above a suitable motor would be an ideal device for storing power. The whole potential energy of it would be available for generating power, and as long as the reservoir and its connections are tight, the energy can be kept in storage for an indefinite period of time.

The hydrostatic storage plant forming part of the Butte power system consists of a number of air receivers connected so the whole forms a receiver of about 66,000 cu. ft. capacity. Some 210 feet above these receivers there is an open water tank 100 ft. in diameter. A pipe leads from the bottom of this tank to the bottom of the air receivers. Air from the compressor plant is piped to the top of these air receivers. Fig. 20 is a general drawing of this plant. Water is run into the elevated tank until the lower tanks and the rising main are filled up to the bottom of the elevated tank. When air is compressed into the lower tanks it drives the water out, and when 66,000 cu. ft. of compressed air have been pumped into these tanks the water is out and nearly fills the upper tank. A long return bend is formed in the water pipe, which bend runs down hill for a considerable distance, so that, if the water is driven out of the air receivers, the air pressure will have to rise very materially before it can force the water out around the bend and before air could enter the water pipe and blow out into the upper tank.

Fig. 18 shows the open tank on top of hill with the substation and compressor house in the background, and Fig. 19 shows the hydrostatic receivers at foot of the hill.

The diagram in Fig. 21 illustrates the capacity of this storage system. When the air receivers are full of air and all the water is in the upper tank we have stored up 66,000 cu. ft. of water under a head of 210 ft. The potential energy of this water is 865,141,200 foot pounds and is represented in the diagram by the shaded area A. In addition to this there is available for power the expansion force of air at 90 lb. pressure represented by the shaded area B in the diagram, the potential energy of which is 730,179,000 foot pounds. We have thus in this plant a total capacity of storage of 1,595,320,200 foot pounds.

[*Editor's Note:* Material has been omitted at this point.]

The advantages of this adopted system over any in which electric motors are directly applied to the hoists are the following:

(1) The existing hoisting engines could be retained and needed comparatively few changes.

(2) They could be operated by steam in case of a serious accident to the power system.

(3) Enough energy is always available in the storage plant to operate the hoists for a short time in case of a short failure of the electric power due to thunder storms, etc.

(4) The capacity of the hoists can be increased, by increasing the hoisting speeds of the loads or the hoisting depth can be increased without changing the motor on the hoist. The compressor capacity only needs to be increased.

(5) A considerable part of the energy liberated in retarding the hoists at the end of a run or when loads are lowered into the mine is returned to the power system and is held in storage without loss for an extended period.

(6) The system can be started with its stored energy without subjecting the electric apparatus to excessive peak loads as is the case when a flywheel of an Ilgner transformer is started.

(7) By this system of compressing, storing and using the air in the hoist cylinders, peak loads on the electric apparatus are absolutely avoided. No matter how the load on the hoist motors may fluctuate, the power to operate the electric motors in the compressor plant is absolutely constant.

(8) Compared with electric hoists the air system has the following electrical advantage: With the air hoist, the electric prime movers are synchronous motors which can be operated with a leading power factor and therefore raise the power factor of the entire electrical system, while with an electric hoist, the prime mover is an induction motor which would have a very low power factor due to the intermittent loads it would have to carry to meet the Butte conditions of hoisting and hence would decrease the power factor of the entire electric system.

As compressed air is used for operating the rock drills it seems consistent and natural, when electric power is used for compressing this air, to enlarge the compressing plant so that the hoists also are operated by air. In most cases of ore mining it will be found that the increase in compressor capacity necessary for hoisting with air is not very large. If proper air storage is provided, the result will be a simpler, cheaper, safer and more economical plant than if the hoists were directly driven by electric motors and motor generators with heavy and fast rotating flywheels were used. The high peak loads resulting from starting these flywheels and from the irregularity of hoisting will be avoided and the electric load will be perfectly constant.

The plant has been a success and the writer believes that it has fulfilled the expectations of the owners. The mechanism used for performing the different functions in both the compressors and the hoists were new, and applied in practice for the first time in the plant. Since the machinery was started there have been no interruptions in its work and very little trouble.

In a new plant operated with compressed air many improvements could be made that were not possible in the Butte plant. Some of these improvements would be of very radical nature and would result in obtaining very high efficiencies without in any way impairing the reliability of the machinery or ease of operation, which, after all, is the most important thing in a mine hoist.

Gas stored in salt-dome caverns

TRANSCONTINENTAL Gas Pipe Line Corp. has completed two solution-mined gas-storage caverns in the Eminence salt dome of Covington County, Miss.

• They are the first salt-dome caverns in the U.S. to be constructed for high-pressure natural gas.

• They have a deliverability rate of 375-MMscfd/cavern and do not require the use of a displacement medium.

• They constitute the first underground storage built to provide a standby supply in case of an emergency, such as a hurricane or freeze-up, in the gathering area.

The Eminence dome is about 20 miles north of Hattiesburg, Miss., and only 1½ miles from Transco's main line. Total cost including development of the storage, the 30-in. connecting line, a 42-in. main-line loop, a 2,000-hp compressor station, equipping the wells, etc., is estimated at $9.7 million.

The project was initiated in Feb., 1967, when Transco engaged Fenix Scisson, Inc., Tulsa, Okla., to investigate Louisiana and Mississippi salt domes that were adjacent to its main line.

Why Eminence. Three reasons lead to the choice of Eminence: (1) It is relatively close to the main line, (2) Top of the salt is relatively shallow, 2,400 ft below the surface, and (3) The top of the dome is large enough to accommodate additional storage if it is ever required.

The Fenix & Scisson study of Eminence reached these conclusions:

1. A seismic survey of the dome indicated there was sufficient salt for the creation of solution-mined gas-storage caverns.

2. Adequate brine-disposal wells could be completed in the Wilcox formation on the flank of the dome.

3. Sufficient land was available for all surface facilities.

4. The optimum condition for storing 2,000 MMscf of usable gas at a pressure of 3,950 psia was determined to be two 13⅜-in. cased solution wells with the 13⅜-in. casing set into the salt at a depth of 5,700 ft with a one-million bbl cavern created from the base of the 13⅜-in. casing to a depth of 6,700 ft in the salt.

Maximum gas deliverability was to be 375 MMscfd. The estimated time for the development of the two caverns was expected to be 25 months at a total cost of $2,600,000. The cost and development time were the principal factors used in selecting the two 13⅜-in. wells cased to 5,700 ft.

Initial operation. In March 1968, Transco authorized Fenix & Scisson to start work. While the leaching plant was being designed and equipment ordered, two fresh-water wells and one disposal well were drilled and tested.

The fresh-water wells were drilled to approximately 600 ft into a massive fresh-water aquifer. They were equipped with Peerless 7-stage downhole pumps driven by 150-hp vertical hollow-shaft electric motors capable of producing 1,000 gpm.

The disposal well had 10¾-in. surface casing set at approximately 1,000 ft. The well was drilled to a total depth of 5,354 ft in the Wilcox, 7-in. casing set and cemented at 5,147 ft. The well was then completed by selectively perforating the casing in the most attractive zones in the Wilcox formation.

The test after acid treatment

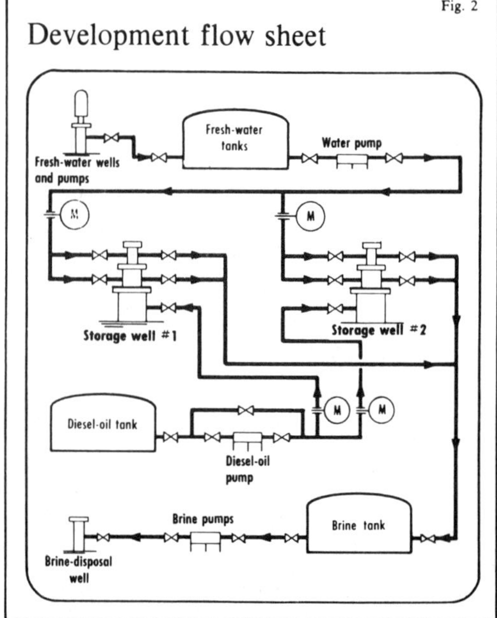

Fig. 1

Typical storage-cavern development

Fig. 2

Development flow sheet

showed the well would accept salt water at a rate of 10,000 b/d at 1,000 psi pump-pressure which was considered adequate. At completion of solutioning, the well had improved such that it would accept salt water at a rate of 35,000 b/d at 850 psi.

Drilling program. Drilling and casing programs for the solution wells follows:

1. 30-in. conductor casing was set and cemented at approximately 50 ft.

2. A 28-in. hole was drilled to 2,700 ft and 20-in. OD casing set and cemented to the surface. Setting casing at this depth served as a secondary seal for the cavern since it sealed off the cap rock and was set approximately 400 ft into the salt stock.

3. A 17½-in. hole was drilled to 5,700 ft in the salt and 13⅜-in. casing set and cemented to the surface. Because of the high operating pressure, the 13⅜-in. casing was tested to 5,000 psi as it was run in the hole.

After the casing was run and cemented and the shoe drilled, it was again tested to 5,000 psi as an additional check on the casing and the cement job.

4. An 11-in. hole was drilled to 6,700 ft, the total depth.

In the first well the salt was cored continuously from 5,700 ft to 6,700 ft, the cavern interval, in order to check the quality of the salt and to determine the amount of insolubles present. Core analyses indicated anhydrite insolubles varying from .09 to 7.42%. An average figure of 4% was used in calculating the net cavern volume since the insoluble material in the form of anhydrite sand was to remain in the cavern.

Casing program. The casing program included running 10¾-in casing to approximately 5,800 ft, or 100 ft below the bottom of the 13⅜-in. casing, and 7-in. casing to total depth. Both strings were suspended from the surface using conventional wellhead equipment. A completed well with the wellhead hookup and the casing in the hole is shown in Fig. 1.

In the leaching or solutioning process fresh water is pumped to the fresh-water tanks. Fresh-water pumps then pump it down the 7-in. casing to the exposed salt. Brine returns to the brine-storage tank at the surface through the 7-in by 10¾-in. annulus. Brine pumps pump brine from storage to the disposal well.

Diesel oil is pumped down the 13⅜-in. by 10¾-in. annulus and the oil-water contact maintained at the bottom of the 10¾-in. casing. This preserves the cavern neck and eliminates leaching above this point. Fig. 2 is a flow diagram of the solutioning process.

Surface facilities. The tank-battery consisted of one 3,000-bbl brine tank, three 500-bbl fresh-water tanks, and one 500-bbl diesel-oil tank. Pumping equipment included one NSCO J60LC horizontal single-acting triplex pump for diesel-oil service driven by a 60-hp electric motor and seven NSCO J250 horizontal single-acting quintuplex pumps driven by 250-hp electric motors.

Three of the J250 pumps were used for fresh water and the others for

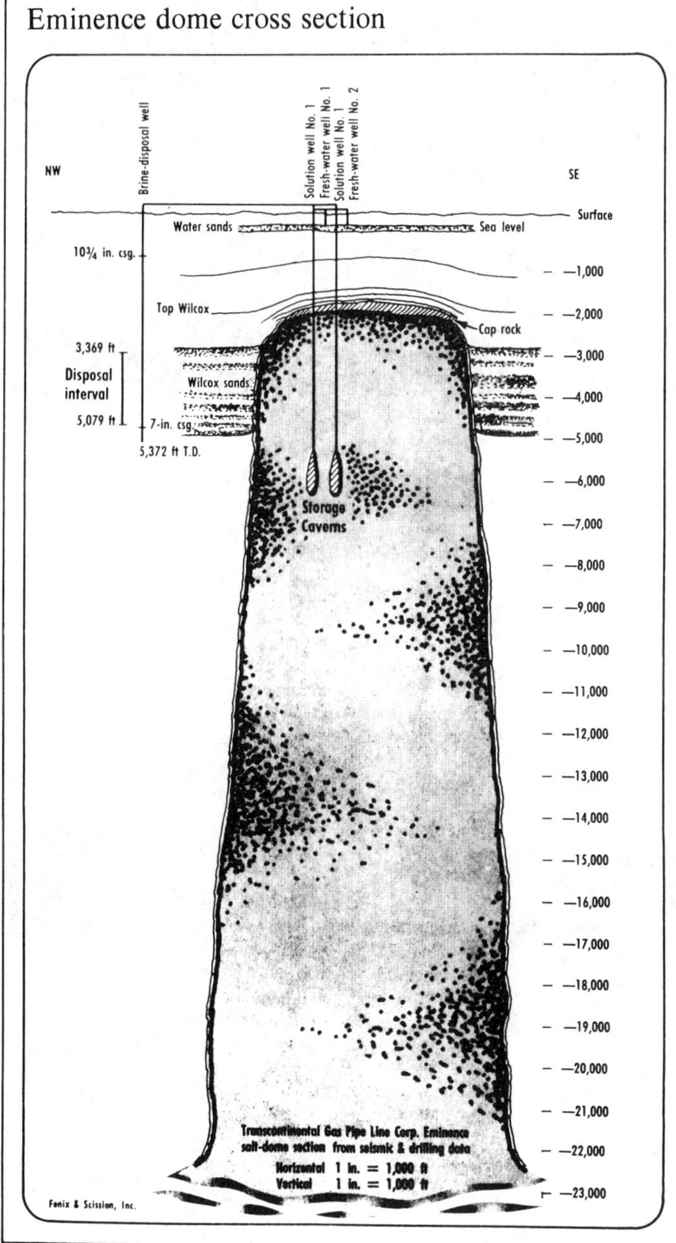

THREE horizontal, single-acting quintuplex pumps were used for fresh-water service and four for brine service in preparation of storage caverns in the Eminence salt dome.

brine service. The tanks were elevated to provide flooded suction to all pumps.

Automatic controls consisted of high-pressure shutdown switches on all units, high and low-level controls from the tanks and time-delay switches in case of power disruptions.

Site development. The plant was built as the solution wells were being drilled and completed. Plant construction began on July 1 and was completed Nov. 20, 1968.

The first solution well was spudded on Aug. 3 and completed Nov. 18, 1968. The second well was spudded Nov. 23, 1968 and completed Jan. 23, 1969. The increased time required for completing the first well was due to coring of the 1,000-ft cavern interval and a nitrogen-gas test to 5,000 psi after drilling was completed.

Solutioning was started in the first well on Nov. 24, 1968, and completed in 393 days with a volume of 1,100,805 bbl. In the second well, it was begun on Jan. 31, 1969 and completed in 423 days with a volume of 1,102,045 bbl.

Although all measurements were based on metered volumes, sonar surveys were run in each cavern as a check on the v o l u m e and cavern shape. These surveys were run twice

WELLHEAD equipment on Transcontinental Gas Pipe Line Corp. gas storage in Covington County, Miss., the first in the U.S. to use salt dome and the first to be located in a gathering area.

AUTOMATIC controls included high-pressure shutdown switches on all units, high and low-elev controls from the tanks and time-delay switches in the event of power disruptions.

in each cavern, when leaching was 50% complete and at completion.

Sand removed. The original plan was to leave the accumulated insolubles in the caverns. Because of the water-retention of the sand an attempt was made to remove as much sand as possible by brine circulation upon completion of leaching. This procedure was reasonably successful.

In cavern No. 1 approximately 37,-000 bbl of sand were removed and in No. 2 approximately 29,600 bbl. In addition to aiding in gas dehydration, additional storage volume was recovered by this procedure.

Upon completion of sand removal, the 7-in. and 10¾-in. casing strings were pulled and the solutioning wellhead equipment removed. Wellhead equipment for 5,000 psi working pressure was installed and 4½-in. tubing run to bottom for displacement.

High-pressure gas was injected into the 13⅜-in. by 4½-in. annulus at a rate of 20 MMscfd displacing the brine through the 4½-in. tubing until the gas-liquid interface reached the bottom of the tubing. The brine was returned to the brine tank and sent to the disposal well on displacing.

Diesel-oil marker. As a precautionary measure, a thin layer of diesel oil was placed on the brine surface in the cavern so that it would be carried to the bottom during displacement. Dielectric sensors to detect diesel oil were placed in the brine return line so that the presence of oil in the return brine would cause the sensor to shut the system in. This indicated that the brine-oil interface was at or near the bottom of the tubing.

This constituted completion of the initial storage system. The caverns are used by injecting or withdrawing gas through the 13⅜-in. by 4½-in. annulus with the 4½-in. tubing shut in.

In the final completion cavern pressure will be reduced to approximately 1,200 psi and the 4½-in. tubing snubbed from the caverns under pressure. The wellhead equipment will be removed and the main gas lines connected directly to the master valves on the 13⅜-in. casing. This will permit full open-flow through the casing with no restriction.

Fig. 3 is a cross section of the Emi-nence dome shown in a NW-SE direction. The drawing is to scale to show the relative positions of the disposal well, the fresh-water wells, and the solution wells.

Advantages. These caverns are the first in the U.S. specifically for the storage of natural gas. Abandoned brine wells have been converted to gas storage but they are relatively shallow and operate at a much lower pressure. Less cushion-gas is required for this than other types of underground storage because of the integrity of the salt.

Probably the most unusual features of these caverns are their potential deliverability rates of 375 MMscfd/cavern and the fact that no displacement is required for their use.

There are storage facilities in which gas is injected into depleted gas reservoirs or other suitable formations. These units function satisfactorily, but are restricted as to deliverability. To provide for deliverability approaching 375 MMscfd from this type of storage would be uneconomical because of the large number of wells required to produce the stored gas.   END

9

HUNTORF—THE WORLD'S FIRST 290-MW GAS TURBINE AIR-STORAGE PEAKING PLANT

WOLFGANG MATTICK
Executive Vice President
Brown Boveri & Co.
Mannheim, West Germany

HANS-GUENTER HADDENHORST
President
Kavernen Bau-und Betriebsgesellschaft mbH
Hannover, West Germany

OTTO WEBER
Manager, Gas Turbine Application Engineering
Brown Boveri & Co.
Mannheim, West Germany

and

Z. STANLEY STYS
Vice President
Brown Boveri Corporation
North Brunswick, New Jersey

INTRODUCTION

Energy storage will become more and more significant in the future. With the number of nuclear power plants growing, the trend of keeping them loaded because of economical reasons will continue. On the other hand, seasonal and daily peaks will traditionally grow sharper. Thus, the possibility of transferring the off-peak, cheaper energy to the peaking period will be the focus of attention of an increasing number of utilities.

Pumped-hydro storage is widely accepted as a significant potential for large-scale energy storage. However, the environmental objections from the siting standpoint seem to limit the possible number of future pumped-hydro storage plants. Besides, the cost of such plants and the time required for their construc-

tion also provide negative factors in their consideration.

Pure gas turbine peaking, so popular only a few years back, makes today's application of this type of prime mover extremely difficult to justify, due to the unavailability and costs of natural gas and oil.

The subject of this paper is a discussion of a system which operates basically on the principle of hydro pumped storage. It provides for the transfer of the off-peak energy into the peaking time, utilizing a modified gas turbine cycle for the purpose, i.e., it has all the inherent characteristics of the gas turbine, such as quick start, low costs, simplicity of operation, etc., but uses only one-third of the fossil energy compared with a pure gas turbine peaking unit.

GAS TURBINE—COMPRESSED AIR STORAGE CYCLE

It is an inherent characteristic of a gas turbine that two-thirds of the generating power is being consumed internally and used for compression of its cycle air. In the plant used for the air-storage peaking application, the compression work is shifted to the off-peak hours, when such energy is generated by the existing plants using more plentiful and less expensive energy sources. The system cycle is presented in Fig. 1.

The ambient air is compressed by an axial-flow compressor, intercooled and boosted up, in a high-speed centrifugal blower, to 1000 lb/in.^{2} Aftercooling follows air discharge, before leading to an air storage facility. The generator is used

122

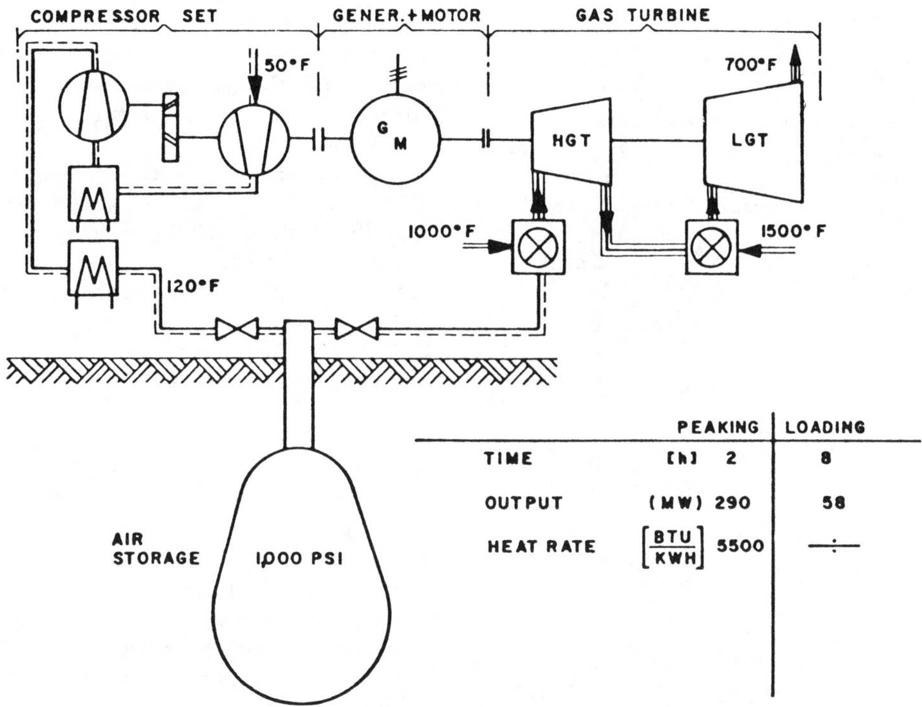

Fig. 1—System cycle, 290-MW air-storage gas turbine plant.

as a motor during the compression cycle.

During the peaking cycle, air is led from the storage through an expansion valve, where the pressure is throttled down to 650 lb/in.² before reaching the HP combustion chamber, where it is heated to 1000 F. After passing through a portion of the expander, the gases are reheated in an LP combustion chamber to 1500 F, and upon passing through the end of the expander, the gases are then exhausted to the atmosphere. It was only possible to justify such a plant by application of a new gas turbine having a pressure ratio of 1:45.

AIR-STORAGE FACILITIES

There are several possible ways to obtain air-storage facilities. Depleted oil and gas wells, defunct iron, potash, salt, and other mines can be used. Aquifers, natural caverns, man-made excavations, leached caverns, etc., are also very suitable for such installations.

The geological structure of the German North Coast, i.e., near Bremen, Hamburg, Oldenburg, etc., bears great similarity to the geological formation of the United States Gulf Coast. Once, in the earth's geological middle ages, these areas were flooded and salt deposits were formed. The subsequent silts, carried by rivers, and other overburden covered these salt deposits to a depth of 20,000 to 30,000 feet. However, the uneven distribution of the overburden masses caused salt in some areas to be squeezed upwards, creating so-called salt domes.

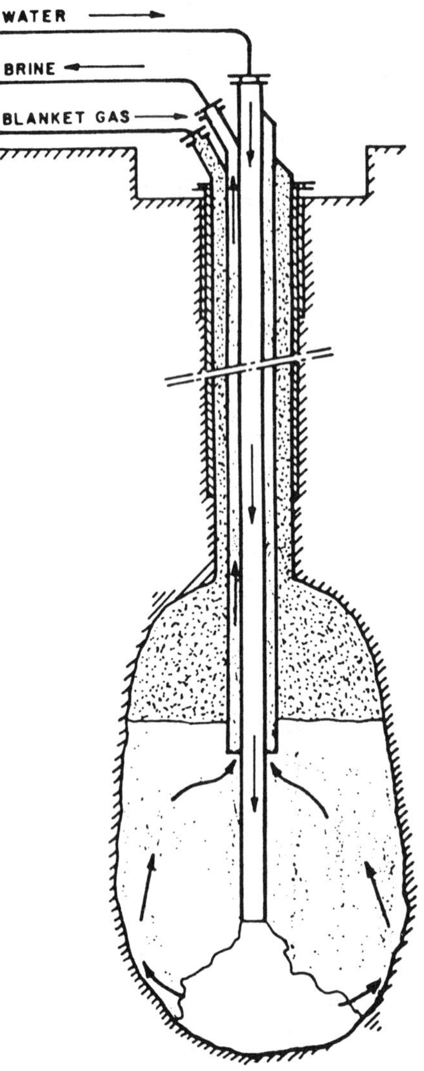

Fig. 2—Cavern leaching process.

They are different in shape and height and some come up close to the earth's surface.

Since usually oil is found around such salt domes, their profile is well known by the geologist specialized in oil exploration.

Salt, being plastic and quite tight as far as gaseous impermeability is concerned, deems itself as very suitable material for the storage of gases under pressure in the cavern, leached in it for this purpose.

In the case to be described, two salt caverns are to be used with a combined capacity of about 10,000,000 cubic feet. These caverns will be leached out in a salt dome, using well known techniques. Sulfur is mined in such a manner on the Gulf Coast. Figure 2 depicts the process used in creating these caverns.

Water is being pumped down through a shaft to the excavation in a salt dome. With the help of a blanket gas, water is circulated and dissolves the salt. Brine is then pumped out and is piped into the ocean.

For control of the shape of the cavern, an instrument is used based on the sound reflection principle. With the help of the blanket gas it is possible to control the leaching process until a more or less cylindrical form of cavern is obtained.

Basically, to leach 1 cubic foot of cavern capacity, 10 cubic feet of water is used. Approximately 14 months will be spent in creating the necessary two 5,000,000-ft^3 cavities.

The top of the salt dome in which these two caverns are located is 1500 feet below the earth's surface. The caverns will have a diameter of about 100 feet and a height of 600 feet and their top will be approximately 1800 feet below surface. They are about 500 feet apart from each other.

HUNTORF POWER STATION

In June, 1974, the world's first compressed-air-storage gas turbine plant was ordered. Its main characteristics are given in *Table I.*

For 1 kWh produced, 0.8 kWh is supplied during charging and 5500 Btu during discharging, added in the form of oil or gas to the gas turbine. A model of the plant is presented in Fig. 3.

MODE OF OPERATION

The unit does not have any starting motor. Startup occurs by leading the air to the HP combustion chamber, where it is heated to 1000 F and exhausted to the gas turbine. The gas turbine shaft starts to rotate and is, at a certain speed, automatically engaged, through a synchro-self-shifting (SSS) clutch, to the generator. The generator is synchronized, the LP combustion chamber is ignited and the plant loaded. It takes only eleven minutes from initial start to full load. (In emergency cases this can be accomplished in six minutes.) The startup can be subdivided as follows:

	Normal	*Emergency*
Preparation for start	.5	.5
Synchronize	3.0	3.0
To Full Load	7.5	2.5
Total startup time, minutes	11.0	6.0

Thus, a residual cavern pressure is always needed to accomplish the startup. For the first loading of the cavern during commissioning of the plant, other provisions have been made. There will be a high-voltage line established between Huntorf plant and a 52-MW gas turbine plant in Emden, about 60 miles away. The units will be electrically interlocked and brought up, by frequency regulation, to synchronous speed.

To pressurize the caverns, the compressor train is engaged to the generator and, after following the same procedures as in the power producing cycle, the combined shaft is brought up to full speed and the generator synchronized. At this point, the gas turbine is shut down,

TABLE I
HUNTORF POWER STATION

Owner: Nordwestdeutsche Kraftwerke AG, Hamburg, Germany

Location: Huntorf in Bremen/Oldenburg area

Cavern made by: Kavernenbau-und Betriebsgesellschaft, Hannover, Germany

Equipment, engineering and design: Brown Boveri & Co., Mannheim, Germany

Clutches: Renk, Augsburg, Germany (Licensee of SSS Gear Works, London, England)

Technical Data:
Capacity during power producing cycle: 290 MW
Power needed to charge the caverns: 58 MW
Duration power cycle: 2 hours/day
Charging cycle: 8 hours/day

Gas Turbine:
Inlet pressure: 650 lb/in.^{2}
Inlet temperature: 1000 F
Reheat temperature: 1500 F
Mass flow: 900 lb/sec
Speed: 3000 r/min

Compressor Set:
LP compressor speed: 3000 r/min (axial)
HP compressor speed: 7600 r/min (centrifugal)
Average power absorption: 58 MW
Mass flow: about 220 lb/sec
Maximum discharge pressure: 1000 lb/in.^{2}

and the compression cycle commences, using the generator as a motor taking power from the grid during the off-peak period.

The starting time of this operation can be subdivided:

	Minutes
Preparation for start	.5
Synchronizing	4.5
Start loading	1.0
Full loading	6.0

Obviously, the generator can also be used as a synchronous condenser. After it is brought to speed and synchronized, the turbine is cut off. The compressors are disengaged during this operation.

Compressor set with generator motor

Fig. 3—Model of Huntorf.

comes to standstill after finishing loading operation in about 15 minutes, turbine alone in about 25 minutes and generator alone in about 35 minutes.

SYSTEM OPERATION

The two hours of peaking at 290 MW and eight hours of charging during off-peak hours is the specific requirement of the grid for which the Huntorf plant was designed. Basically speaking, coverage of any peak, or for this matter, any mode of operation, can be accomplished if the proper energy balance between input and output is established. Once the basic parameters for such a plant are fixed, such as charging and discharging times, other free variables can be chosen to match these conditions. Thus, air storage

pressure, cavern size, etc., can then be determined.

In the case of an available storage facility, such as a salt or potash mine, aquifer, defunct iron mine, depleted oil or gas well, the situation is slightly different. After geological soundness is established, corresponding storage pressure is chosen. This pressure might be the imposing and limiting factor influencing the design of a particular plant.

EQUIPMENT

To ensure high reliability of the plant from the start, the choice of equipment was limited to the components already proven in service. Where a new design was called for, known technologies were applied to limit the unknown factors to the absolute minimum. Thus, no basic new additions to the art of engineering were made in designing the machines for the Huntorf plant. However, the proper selection was made to ensure compliance with the plant design parameters.

Turbine

The pressure ratio of the existing gas turbines was not suitable for the economic assumptions of such plants, since the storage pressure must be kept as high as possible. Thus, a two-bearing machine was designed in which the front end complied with the basic design parameter of a steam turbine, such as low temperature and high pressure. However, at the back end, standard gas turbine technology was used: high temperature and low pressure. Each part has its own combustion chamber.

Combining these two technologies (steam and gas turbine) was only possible because in the case of Brown Boveri they are inherently compatible. The same builtup rotor construction and the same welding technique were used that apply in the manufacturing of both types of machine.

Combustion Chambers

The single-type combustion chamber used allows combustion not only of high grade fuels such as natural gas and oil, but also of low-Btu value fuels such as blast furnace gas or products of coal gasification. Blast furnace gas experience is available from over 20 years operation of such combustion chambers in a dozen units.

Generator—Motor

The standard 341-MVA, 21-kV, 50-Hz, 3-phase generator was chosen for this duty. The stator is water cooled and the rotor hydrogen cooled. Excitation current is fed in from the compressor side of the generator rotor. The output of this side is thus limited to 200 MW which is still sufficient to reach the loading ratio of 1:1.

Compressor Train

To achieve compression of a large volume of air to the comparatively high pressure, an axial-flow compressor of a special industrial design was chosen, followed by a six-stage centrifugal booster. The gearbox, having a ratio of 3000/7600, was chosen to provide the optimum speed for these machines. Intercooling lowers the input of the second compression stage.

Experience gained from operation of such compressor trains for industrial applications, such as oxygen plants for instance, was directly applied to the design of this set. The compressor mass flow is about 220 lb/h and discharge pressure 1000 lb/in.^{2}

Clutches

Self-synchro-shifting (SSS) clutches are provided at both ends of the generator-motor. The clutches are designed to connect these machines (generator and gas turbine) at the instant a predetermined speed differential is reached at the corresponding shafts. Engagement and disengagement of the clutches is completely automatic. A separately controlled oil supply is used to hold the clutch in the engaged position by means of a hydraulic lock when so required. Figure 4 shows the basic elements of such a clutch.

The clutch's lubricating and hydraulic locking oil is supplied from the main lubricating system and delivered to the clutch from the oil catchers incorporated in the turbine/motor-generator flanges. The lubricating and locking oil is caused

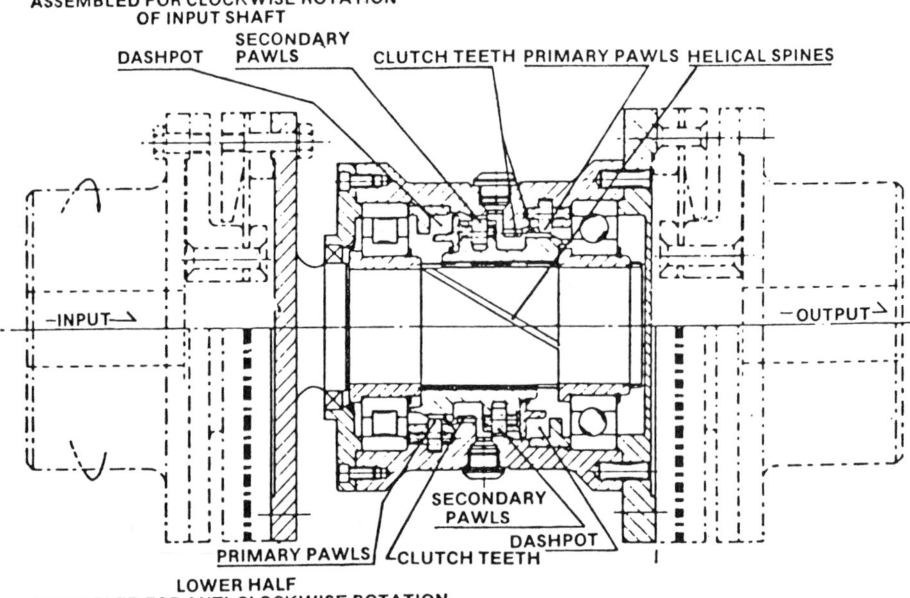

Fig. 4—Basic elements of SSS clutch.

to flow through the clutch by centrifugal force. The clutch does not require any special high-pressure oil supply.

Regulation and Governing

The output of the gas turbine is regulated by the mass flow and temperature regulation of the LP section of the turbine, as shown in Fig. 5.

The excitation system is static. Thyristors, the generally accepted element in this type of excitation, are used. Built-in redundance makes changing of individual elements possible during operation. The whole plant is remotely controlled from a distance of about 100 miles.

PHILOSOPHY OF THE PLANT DESIGN

The Huntorf plant is the world's first of this type, i.e., first compressed-air gas turbine peaking plant. In its design the following basic assumptions have been made: simplicity, lowest cost, dependability. The sacrifice of comparatively low efficiency has thus been taken into account. The throttling loss in the pressure regulating valve between the cavern and the inlet to the gas turbine is a considerable one but it can be eliminated if the cavern is put under pressure, i.e., by means of a hydrostatic water column from a reservoir located on the earth surface.

Another considerable loss can be avoided by installation of a regenerator: The air from the cavern to the combustion chamber would be preheated by the exhaust gases now leaving the stack at a temperature of 700 F. Such a regenerator would lower the heat consumption from 5500 Btu/kWh down to about 4000 Btu/kWh.

The intercooling and aftercooling of

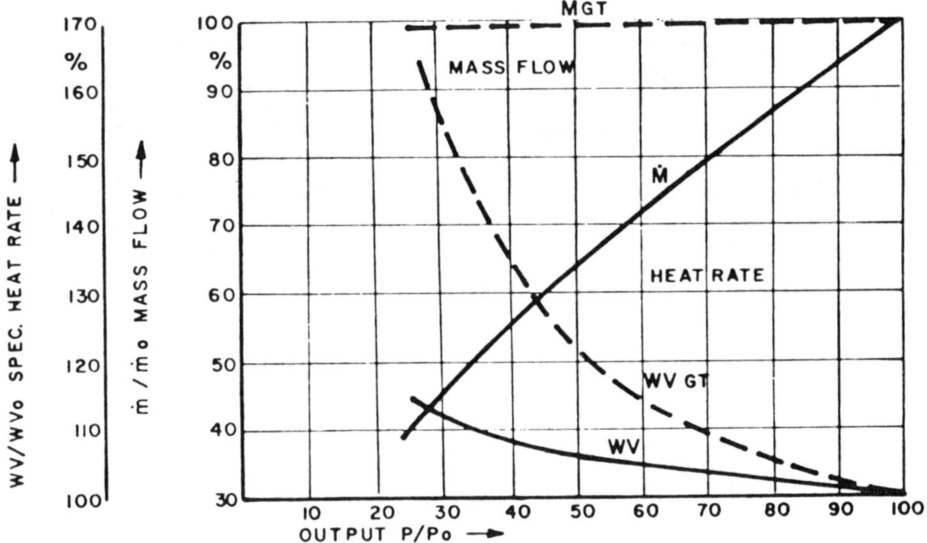

Fig. 5—Partial load of the air-storage gas turbine and open-cycle gas turbine.

the air also represents a loss: its incorporation into the cycle would represent another improvement, although comparatively small, compared with those mentioned above.

Thus, for simplicity and cost saving, it was decided to forego these improvements at Huntorf.

A high degree of noise intinuation was, however, desired, because of the plant location and it was achieved by the installation of silencing equipment.

CONCLUSIONS

The idea of the compressed-air storage, gas turbine peaking plants is basically not a new one. Although there were several studies made, some of them still active, there is basically no project close to realization in this country yet. Several papers were presented in the past on this subject, theoretical cost analyses and comparisons were made, and some of them were quite conclusive.

However, the fact that there is a Huntorf plant being built that will be commissioned by the middle of 1977 resurrected a great interest among utilities and architect-engineers in this country. With the availability of salt domes in the Gulf area, aquifers in the Midwest, depleted oil and gas wells in Texas and California, defunct salt, potash, iron and other mines in other parts of the country, there is a natural potential to build such plants all over the country. Even a man-made cavern, especially excavated for air storage, is not too exhorbitantly high in cost, compared with today's extra additions to fossil and nuclear plants, to comply with environmental and safety requirements only.

The basic idea to be able to transfer off-peak power to the peaking period, to have a better load factor on the machines, or even defer new capacity investments, is appealing more and more to many utilities.

Building an energy storage facility,

similar to pumped-hydro storage in a flat country, certainly presents definite advantage from an environmental standpoint of view.

Thus, it is hoped that the Huntorf plant fulfills industry expectations and its example will be followed by other utilities in this country.

REFERENCES

1. Herbst, H. C., "The Air Storage Gas Turbine: A New Peaking Concept in Status Nascendi," Paper presented at the NATO Symposium on Energy Storage, 1974.

2. Weber, O., "Air Storage Gas Turbine Plant," Internal Rep. Mannheim, Brown Boveri Corp., n.d.

10

GAS TURBINE SYSTEMS USING UNDERGROUND COMPRESSED AIR STORAGE

D. L. AYERS
Manager
Fluid Systems Laboratory
Westinghouse Electric Corporation
West Lafayette, Indiana

and

D. Q. HOOVER
Generation Consultant
Power Generation Systems
Westinghouse Electric Corporation
East Pittsburgh

INTRODUCTION

Energy storage systems are not new to the power generation industry. In this country, we have seen in recent years the advent of pumped hydroelectric power systems as a means of storing off-peak base power to be used subsequently during peak power periods when demand exceeds base-plant capacity. Even more recently, the concept of storing excess base-plant energy in the form of compressed air in mined caverns, for subsequent use with peaking gas turbines, has drawn considerable utility interest, especially in Europe. Other more advanced energy storage concepts are under development in the laboratory.

There are disadvantages which tend to detract from pumped hydroelectric and cavern-stored air systems. The investment cost for a pump hydroelectric plant is quite high compared with a system utilizing a gas turbine. In addition, suitable sites for the required large elevated water reservoirs are not necessarily found near major load centers. The cavern-stored air system suffers from the high cost of excavating the storage cavity, which can be as much as or even

more than the above-ground equipment cost. Also, the cavern-stored air system requires proximity to a body of water which hydrostatically maintains isobaric conditions in the cavern.

This paper examines an energy storage system similar to the cavern-stored air concept, except that the storage volume is an aquifer—a porous, underground stratum normally filled with ground water. A mathematical model covering the thermodynamics, fluid dynamics, and economics of such a system is described. The economics of the plant are compared with those of a standard gas turbine.

AQUIFER AIR STORAGE

Aquifers are naturally-occurring sedimentary geologic formations with interstitial pore spaces normally filled with ground water. They can be found over broad areas of this country, as shown in Fig. 1, and at a variety of depths below the earth's surface. From the standpoint of fluid dynamics, porous aquifers are characterized by their porosity ϵ, the ratio of pore volume to total volume, and permeability k, the flow rate of a given fluid through the porous structure per unit of pressure gradient across the structure.

The type of aquifer which is ideal for air storage has three naturally occurring properties. These are: (1) A dome or anticline trap which prohibits lateral air bubble migration, as pictured schematically in Fig. 2; (2) An impermeable caprock covering the porous storage volume; and (3) A porous stratum of sufficient volume under the caprock where the air can be stored, with suitable void volume and permeability to

131

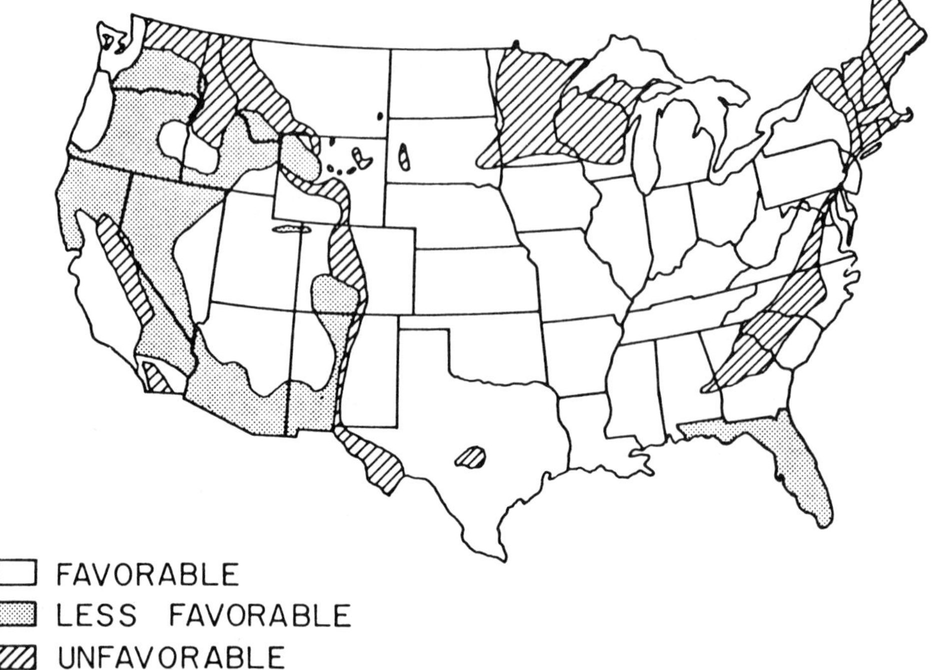

Fig. 1—Prospective areas for aquifer storage in coterminous United States

the flow of air. As Fig. 1 shows, the geologic structure beneath a large portion of this country is sedimentary and thus it is probable that aquifers would be discovered there. The anticlines or domes are local features and must be found by exploration.

To store air in an aquifer, wells are sunk near the apex of the trap, through the caprock, and into the porous bed. The air is compressed to a pressure above water pressure in the aquifer and as it is injected it forces the water away from the well bore. This forms a stable isobaric bubble which is confined around the bore by the caprock overburden and by the shape of the trap. When air is to be withdrawn, the water surrounding the bubble drives the air out the well and reoccupies the void volume vacated by the air. The aquifer thus acts like a large elastic chamber when air is injected or withdrawn.

There may be skepticism on the part of some who feel that energy storage by means of compressed air in aquifers is either technically or economically unfeasible. We hope to demonstrate later in this paper that the idea is quite feasible economically. As for technology, the natural gas utilities have for years been using this concept for annual peak demand shaving. During the summer months, when gas transmission line capacity exceeds demand, the surplus gas is compressed and stored in aquifers near load centers, as described above. During the subsequent winter months, when demand exceeds transmission line capacity, the aquifer-stored gas is withdrawn to meet demand. Today, there are over 50 such underground natural gas storage systems in the United States. Twenty-eight of them are located in Illinois alone. Their storage volumes range from the hundreds of millions

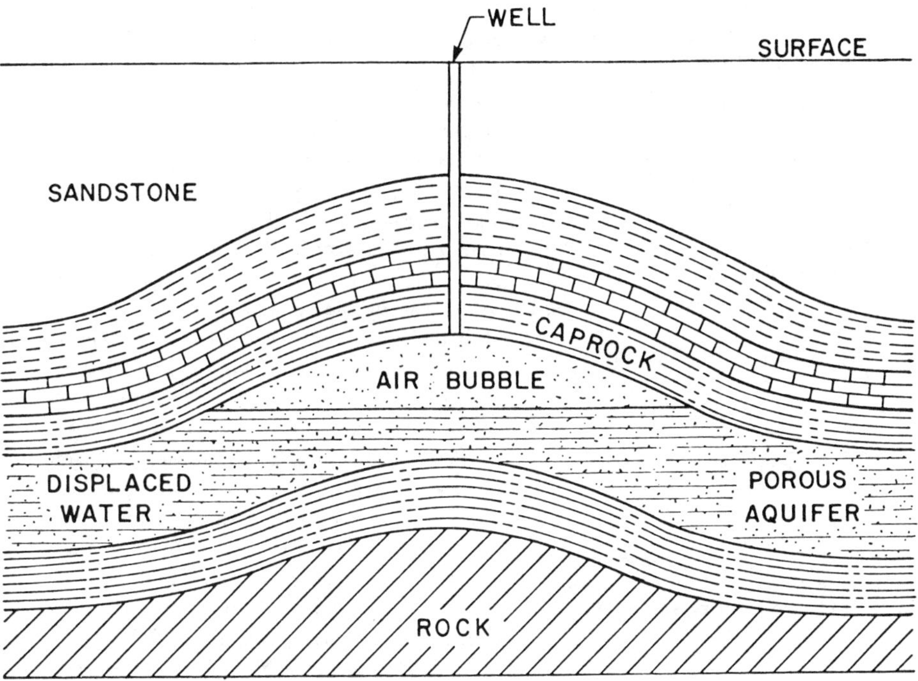

Fig. 2—Sedimentary geologic structure suitable for underground air storage.

to the tens of billions of cubic feet. As will be seen later in this paper, this is the range of storage volume that would be required for an energy storage system using compressed air.

THE SYSTEM

The aquifer-stored air peaking system is shown schematically in Fig. 3. It is considerably different from a standard gas turbine system in that the compressor is not integral with the turbine. Interposed between the turbine and the compressor is a motor/generator which is separated from each by a clutch. By engaging or disengaging the clutches, the turbine can either drive the motor/generator as a generator, or the motor/generator running as a motor can drive the compressors.

The system works in the following manner. During evening and night periods, when electrical demand is less than a utility's base-plant capacity, the surplus base power is used to run the motor which drives the compressor. The right-hand clutch is disengaged so that the turbine is idle. Air is ingested into the LP compressor and is then intercooled and compressed again in the HP compressor. The air is aftercooled (possibly optional) and is conducted through feeder lines and wells to the aquifer, where a stable air bubble exists. During the injection, incoming compressed air displaces water which normally occupies the porous aquifer stratum. During the next period in time (probably the next day) when electrical demand exceeds base-plant capacity, the compressed air, driven by the displaced ground water, is allowed to pass up the wells into the combustor where fuel

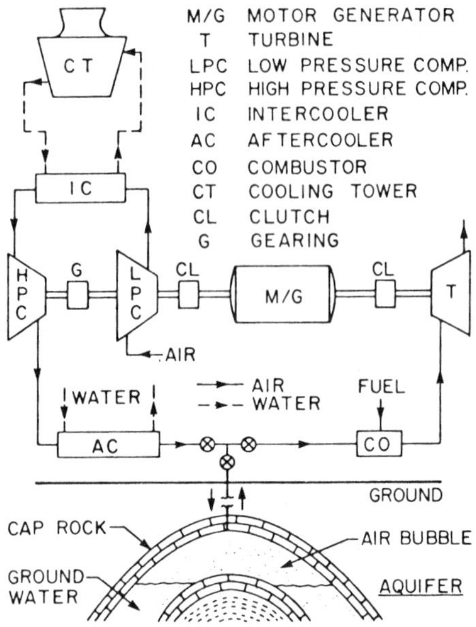

Fig. 3—Aquifer-stored air peaking plant.

is injected and burned. The combustion products drive the gas turbine. In this mode of operation, the right-hand clutch is engaged and the left is disengaged, leaving the compressor idle. In an emergency, if sufficient excess air has been stored, the system can be run much longer than the daily peaking period. If the emergency persists and the air in the aquifer is finally exhausted, both clutches can be engaged and the system run with the compressor exhausting directly into the combustor. In this configuration, the system capacity is somewhat less than maximum.

The aquifer-stored air peaking system has several advantages. First, it allows a utility to store base-plant energy during slack periods. The base plant can run at or near full capacity for the entire day—at its most efficient operating point. Second, the system has a net output more than twice that of a standard gas turbine system with the same size turbine since, when the turbine is oper-

ating, all of its output is used to drive the generator. As the equipment that is used costs somewhat less than twice that of a standard gas turbine system, the equipment investment cost ($/kW) is less than for a standard gas turbine. Third, the aquifer-stored air peaking system conserves gas turbine (natural gas and oil) fuels. Because base-plant power (and base-plant fuel) is used for compression, all the energy derived from gas turbine fuel goes to net output, with the result that system heat rate based on gas turbine fuel is about one-half that of a standard gas turbine system operating under the same conditions. As will be seen later in this paper, however, the overall heat rate is greater than that of a comparable standard gas turbine system. Thus, the aquifer-stored air peaking system is not conservative of energy in an absolute sense. The fourth advantage of this system is that, with reduced capital investment/kW capacity and with cheaper base-plant power for compression, the generating cost in mills/kWh is lower than that of a comparable standard gas turbine system.

In our studies of the system shown in Fig. 3, we have not included means for recovery of energy in the turbine exhaust, i.e., recuperators or bottoming steam cycles, as our intent here was to focus on the air storage aspects of the system. Practical systems would include such devices if design calculations show cost improvement.

MATHEMATICAL MODEL

A rigorous system thermodynamic, fluid flow, and heat transfer analysis has been performed and programmed for digital computer solution. The program also includes a system economic model which determines equipment first cost and generating cost. The purpose of the modeling and analysis was to determine if the aquifer-stored air peaking concept is feasible economically, and to determine how the various parameters describing the system affect performance and costs.

The analysis determines the state points (temperatures and pressures) throughout the system and sizes the various components shown in Fig. 3. The physical modeling of the rotating equipment and heat exchangers comes from classical thermodynamics and basic heat transfer. Modeling for pressure drop of air during injection and withdrawal in the aquifer is similar to that given in Reference 1. Heat transfer in the aquifer is determined by a rather complex regenerative heat exchanger model. Air pressure loss and temperature change in wells and feeder lines are also considered. Attainable efficiencies are used for all thermodynamic and electrical equipments.

The cost model in the analysis is divided into two parts: capital investment and operating costs. The investment costs are further divided into those above or below ground. The aboveground items include all major equipment (turbine, compressors, auxiliaries, controls, combustors, motor/generator, heat exchangers, cooling towers, clutches, gearing, and pumps) and erection. Cost figures for these equipments are well-established late 1973 figures. Belowground costs include land, wells, and feeder lines. Natural gas transmission line costs are used for feeder lines and were taken from Reference 2. They may be slightly high. Well costs are from Reference 3 and (as in the case of feeder line costs) are typical total installed prices.

Generating costs are those due to capital at 18 percent/annum, base-plant fuel at 110 percent fuel replacement cost, gas turbine fuel, and 1.0 mill/kWh for O & M.

Not included in the economic model are any charges for engineering (unique to each job) necessary for putting together the system. Investment costs also do not include aquifer exploration and electrical transmission lines which also are unique to each job.

To give one a feel for how these added costs might affect our findings, for a typical system, if the added costs are \$1,000,000, investment cost would increase about \$6/kW and generating costs would increase about 1.2 mills/kWh.

RESULTS OF SYSTEM STUDY

Our main aim in this study was to determine whether a single turbine design, with fixed operating parameters, namely pressure ratio of 10:1, inlet temperature of 1850 F and an air flow of 650 lb/sec. (with 8.5 percent bypass cooling air), could be used with aquifers having disparate porosity and permeability and located at varying depths. The rationale here was that if a unique turbine design were to be necessary to match each aquifer system, the concept would not be feasible because of excessive equipment design cost. Our motive for selecting these particular turbine parameters was that accurate costs for a standard gas turbine system of this description were available and valid comparisons could be made.

Figure 4 is a pressure and temperature map for a typical aquifer-stored air peaking system. The aquifer has 50 wells, 12 inches in diameter, and is at a 520-ft depth. Its porosity is 0.12 and permeability is 1.13 darcy. The turbine is considered to have a load factor of 0.1 and the compressor 0.15. Ground water pressure is the hydrostatic head at 520 ft, i.e., 225 lb/in.^{2} The aquifer air bubble is five times larger than daily demand. As Fig. 4 shows, an overpressure of 50 lb/in.^{2} is required to inject the air while a drop of 75 lb/in.^{2} below hydrostatic head is experienced during withdrawal. The net result is that the compressors must have an overall 18.6:1 pressure ratio compared with 10:1 for the turbine. Another point to note in Fig. 4 is that the analysis assumes many cycles of operation where the air bubble expands and contracts over the daily cycle, during which water does not reflood the

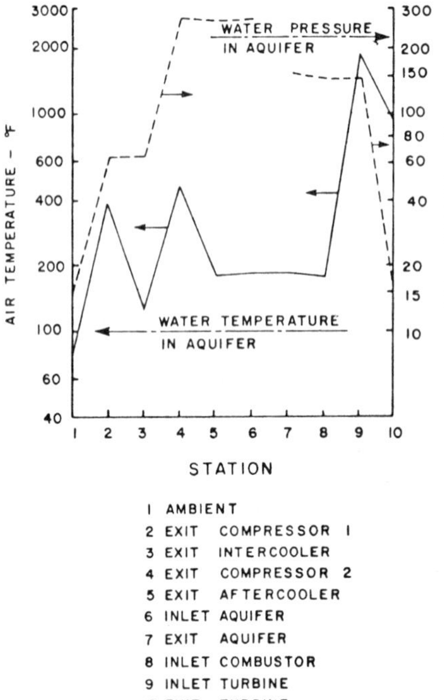

Fig. 4—Air pressure and temperature map for an aquifer-stored air peaking plant.

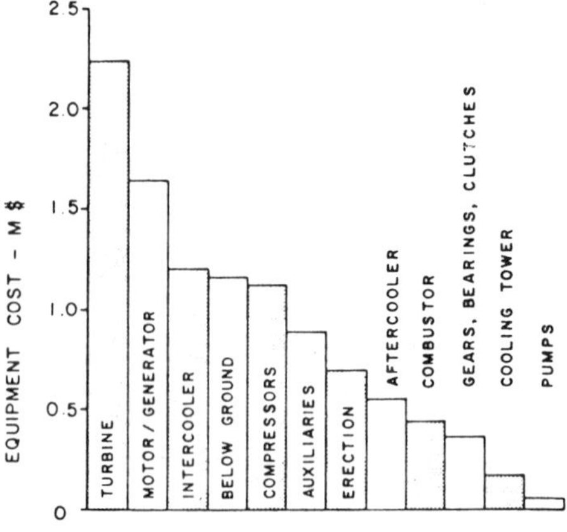

Fig. 5—Equipment costs for an aquifer-stored air peaking plant.

aquifer near the well bores. The effect is a gradual increase in stored air temperature as, each day, injected air heats the solids slightly. The temperature level shown for the air in the aquifer in Fig. 4 is the final equilibrium air-solid temperature near the well bore.

The net electrical output for this configuration is 168 MW. If the same turbine had been employed in a standard gas turbine system operating under the same thermodynamic conditions, the compressor would require about 90 MW and net output would be 78 MW.

Figure 5 shows the distribution of equipment costs for the system described above. Total capital cost is $10.6 x 10⁶ or $63/kW capacity. The comparable standard gas turbine would cost around $76/kW.

A point to note here is that belowground costs are $1.18 x 10⁶ or $7/kW. This compares well with the estimate of $5 to $15/ kW.[4] In addition, in terms of the amount of air stored (about 650 x 10⁶ standard cubic feet), the belowground cost is $1.80/1000 ft³, which is well above the natural gas storage experience. Thus, even though belowground costs are quite low compared with published estimates for comparable air storage systems using mined cavities,[5] the figure for aquifer storage is realistic, if not conservatively high.

The generating cost for the aquifer system, assuming gas turbine fuel cost of $1.00/10⁶ Btu and base-plant fuel at $.20/ 10⁶ Btu (a nuclear plant with 10,000 Btu/ kWh heat rate), is 2.2¢/kWh. The standard gas turbine system shows a comparable cost of 2.6¢/kWh. Another interesting comparison is heat rate. For the aquifer system, the heat rate is 13,300 Btu/kWh, of which 6200 Btu/kWh is from gas turbine fuel and 7100 Btu/kWh from base-plant fuel. For the comparable gas turbine system, the heat rate would be slightly more than 11,000 Btu/kWh, all of which is from gas turbine fuel. Thus, the aquifer system is seen to

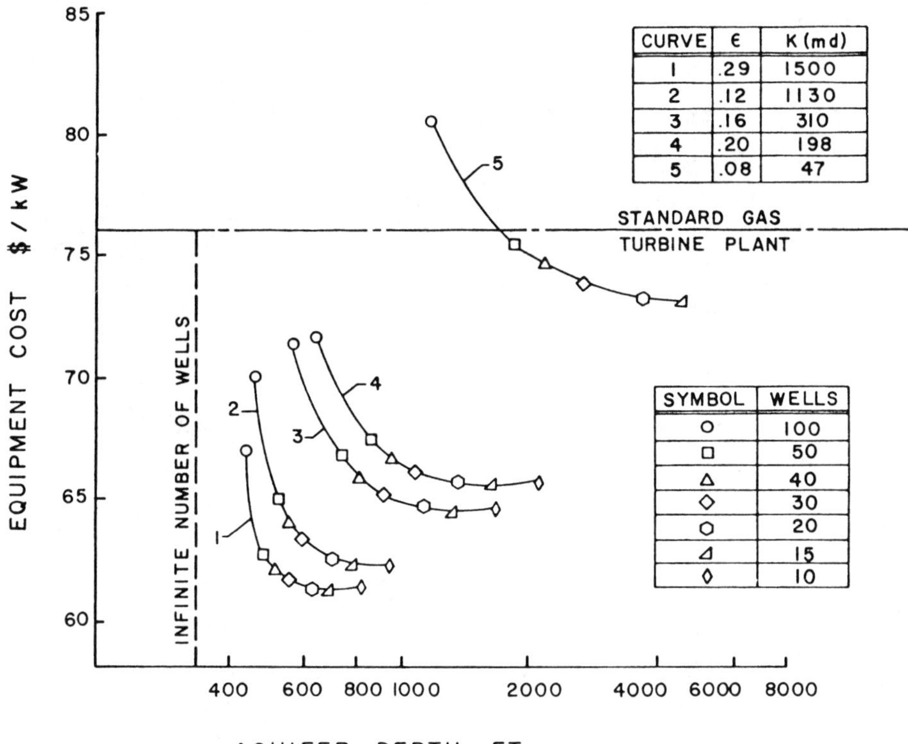

CURVE	ϵ	K (md)
1	.29	1500
2	.12	1130
3	.16	310
4	.20	198
5	.08	47

SYMBOL	WELLS
○	100
□	50
△	40
◇	30
○	20
◁	15
◇	10

Fig. 6—Effect of aquifer depth on equipment cost.

conserve gas turbine fuel, even though more total fuel is used for a unit of electrical output. The added heat rate for the aquifer system over the standard gas turbine is caused by the large compressor pressure ratio needed to overcome air pressure losses during aquifer injection and withdrawal, and by the fact that the air enters the combustor at a relatively low temperature.

Figure 6 shows expected equipment investment cost of an aquifer-stored air peaking system as a function of aquifer depth for five different porosity and permeability combinations. These porosity/permeability pairs are representative values in existing natural gas storage systems.[6] Curve No. 1 is a reservoir which is quite porous and permeable to flow, while No. 5 is the other extreme. For all cases here the turbine system is as described above: namely, 168 MW output, 10:1 pressure ratio, 1850 F inlet temperature, inlet air flow of 650 lb/sec., a turbine load factor of 0.1, and compressor load factor of 0.15. Each curve in Fig. 6 shows the number of wells required at certain aquifer depths. The number of wells is determined by dividing the total required flow by the amount of air each well can deliver at 10 atmospheres pressure at the combustor inlet. Three interesting points can be seen here. First, for a given aquifer porosity/permeability, as aquifer depth increases the required number of wells decreases. This tends, at the smaller depths, to reduce

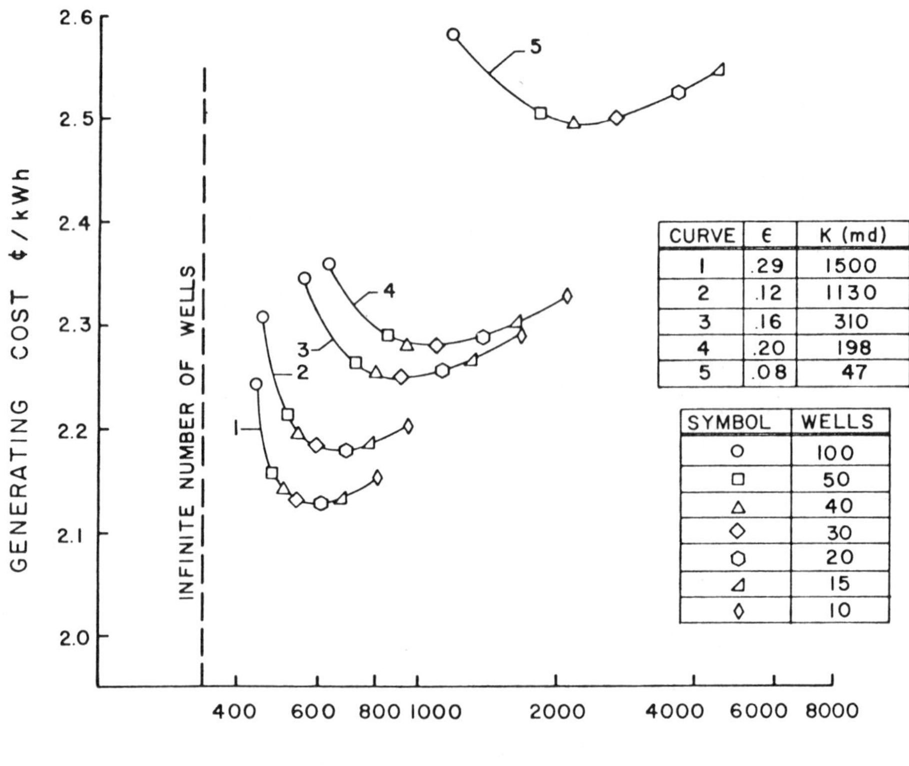

CURVE	ε	K (md)
1	.29	1500
2	.12	1130
3	.16	310
4	.20	198
5	.08	47

SYMBOL	WELLS
○	100
□	50
△	40
◇	30
○	20
◿	15
◊	10

Fig. 7—Effect of aquifer depth on generating cost.

equipment cost. However, at greater depths the increased cost of the compressors offsets reduced well costs and equipment cost is found to be insensitive to depth. The second point to be noted is that at the smaller depths, costs increase drastically and asymptotically approach the line through a depth of 340 feet. At this depth, the hydrostatic head is 10 atmospheres and it would theoretically take an infinite number of wells to deliver air at 10 atmospheres above ground. The third point of interest is that, at a given depth, equipment cost increases as permeability decreases. The increase is caused by the requirement of more wells and higher pressure ratio compressors.

Overall, one should note from the findings in Fig. 6 that, for a given porosity/permeability, there is an optimum range of aquifer depth which tends to minimize equipment cost. One should also note that only in the case of the most impermeable aquifer does equipment cost approach that of a standard gas turbine system.

Figure 7 presents generating costs for the same conditions in Fig. 6, with \$1.00/10⁶ Btu gas turbine fuel, base-plant heat rate of 10,000 Btu/kWh with \$0.20/10⁶ Btu fuel charged at 110 percent replacement cost, and the cost of capital at 18 percent/annum. For the comparable standard gas turbine system, generating cost would be about

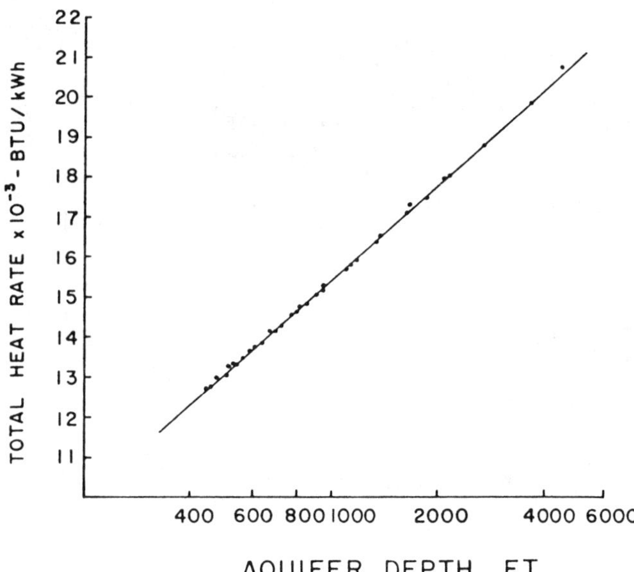

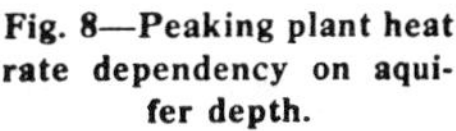
Fig. 8—Peaking plant heat rate dependency on aquifer depth.

2.6¢/kWh. The same conclusions as drawn in Fig. 6 still apply; namely, that costs are reduced for the more permeable aquifers (at a given depth) because of reduced investment cost and because less energy is required to inject the desired amount of air. Again, there appears to be an optimum range of depth for each curve in which generating costs are at a minimum and are insensitive to depth.

Figure 8 gives the heat rate for each point shown in Figs. 6 and 7. Heat rate has been found to be a function of aquifer depth only. There appears to be no direct relationship between porosity or permeability and heat rate, but there may be an indirect one in that the less permeable aquifers must be found at greater depths to give the required flow of air. For all cases in Fig. 8, the gas turbine fuel heat rate is 6200 Btu/kWh and the remainder is in base-plant fuel.

In all of the systems considered previously, the analysis assumed a turbine load factor of 0.1 and compressor load factor of 0.15. The compressor load factor is probably conservatively low, since a load factor of 0.15 is roughly equivalent to a six-hour, full-capacity compression each night. When one considers longer nightime and weekend compression periods, this figure might well be 0.30. Figures 9 and 10 show the effect of varying load factors on equipment and generating costs, respectively. In Fig. 9, we see that increasing compressor load factor while maintaining a constant turbine load factor will reduce equipment costs due entirely to reduced compressor, intercooler, and aftercooler size. When one increases turbine load factor at constant compressor load factor, equipment costs increase because the sizes of compressors and coolers must increase and the required number of wells increases slightly. In Fig. 10, the same conclusions apply to increasing compressor load factor, i.e., increased factor decreases cost. This is true when one increases turbine load factor as well. In both cases, the improvement occurs because equipment

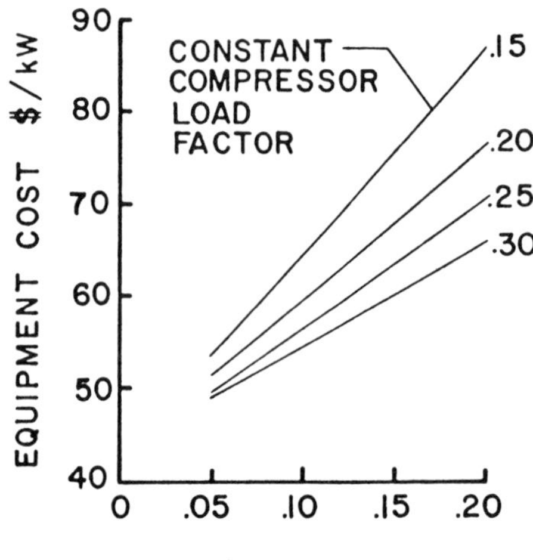

Fig. 9—Effect of load factors on equipment cost.

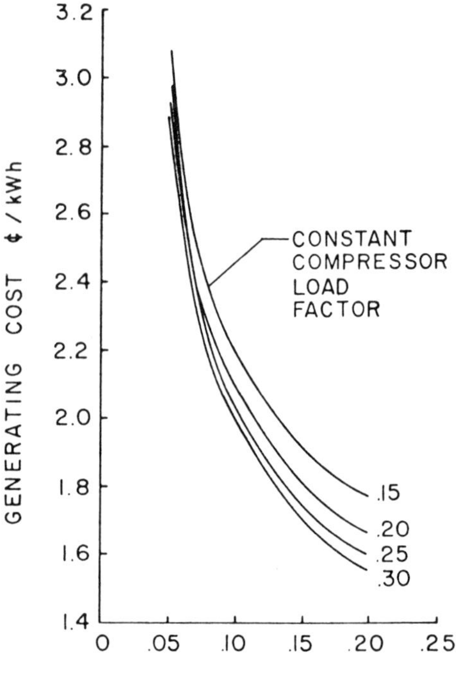

Fig. 10—Effect of load factors on generating cost.

costs are written off over more operating hours.

The net effect of increasing gas turbine fuel costs above the $1.00/10^6 Btu used throughout this paper is, of course, an increase in generating cost. However, since only a part of the heat rate is supported by gas turbine fuel, the generating cost advantage relative to a standard gas turbine system would increase.

CONCLUSIONS

Large portions of our country possess a sedimentary geologic structure which is favorable to discovery of porous water-bearing strata which are suitable for storage of large volumes of air. Over 50 such systems are in use today for the storage of natural gas. The technology necessary for discovery and exploitation of aquifers for such gas storage is well advanced and is proven.

Large blocks of a utility's surplus generating capacity can be used during slack demand periods for compressing air for storage in aquifers, to be withdrawn during subsequent peak-demand periods when base-plant capacity cannot meet demand. The air can be used to feed a gas turbine which, running without an intergral compressor, delivers more than twice its normal output. Such systems cost less in initial investment/unit of output and deliver power at less cost than standard gas turbines operating under comparable thermodynamic conditions. And the heat rate for gas turbine fuel is slightly more than one-half that of the comparable standard system. Aquifer-stored air peaking systems can also be used for emergency power, either by utilizing the excess stored compressed air until it is depleted or by running the compressors directly connected to the turbine.

Analysis has shown that it is not necessary to tailor the turbine to each aquifer. Off-the-shelf designs which are similar to standard gas turbine designs on the market today can be developed if there is demand for them. However, air compressors which are driven separately from the turbines must be matched to each aquifer system, their capacity and pressure ratio being dictated by design duty cycle and the physics of the aquifer.

11

Reprinted from *Report ORNL-NSF-EP-11*, Oak Ridge National Laboratory (operated by Union Carbide for the Dept. of Energy), 1972, 17pp.

UNDERGROUND AIR STORAGE AND ELECTRICAL ENERGY PRODUCTION

R. B. Korsmeyer

ABSTRACT

With the trend toward very large power plants, the problem of meeting peak power demands becomes acute, and large utilities have been turning to energy storage as a means of leveling out demand into a nearly steady base load. The only way that energy of such magnitude can be stored is either by pumping water to a higher elevation to take advantage of gravity or by compressing air in large underground caverns. Pumped-water storage is the more flexible method and can be fitted into various demand patterns with both hydro- and thermal-electric generating systems, but it generally must be situated in at least hilly terrain to obtain adequate differences in elevation and it uses considerable land surface.

Underground air storage, on the other hand, can be located almost anywhere that firm rock can be found at depth, uses only about one-tenth of the land surface that equivalent pumped-water storage does, and uses the stored energy more effectively; but it can be utilized only in combination with gas turbine generators. At present air energy storage requires equipment development and, therefore, there are uncertainties in equipment cost estimates. There are also large uncertainties in excavation cost estimates and in possible environmental effects. In this review we examine the potential advantages of pumped-air storage and develop recommendations for research required to clarify the options.

I. INTRODUCTION

With the trend toward very large power plants, and especially nuclear-electric generating stations, the problem of meeting peak power demands becomes acute because it is not economical to change the plant load either rapidly or frequently. Increasingly, large utilities have been turning to energy storage as a means of leveling out the demand into a nearly steady base load. The peak-to-average power demand varies most severely on a daily basis in industrial and urban areas, although weekly and seasonal variations are often important too. In large American cities the daily peak demand generally runs from 20 to 30% above the average for a period of 5 to 8 hours.[1] This is equivalent to a storage capacity of roughly 1.5 MW hours per MW average load.

An alternative to energy storage for mitigating the peaking problem is the use of interconnections among utility systems that span long distances across the country. Diurnal peaking can be alleviated by east-west interconnections sufficiently extensive to cover several time zones, and seasonal peaking between winter heating and summer air conditioning can be reduced by north-south interconnections. A considerable increase in transmission capacity with attendant capital and environmental costs would no doubt be required. A number of utilities are meeting the variable load problem by installing gas-turbine peaking plants which are operated only during periods of heavy demand because of their ease of rapid startup and shutdown.

II. METHODS OF ENERGY STORAGE - PUMPED WATER

Pumped-water storage is a flexible method and can be fitted into various demand patterns with both hydro- and thermal-electric generating systems. The simplest scheme utilizes two reservoirs located at different elevations. Water is pumped uphill during off-peak periods and then is allowed to flow downhill to generate additional power during the peaks. The difference in elevation need not be great (say, 50 ft) but the reservoir size and pumping losses are inversely proportional to the height for a given energy storage. The maximum practical elevation difference for single station operation is about 1700 ft as this is the maximum head against which the reversible Francis turbine can work--as a pump during off-peak periods and as the generator drive during peaking. There are other advantages to pumped-water storage, such as easy provision of "spinning reserve," etc.[2,3,4]

Whereas United States installations have been entirely on the surface, the Swedes have utilized the tail raceways of underground hydroelectric plants of very high head as reservoirs from which water is returned to an upper elevation during off-peak hours for reuse in peaking periods. These tunnels, however, entail excavation of enormous volumes of rock.[5] Even so, Swedish engineers believe that it is possible to mine very deep caverns almost anywhere as catch basins for water that is used to generate power on the way down and is pumped back up again during off-peak hours.[6] In fact, this scheme has been proposed in connection with a pumped-storage system as part of Chicago's storm water and pollution control systems.[7]

The extent of pumped-storage system development in the United States is shown in Table 1.[8] Predictions have been made of very large numbers of pumped-water storage installations in this country in the next thirty years.[9] Table 2, taken from ref. 9, shows the projected pumped-water storage capacity as 24 million kW by 1980 and 65 million by 1990. However, since there is no overall siting plan for the country, it is not possible to say whether the construction of all these installations is feasible or whether there may be an urgent need for the development of alternatives such as underground air. Already the conflict over land use for pumped-water storage has arisen and litigation has forestalled construction of at least one large project for a number of years (the Storm King project at Cornwall, N. Y.). We expect that the conflict will increase greatly as time goes on, particularly in the absence of a well thought-out plan that optimizes land use with growing power and recreational requirements.

III. METHODS OF ENERGY STORAGE - COMPRESSED AIR

When large thermal-electric power plants are situated in areas unfavorable for construction of pumped-water surface systems, it is possible to store the water in underground caverns. However, very large caverns are required to accommodate these pumped-water systems. A possibly more attractive alternative is compressed-air storage in combination with gas-turbine peaking generators. The use of such a system is particularly applicable in this country where utilities are installing multiples of the largest gas-turbine generators available today (about 60 MW in the U. S.,[10] 100 MW outside[11]) for supplying peak power. These gas turbines, fueled with natural gas or oil, contribute more heavily to atmospheric pollution than similarly fueled steam power plants because

Table 1. U. S. Pumped-Storage Projects

Name	State	Owner	Date	Head, ft	Reversible Capacity, kW
		In Operation			
Rocky River	Conn.	Conn. Lt & Power	1928	230	7,000[1]
Buchanan	Tex.	Lower Colo. River Authority	1951	135	11,000[1]
Flatiron	Colo.	BuRec	1954	1,115	9,000
Hiwassee	N. C.	TVA	1956	254	59,500
Lewiston	N. Y.	Power Auth. N. Y.	1961	85+300	240,000
Taum Sauk	Mo.	Union Elect.	1963	830	350,000
Smith Mtn.	Va.	Appalachian Power	1965	227	132,000
Yards Creek	N. J.	Jersey Cent. P. & L. Pub Service E. & G.	1965	760	338,000
Senator Wash	Calif.	BuRec	1965	65	7,200
Cabin Creek	Colo.	Pub. Serv. of Colo.	1967	1,226	350,000
Muddy Run	Pa.	Philadelphia Elec.	1967	411	800,000
Oroville-Thermalito	Calif.	State of Calif.	1968	675 102	345,000 91,200
Salina	Okla.	Grand River Power Authority	1968	251	520,000
San Luis	Calif.	BuRec	1968	320	424,000
		Under Construction			
Jocassee	S. C.	Duke Power Co.		310	450,000
Carters	Ga.	Army Engrs.		400	250,000
Kaysinger Bluff	Mo.	Army Engrs.		46	160,000
Kinzua	Pa.	Penn Elec. & Cleveland Elec.		798	330,000
Northfield Mtn.	Mass.	Northeast Utilities		800	1,000,000
Clarence Cannon	Mo.	Army Engrs.		75	27,000
DeGray	Ark.	Army Engrs.		175	28,000
		Pending			
Cornwall	N. Y.	Con Edison		1,160	2,000,000[2]
Blue Ridge	Va.	Appalachian Power		210	1,600,000
Laurens Shoals	Ga.	Georgia Power		95	216,000
Montezuma	Ariz.	Ariz. Power Auth.		1,690	500,000[3]
Longwood	N. J.	Jersey Central Power & Light		390	122,000
Blair Mtn.	Colo.	Colo. River Water Cons. Dist.		2,200	525,000
Blenheim-Gilboa	N. Y.	Power Auth. N. Y.		1,000	1,000,000
Raccoon Mtn.	Tenn.	TVA		1,000[4]	1,400,000
Ludington	Mich.	Consumers Power		362	1,900,000
Bear Swamp	Mass.	New England Power		730	600,000

Table 1. (Continued)

Name	State	Owner	Date	Head, ft	Reversible Capacity, kW
Marble Valley	Va.	Virginia Elect.		800	1,000,000 [5]
Mormon Flat	Ariz.	Salt River Project		129	47,000
Horse Mesa	Ariz.	Salt River Project		264	70,000
Castaic	Calif.	Los Angeles Dept of Water & Power		1,000 [4]	1,200,000
Tocks Island	N. J.	Jersey Cent. P. & L. Pub Service E. & G.		1,200	950,000
Jocassee No. 2	S. C.	Duke Power Co.		1,180	2,300,000
Dirty Face Mtn.	Wash.	PUD No. 1 of Chelan County		670	65,000
Merrill Lake	Wash.	PUD No. 1 of Cowlitz County		1,060	500,000

[1] Separate pumps and turbines.

[2] May be increased to 3,000,000 kW.

[3] May be increased to 1,000,000 kW.

[4] Tentative head.

[5] May be increased to 1,500,000 kW.

Table 2.* Electric Utility Requirements and Supply,[1] 1965-90

	1965	1970	1980	1990
Energy requirement (trillion kWhr)	1.06	1.52	3.07	5.83
Peak demand (million kW)	188.0	277.0	554.0	1,051.0
Total installed capacity (million kW) [2]	236.1	344.0	668.0	1,261.0
Hydroelectric capacity (million kW)	41.7	51.4	68.0	83.0
Pumped storage capacity (million kW)	1.3	3.6	24.0	65.1
Internal combustion and gas turbine capacity (million kW)	4.9	16.2	27.0	42.0
Fossil steam capacity (million kW)	187.5	261.2	399.0	562.0
Nuclear capacity (million kW)	.7	11.6	150.0	509.0
Capacity dependent on cooling water (million kW)	188.2	272.8	549.0	1,071.0

*Taken from ref. 9.

[1] Excludes Alaska and Hawaii.

[2] Does not add up due to rounding.

they are less thermally efficient. For the same reason, they more
rapidly consume our dwindling natural gas and petroleum reserves.

Since a major part of the shaft power developed by gas turbines is
required to compress the air for the burner, the electrical output, and
hence efficiency, of gas-turbine generators is low. By compressing the
air with off-peak, base-load power and storing it underground, the gas-
turbine shaft power may be applied entirely to electric power generation.
In this way the gas turbine is able to produce about 2.5 times the
electrical power directly obtainable otherwise.[12,13]

Containment for the air storage is provided by underground caverns
connected to a hydrostatic leg terminating in a suitable body of water
at the surface. The hydrostatic leg provides an essentially constant
air pressure during withdrawal from the cavern. The system is shown
schematically in Figure 1. It appears that the "breathing" of the cavern,
as represented by water flow, is about 10 to 12% of the water flow in a
pumped-water storage system of the same head and energy storage capacity.
Thus, a much smaller lake is required, and less land surface is used. In
fact, almost any body of water, such as a river or the ocean, could be
used without land damage, and site selection would no longer be dependent
on mountainous terrain.

The pressures to which air is usually compressed for gas-turbine
combustion is 20 atm or more. Stal-Laval[13] believes the economic optimum
is about 40 atm, taking into account the cost of mining the cavern. The
cavern must be located at a depth of 1400 ft below ground to develop
this pressure with a hydrostatic leg, and it must be excavated from
substantially impervious rock to minimize sealing costs.

IV.   CURRENT STATUS OF DEVELOPMENT

The compressed-air energy storage concept has not been put into
practice at the present time. Considerable attention has been given to
it in Sweden, Finland, Denmark, Yugoslavia, and France.[14] Electricité
de France has had some interest, but currently their interest is dormant.
No public comments either way have been made by U. S. utilities.[15] TVA
is purchasing considerable gas-turbine generating capacity for peak power

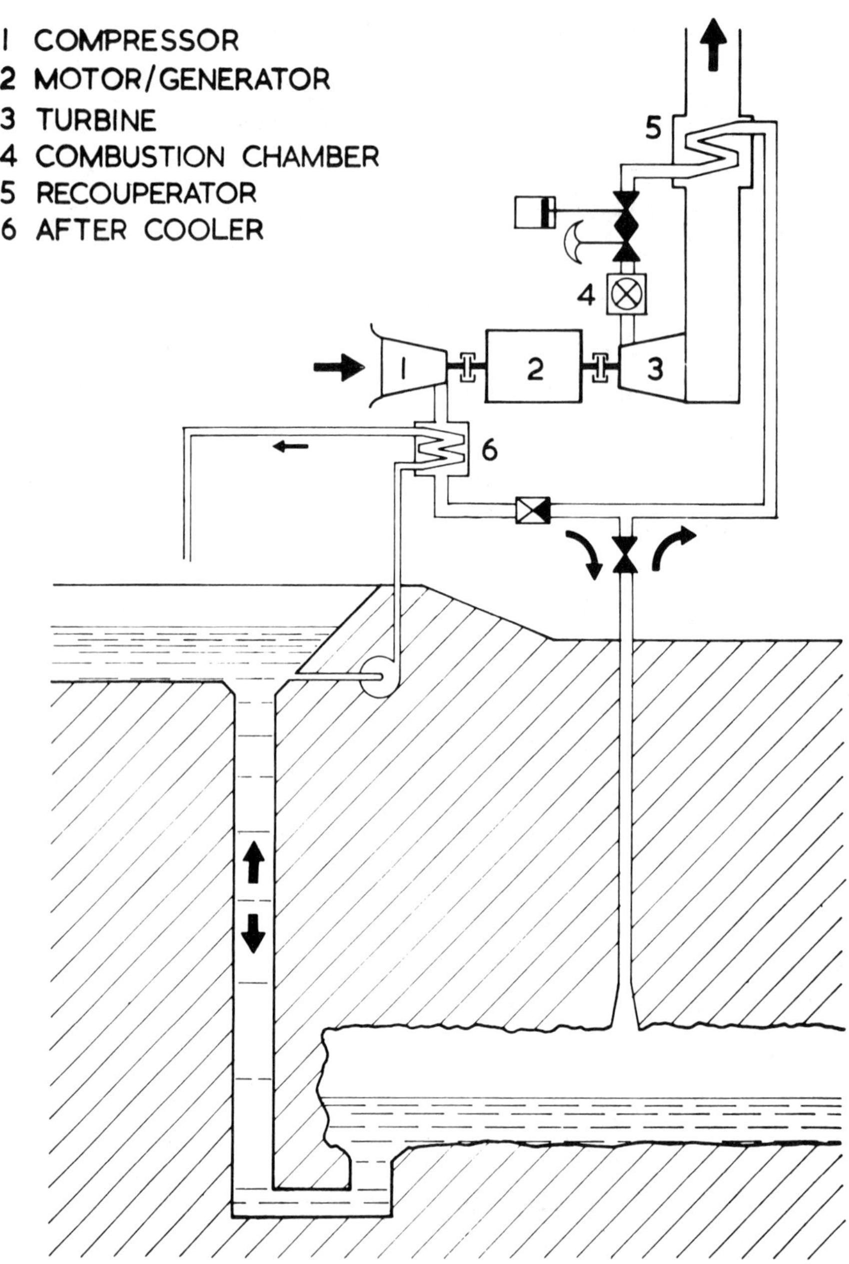

Fig. 1. Schematic View of Underground Air Storage Power System (taken from ref. 22).

needs and is currently designing a pumped-water storage system. They currently believe that geological problems (if not others) preclude air storage consideration in their area, but they are keeping an open mind.[16]

Probably more attention is devoted to the study of underground air storage in the U. S. than the paucity of literature would imply. The mining and tunneling specialists, Fenix and Scisson, collaborating with Stal-Laval "and others" in connection with mining costs, find that the idea has considerable merit but that "mechanical designs are about five years from realization." They have not studied the problems in sufficient detail to produce meaningful cost curves for cavern size (and depth) and do not plan to do so in the immediate future.[17]

V.   ENGINEERING AND ECONOMIC PROBLEMS

Efficiency and Effectiveness

Simple, open-cycle, conventional gas-turbine plants have much lower thermal efficiency than modern steam plants. These gas-turbine units use about 12,000 Btu/kWhr[11], which is equivalent to about 29% thermal efficiency. This compares to 39% efficiency (8800 Btu/kWhr) of, for example, the TVA Bull Run Plant.[18] Because of this relatively poor fuel economy, gas-turbine plants are used mainly for supplying peaking power in combination with large, modern base-load plants. On the other hand, a gas-turbine plant with compressed-air storage can operate in the generating mode for eight or more hours a day, and unit size can be from 100 to 250 MW. Fuel consumption is reduced to about 3960 Btu/kWhr to yield a generating efficiency of 86%, exclusive of the contribution of the stored air energy. In summary, to generate 1.00 kWhr of peak-load energy, the compressed-air storage plant requires an energy input of 0.82 kWhr in the form of off-peak electric energy to compress air, and 3960 Btu in the form of fuel.[12] (Olsson[13] takes a slightly more conservative view.)

Overall effectiveness of storing off-peak energy can be calculated in various ways. For example, with steam units having heat rates around 9000 Btu/kWhr, the 3960 Btu/kWhr thermal energy used by the gas turbine could generate off-peak electric energy of 0.44 kWhr. The effectiveness

149

(not thermal efficiency) of storing power therefore is 1.00 kWhr/(0.82 + 0.44) kWhr, or 79%. By comparison, a pumped-water storage system has an effectiveness of about 67% because about 1.5 kWhr of pumping power is required for every kWhr returned to the grid as peak power. Another way of comparing effectiveness that is meaningful from the standpoint of air pollution is to estimate the total fuel consumption per unit of peak power produced. For the gas-turbine system which requires 0.82 kWhr of off-peak energy and 3960 Btu/kWhr of thermal energy, the total fuel consumed is 0.82 x 9000 + 3960, or 11,340 Btu/kWhr. The total fuel consumed by the pumped-storage plant is 1.5 x 9000, or 13,500 Btu/kWhr (19% more than the gas-turbine plant). An inventory of thermal discharges from the 2 plants is given in Table 3.

Table 3.   Inventory of Thermal Discharges from Pumped-Water
and Compressed-Air Energy Storage Plants
(Btu/kWhr peak energy)

	Pumped-Water	Compressed-Air
Base-load plant stack	1,350	738
Base-load plant condenser	7,031	3,844
Upper reservoir	853	–
Lower reservoir	853	–
Turbine exhaust	–	546
Aftercooler	–	2,799
Total Thermal Discharge	10,087	7,927
Power output	3,413	3,413
Energy Input	13,500	11,340

Possibilities for improving the compressed-air storage system shown in Figure 1 include reclamation of the aftercooler energy or the addition of intercoolers. After compression to 40 atmospheres, the air temperature would be approximately 1200°F if intercoolers are not used. There may be worthwhile purposes to which this thermal energy may be applied, even though it is only intermittently available. Heat removal from the air during the compression process by intercoolers would reduce the power required to drive the compressor and improve the effectiveness of the plant. A compressor power reduction of 20% or more should be achievable.

Turbine Generator Costs

The largest modern gas-turbine plants made in the U. S. cost about $83/kW. In these plants the compressor, turbine, and generator are all mounted on a single shaft. In the stored energy plant, however, the three components must be separated by clutches so that the compressor and turbine may be individually coupled to the generator, depending on whether the generator is supplying power to the grid (turbine-coupled) or whether, as a motor, it is driving the compressor (compressor-coupled). Because the compressor in a normal gas-turbine plant may absorb 60% or more of the turbine output, the generator in a decoupled plant would be at least 2.5 times larger than in the normal plant having the same compressor and turbine. The estimated cost of the gas-turbine unit for use in an energy storage plant, including clutches and based on the larger generator rating, is in the range of $50 to $70/kW.[18]

Cavern Problems

According to Olsson[13] a 220-MW gas-turbine stored air plant would have an optimum cavern size of 4.8 million cubic feet at a depth of 1400 feet. The cavern would probably have to be mined in the form of a series of excavated tunnels, as the likelihood of finding a natural cavern of suitable size, shape, and depth is extremely small. While one might form such a cavity in a salt dome without mining by dissolving away many tons of salt, there is the problem of handling extremely saline water in the hydrostatic leg and perhaps in other equipment. A somewhat different concept, with compressed air stored in solution cavities in salt domes, has been proposed by Whitehouse.[19] In this case there was no hydrostatic leg, and energy recovery was through a rotary screw air compressor. Depleted oil and gas wells would not ordinarily be suitable since the fluid flow required implies passages of large cross-section.

In all of the published studies the availability of nearly impervious rock for the cavern walls is assumed. Costs could be increased significantly if sealing were found necessary, and the cavern, air shaft, and hydrostatic leg must certainly be established as substantially impervious to outward seepage to avoid the possibility of contamination of the surrounding aquifer, particularly if salty or polluted water is utilized.

One might reasonably assume that a well-made cavern in good rock should undergo little or no measurable degradation over the years, and hence its expected lifetime would exceed that of the best power plants. Nonetheless, structural components and hardware (such as the lower end of the hydrostatic leg) will deteriorate, and it seems only prudent to provide some means of periodic inspection and checking leak-tightness-- at least a few times during the life of the power plant.

Cavern Costs

The cost of mined caverns, and particularly special cavern and structural features that may be required, is essentially unknown because the figures that have been published are considerably at variance with other mining estimates. The attitude of Fenix and Scisson mentioned above throws considerable doubt on the accuracy of cavern mining figures published abroad[6,13] and also of those indicated by, for example, the Harza engineering firm of Chicago.[20] Some of the estimates have apparently been based on a rock removal cost of about \$0.04/ft^3 (Olsson estimates \$0.50/ft^3), but current tunneling costs in the U. S. are much closer to \$1.00/ft^3 under favorable conditions and could easily approach double that in unfavorable geologic formations.[21] In this connection it can be noted that 1967 Fenix and Scisson figures for 25 ft diam unlined tunnels in moderately stratified rock ran between \$700 and \$900 per linear ft. This corresponds to about \$1.60/ft^3, but it includes support with steel horseshoe ribs. The existence of more recent estimates from Stal-Laval can be inferred from the paper by Harboe,[22] where costs appear to be between \$0.70 and \$0.90/ft^3.

Effects on Natural Resources

A well-sealed cavern, hydrostatic leg and shaft would probably have no adverse effect on underground resources, but the problem of aquifer lowering or contamination is important, for it may involve damage to urban structures and services as well as potable water supplies.[23]

Land surface degradation is potentially less than with present pumped storage, but each situation must be independently examined.

Cavern Degradation

The constant leaching and drying of the cavern walls with change in water level is probably unimportant in granite or even limestone, but if veins of some minerals are encountered, one can imagine rock deterioration leading to hydraulic flow problems in the course of years and to loss of imperviousness for unlined caverns and hydrostatic legs. Even concrete linings may be attacked if sulfate is present.[24,25]

VI.   TENTATIVE RELATIVE COSTS OF COMPRESSED-AIR AND PUMPED-WATER STORAGE

If one accepts the published estimated costs of compressed-air storage energy systems and their counterparts in pumped-water storage, it is possible to see the relative effects of estimating errors in terms of the cost of peaking power. For example, both Kalapasev[12] and Olsson[13] estimate the capital investment for a 220 to 250 MW compressed-air storage plant at $50/kW. With cost of capital calculated at 7% interest and write-off periods of 28 years aboveground and 40 years underground, plus an annual charge of 30¢/kW for miscellaneous costs and services, the total fixed cost is $3.75/kW-year excluding taxes.[13] If a return on capital of, say, 13% (including taxes), more commonly used in U. S. accounting practices, is applied, the fixed cost becomes $6.80/kW-year. Using the latter figure plus fuel cost of 83¢/MBtu* (3.3 m/kWhr), running time of 2000 hr/year, and a base-load power cost of 4.0 m/kWhr, the cost of peaking power is 10.2 m/kWhr.

For an equivalent size pumped-water storage system, Hubbard[26] states that the capital investment is about $110/kW of which 43% is for reservoirs, dams, etc., that are amortized over long periods. Using a 13% return on the power plant and 6% for the dams and reservoirs, the annual charge becomes $11/kW-year, or 5.5 m/kWhr for 2000 hr/year. Allowing 0.3 m/kW for maintenance and an effectiveness of 67% with base-load power at 4 m/kWhr, the cost of peaking power is 11.8 m/kWhr.

---

*Cost used by Consolidated Edison for fuel oil.

Since gas-turbine generators are not available in larger units, greater plant capacity can be obtained only by multiplying the number of units, and it is not clear whether larger sized caverns have lower unit construction costs or not. Therefore, one is forced to assume that, say, a 1000 MW compressed-air storage system would cost essentially four times that of a 250 MW plant. On the other hand, Hubbard[26] states that 1000 MW pumped-water storage plants cost about $70/kW, or only about 64% of the unit cost of 250 MW plants. In this size, therefore, the pumped-water plant would be more economical than compressed-air storage. These three cases are compared in Table 4.

A point not clarified by Hubbard is the effect of elevation (between reservoirs) on power cost. He states merely that his cost estimates are for elevations equal to or greater than 350 ft. However, it is obvious that the quantity of water that must be pumped is inversely proportional to elevation, and therefore reservoir size decreases with elevation for given storage capacity. Economies in reservoir size may, however, be largely offset by other construction costs.

Table 4. Comparison of Compressed-Air and
Pumped-Water Energy Systems

	250 MW Compressed-Air	250 MW Pumped-Water	1000 MW Pumped-Water
(m/kWhr @ 2000 hr/year)			
Fixed costs	3.4	5.5	3.5
Fuel cost	3.3	–	–
Maintenance	0.2	0.3	0.3
Base-load pumping @ 4 m/kWhr	3.3	6.0	6.0
Total peaking power cost	10.2	11.8	9.8

Effect of Cavern Mining and Equipment Costs on Air Storage Power

If mining costs are $1.00/ft^3, instead of Olsson's $0.50/ft^3, the cavern cost (in his estimate) is \$22/kW instead of \$11/kW. Fixed cost would then be increased from \$6.80 to \$8.23/kW-year, and power costs would increase by 0.7 m/kWhr, to 10.9 m/kWhr, for a 250 MW plant. Undoubtedly, one would consider going to deeper, smaller caverns for minimum power costs, but the overall effect on the power cost is complicated.

If the high estimate for the gas-turbine equipment (\$70/kW) is combined with the high estimate for cavern mining (\$22/kW), the fixed cost becomes \$12.26/kW-year and the cost of peaking power becomes 12.9 m/kWhr.

VII.   CONCLUSIONS

1.   Costs of modified gas-turbine generators for stored air utilization are not firm. Present estimates range from \$50/kW to about \$70/kW. Pumped-water storage systems cost about \$70 to \$110/kW.

2.   Cost of mined caverns, and particularly special cavern structures that may be required are essentially unknown because the figures that have been published are at variance with known U. S. tunneling costs in hard rock.

3.   The air-storage concept appears to be a more effective method of storing energy than the pumped-water concept in that less energy is wasted per kWhr of peaking energy produced. However, the incremental energy required over base-load energy must be supplied by gas or liquid fuel, whereas all of the peaking energy derived from pumped-water storage comes from whatever fuel is used for the base-load power generation.

4.   All published proposals assume the cavern to be located in impervious rock such as granite, limestone or well-compacted shales, as the cavern leakage must be quite small to be economical. Leakage also needs to be negligible from the standpoint of ground contamination if polluted or salty water is used for the hydrostatic leg. Land surface degradation is potentially less than with pumped-water storage on account of the much smaller water movement (one-eighth to one-tenth as much as with comparable pumped-water storage).

5. Further work is needed in a number of areas in order to assess the economics and potential environmental effects of compressed-air energy storage compared to pumped-water storage. The work recommended is to:

  a. Obtain firmer estimates of modified gas-turbine costs.

  b. Develop reliable methods for estimating cavern mining and preparation costs for different types of geological formations.

  c. Conduct a geological survey to determine as nearly as possible what areas of the country are potentially suitable for underground air storage and what areas are not. Information developed by current ORNL seismic studies relating to siting of nuclear power plants is pertinent.

  d. Compare the geological information with the related topography and land development to assess the relative merits of pumped-water storage and compressed-air storage for given locations.

  e. Study in detail the comparative overall environmental effects of both systems of storing energy for a few selected sites in the country.

VIII.   REFERENCES

1. A. J. Miller et al., Use of Steam-Electric Power Plant to Provide Thermal Energy to Urban Areas, ORNL-HUD-14, pp. 22-23, January 1971.

2. H. Susskind and C. J. Roseman, Combined Hydroelectric Pumped Storage and Nuclear Power Generation, BNL 50238, April 1970.

3. S. R. Knapp, Pumped Storage - the Handmaiden of Nuclear Power, *IEEE Spectrum*, pp. 46-52, April 1969.

4. John Pett, How to Evaluate Pumped Storage for Peaking Generation, *Power Engineering*, pp. 28-32, July 1964.

5. Op. cit., Ref. 2, p. 45.

6. Engineering News-Record, p. 36, August 29, 1968.

7. Op. cit., Ref. 2, p. 46.

8. Engineering News-Record, p. 24, January 2, 1969.

9. Environmental Effects of Producing Electric Power, Hearings before the Joint Committee on Atomic Energy, Ninety-first Congress, Part I, pp. 55-56.

10. V. V. Schloesser, Gas-Turbine Peaking Plant Provides 58 MW in a Modular Package, paper given at 31st Annual Meeting of the American Power Conference, April 22, 1969.

11. P. S. Chernishev et al., Russia's 100 MW Gas Turbine, *Mech. Eng.*, p. 26, December 1970.

12. S. D. Kalapasev, Constant Pressure Compressed-Air Storage Extends Peak Periods for Gas Turbine, *Power*, pp. 90-93, March 1970.

13. E.K.A. Olsson, Air Storage Power Plant, *Mech. Eng.*, pp. 20-24, November 1970.

14. *Business Week*, p. 76, August 1, 1970.

15. B. Flanagan, *Business Week* editorial staff, personal communication.

16. H. Falkenberry, TVA, personal communication.

17. C. T. Brandt, Vice-President of Fenix & Sisson, letter of February 10, 1971.

18. M. L. Myers, personal communication.

19. G. D. Whitehouse, M. E. Council, and J. D. Martinez, Peaking Power With Air, *Power Engineering*, January 1968, pp. 50-52.

20. Op. Cit., Ref. 2, p. 47.

21. G. A. Cristy, personal communication.

22. Henrik Harboe, Economical Aspects of Air Storage Power, presented at the American Power Conference, Chicago, Illinois, April 20, 21, and 22, 1971.

23. C. O. Morfeldt, Significance of Groundwater at Rock Construction of Different Types, p. 306 in Proceedings of the International Symposium, Oslo, September 23-25, 1969, *Large Permanent Underground Openings*, (Experience at Stockholm), (see also pp. 316-317), edited by Brekke and Jorstad, Universitetsforlaget 1970.

24. J. Martna, Engineering Problems in Rocks Containing Pyrrhotite, p. 87 in Proceedings of the International Symposium, Oslo, September 23-25, 1969, *Large Permanent Underground Openings*, edited by Brekke and Jorstad, Universitetsforlaget 1970.

25. T. H. Hagerman, Groundwater Problems in Underground Construction, pp. 319-321 in Proceedings of the International Symposium, Oslo, September 23-25, 1969, *Large Permanent Underground Openings*, edited by Brekke and Jorstad, Universitetsforlaget 1970.

26. C. W. Hubbard, These Factors Must be Weighed in Developing Pumped-Storage Plants, *Power Engineering*, February 1966, p. 42.

Part III

CHEMICAL ENERGY STORAGE

Editor's Comments
on Papers 12 Through 18

For this volume the term *chemical energy storage* has been restricted to include only those substances that are produced specifically for storage and that are consumed in combustion or some other oxidation process. Compounds included are hydrogen, ethanol, methanol, and synthetic gases. Although processes, they are generally used to produce electrical or mechanical energy or heat. Several conventional and synthetic fuels are compared in table 1.

There are many actual and proposed uses of these chemicals. Alcohol is now being mixed with gasoline to produce gasohol, a mixture of approximately 90 percent gasoline and 10 percent alcohol, which functions quite well in internal combustion engines. Unfortunately, if a large percentage of alcohol is used in

Table 1. A Comparison of the Physical Properties of Several Chemical Fuels

Chemical	Density (g/cc)	Energy Density (MJ/m³)	Flame Temperature in Air (K)	Molecular Weight	Heat of Combustion (MJ/kg)
Gasoline	0.74	33×10^3	2470	107	44
Liquid hydrogen	0.071	8.5×10^3	2318	2	120
Metal hydrides	2.5–5	5–20×10^3	2318	30–100	2–4
Ethanol	0.79	17×10^3	2400	44	22
Methanol	0.80	16×10^3	2370	32	20
Methane gas	0.72×10^{-3}	37	2065	16	50
Natural gas	0.72×10^{-3}	40	2060–2120	18	57
City gas	0.9×10^{-3}	21	1680–2170	25	23
Propane	1.52×10^{-3}	92	2115	44	51

the mixture, the compression ratio of the engine must be increased to maintain satisfactory operation.

In many areas a mixture known as "city gas," which is produced through the partial combustion of coal, is stored and used in place of natural gas. Less energy can be produced per unit volume of city gas than from natural gas because the former contains some N_2 and some gaseous combustion products such as CO and CO_2.

Hydrogen, which does not appear freely in nature, can be produced by the electrolysis of water and by other techniques. Its potential for use as an energy-storage medium is very great. Proposed applications of hydrogen range from aircraft fuels to an additive to the natural gas distribution system. Gases such as methane, which can be produced from wood, for example, can be used in many of the same applications as hydrogen. These gases have not been suggested for widespread use, however, because their production is generally rather complex and they have had no special advantages over natural gas, which in the past has been widely available.

Hydrogen has been proposed as a solution to many of our future energy problems even though its widescale application will introduce many new problems and it will not be especially cheap compared to fossil fuels. The use of liquid hydrogen by NASA in the U.S. space program in the early 1960s showed, however, that the potential problems can be overcome and that the major task in implementing at least a minihydrogen economy is retraining people to accommodate the idiosyncrasies of this material. Contrary to popular opinion, it does not appear that hydrogen will be any more hazardous than fuels such as natural gas, methane, or gasoline, but it must be handled differently.

There are three ways to store hydrogen: as a gas at high pressure, as a liquid at 20 K, and as metal hydride. Hydrogen was one of the last gases to be liquified (1898) because of the very low temperature, 20 K (–423°F), at which liquid and vapor are in equilibrium at 1 atmosphere pressure. Storage as a liquid has been investigated thoroughly for the space application where light molecules provide the most effective means of propelling a vehicle. The specific thrust of a fuel, a measure of its effectiveness as a rocket propellant, increases as the molecular weight of the reaction products decrease; thus, hydrogen, which has the smallest possible atomic weight, 1, is the most effective rocket exhaust gas. Nuclear rocket propulsion, in which the exhaust gas is heated but not burned, works most efficiently with hydrogen. Liquid hydrogen has been used on an experimental basis as an automotive fuel and has been proposed for airplane and ship fuels.

Metal hydrides have been studied for hydrogen storage because they are produced easily and because more hydrogen can be stored in a given volume as a hydride than as a liquid. The propensity of hydrogen to form hydrides, which is an advantage in hydride storage, makes the storing of hydrogen as a compressed gas in a metal container difficult because the strength of the hydride is considerably less than that of the fabricated metal. This characteristic is referred to as hydrogen embrittlement. Recently there has been a renewed interest in storing hydrogen as a compressed gas. The interest is in using permeable geologic formations, usually exhausted natural gas fields or caverns, as the storage vessels.

There has been considerable discussion of the relative merits of liquid hydrogen or metal hydrides for various applications. Some of this discussion appears in the papers reprinted here. The first applications of hydrogen will likely be as an additive to natural or synthetic gases for residential, commerical, and industrial applications and in fuel cells for the production of electrical energy. Future uses will be as a fuel for the propulsion of automobiles, trucks, buses, airplanes, and ships.

It is not the purpose of this book to describe the methods of storing natural fuels; however, because the technology that developed around the liquefication, storage, and transportation of fuels that are gaseous at ambient temperatures and atmospheric pressure will be applicable to similar synthetic fuels in the future, a brief summary of liquefied gas technology is presented here.

Liquefied natural gas (LNG) is a major storage and transportation medium for large quantities of natural gas, which is mainly methane. The boiling point of LNG is about 110 K, which demands a well-insulated vessel for storage and transport. There are three major potential uses of LNG: to augment the local natural gas supply where there are shortages, as a fuel for peaking turbines on electric utilities, and as a fuel for automotive propulsion. Because of the limited storage time possible resulting from the small size of the cryogenic storage vessels that can be used for automobiles, liquefied petroleum gas (LPG), which can be stored as a liquid at ambient temperature, and compressed natural gas (CNG) are preferred for automotive applications.

The main countries importing LNG are France, Italy, Japan, Spain, and the United Kingdom; much of the LNG used in the United States comes from Algeria, but some is from gas fields in the Middle, East, Alaska, and the North Sea. Modern liquefaction plants have a production capacity of about 150×10^6 standard cubic feet (SCF) of gas per day. Often several of these plants are sited together in a liquefaction and storage complex having a liquefaction capacity of 10^9 SCF or more and a storage capacity of 1.2 million barrels of liquid. (One barrel of LNG is equivalent to 3,460 SCF of natural gas.) The LNG is usually shipped from the production site in tankers such as the *Arctic Tokyo*, which has a capacity of 71,500 cubic meters or 450,000 barrels.

There are four different types of LNG storage tanks; metallic tanks of double-wall construction, prestressed concrete tanks, frozen-ground tanks, and, for smaller applications, vacuum-insulted tanks. Double-wall metal tanks have been built in capacities up to 600,000 barrels, the concrete tanks up to 900,000 barrels, and frozen-ground tanks, which have had some operating problems, up to 300,000 barrels.

Although the United States did not import significant amounts of LNG until the mid-1970s, it has considerable experience with the handling of LNG through interstate transport. The United States is expected to become a major importer of LNG in the future, with estimates of 3.8×10^{12} to 5.7×10^{12} SCF per year imported into the lower forty-eight states by 1990.

Considerable data on hydrogen, methane, and gasoline in the form of a rather lengthy but very useful table is contained in Paper 12. The only other major part of this paper that is included is the introduction, which contains the important conclusion that, for many applications, hydrogen is as safe as natural gas and gasoline.

The Nuclear Rocket Development Station near Las Vegas, Nevada, was created to test and develop nuclear rocket engines using high-temperature hydrogen gas for thurst. Paper 13 describes the actual operation of the first large-scale liquid-hydrogen storage system, which consisted of two liquid-hydrogen storage tanks having a capacity of 500,000 gallons and two with capacities of 50,000 gallons.

The application of hydrogen to automotive transportation is the subject of Paper 14.

Paper 15 is an early treatise on the adsorption of hydrogen on the surface of iron, which was in fact the formation of a hydride. The proposed use of hydrogen as a fuel in commerical jet aircraft is explained in Paper 16.

Paper 17 looks at the possible applications of methanol. Paper 18 presents excerpts that summarize the present status of alcohol fuels and gives a projection for future applications.

BIBLIOGRAPHY

Alcohols and Hydrocarbons as Motor Fuels. 1974. Conf. Soc. Automotive Eng. Proc.

Bernstein, I. M., and A. W. Thompson, eds. 1974. *Hydrogen in Metals.* Int. Conf. Effects of Hydrogen on Materials, Properties and Selection and Structural Design, Proc. Am. Soc. Metals.

Bibliography of Reference for the Transport Properties of Liquid and Gaseous A, Kr, N_2, O_2, CH_4. 1963. National Bureau of Standards, Boulder, Colo.

Bockris, J. O'M. 1975. *Energy: The Solar-Hydrogen Alternative.* Wiley, New York.

Brunshteyn, B. 1969. *Production of Alcohols from Petroleum and Gas Raw Material* (translated from Russian). NSF publication, NTIS No. AD 689793.

Cosci, C., ed. 1964. *Fuels and New Propellants.* Pergamon, Oxford.

Cox, K. E., ed. 1974. *Hydrogen Energy: A Bibliography with Abstracts, Cumulative Volume, 1953 Through 1973.* Univ. New Mexico, TAC-H-74-500.

Cox, K. E., and K. D. Williamson, Jr., eds. 1977. *Hydrogen: Its Technology and Implications,* vols. 1, 2, and 3. CRC Press, Cleveland, Ohio.

Dickson, E. M., J. W. Ryan, and M. H. Smulyan. 1977. *The Hydrogen Energy Economy: A Realistic Appraisal of Prospects and Impacts.* Praeger, New York.

Donner, G. A., ed. 1970. *Methanol Technology and Economics.* American Institute of Chemical Engineering, New York.

Hydrogen Congressional Record. 1975. Hearings before the Subcommittee on Energy Research, Development and Demonstration of the Committee on Science and Technology, U.S. House of Representatives, 94th Congress, June 10 and 12.

The Hydrogen Economy Miami Energy Conference, Proc. 1975. Vols. 1 and 2. Plenum, New York.

Hydrogen Fuels: A Bibliography. 1976. Energy Res. Dev. Admn. Rept. TID-3358.

Ishimasa, Y. 1979. The Tokyo Gas Company. *Cryogenics* **192**:67–72.

Liebhafsky, H. A., and E. J. Cairns. 1968. *Fuel Cells and Fuel Batteries: A Guide to Their Research and Development.* Wiley, New York.

Mathis, D. 1976. *Hydrogen Technology for Energy.* Noyes Data Corp., Park Ridge, N. J.

Plaeth, S. T. W. 1949. *Alcohol, a Fuel for Internal Combustion Engines.* Chapman and Hall, London.

Rapoport, I. B. 1962. *Chemistry and Technology of Synthetic Liquid Fuels* (translated from Russian). NSF publication.

Ring, E., ed. 1963. *Rocket Propellant and Pressurization.* Prentice-Hall, Englewood Cliffs, N. J.

Roder, H. M., C. E. Childs, R. D. McCarty, and P. E. Angerhofer. 1973. Survey of the Properties of the Hydrogen Isotopes below Their Critical Temperatures. *Natl. Bur. Stand. Tech. Note 641.*

Seglin, L., ed. 1974. *Symp. Methanation of Synthesis Gas, Proc.* Am. Chem. Soc.

Vielstich, W. 1970. *Fuel Cells: Modern Processes for the Electrochemical Production of Energy* (English). Trans. D. J. G. Ives. Wiley Interscience, London.

Von Ottingen, W. 1943. Aliphatic Alcohols: Their Toxicity and Potential Dangers in Relation to Their Chemical Constitution and Their Fate in Metabolism. *Public Health Service Bull. 281.*

Wilhoit, R., and B. Zwolinski, eds. 1973. *Physical and Thermodynamic Properties of Aliphatic Alcohols.* Am. Chem. Soc.

Williams, A. F., and W. L. Lom. 1973. *Liquified Petroleum Gases: A Guide to Properties, Applications and Usage of Propane and Butane.* Ellis Horwood, Chichester.

Williams, D. A., and G. Jones. 1963. *Liquid Fuels.* Pergamon Press, Oxford.

12

Reprinted from pages 1 and 5–6 of *Natl. Bur. Stand. Tech. Note 690, 1976, 35pp.*

IS HYDROGEN SAFE?

J. Hord

Cryogenics Division
Institute for Basic Standards
National Bureau of Standards
Boulder, Colorado 80302

The safety aspects of hydrogen are systematically examined and compared with
those of methane and gasoline. Physical and chemical property data for all three
fuels are compiled and used to provide a basis for comparing the various safety
features of the three fuels. Each fuel is examined to evaluate its fire hazard,
fire damage, explosive hazard and explosive damage characteristics. The fire
characteristics of hydrogen, methane and gasoline, while different, do not largely
favor the preferred use of any one of the three fuels; however, the threat of fuel-
air explosions in confined spaces is greatest for hydrogen. Gasoline is believed
to be the easiest and perhaps the safest fuel to store because of its lower volatility
and narrower flammable and detonable limits. It is concluded that all three fuels
can be safely stored and used; however, the level of safety risk for each fuel will
vary from one application to another. Generalized safety comparisons are made herein
but detailed safety analyses will be required to establish the relative safety of different fuels for each specific fuel application and stipulated accident. The technical
data supplied in this paper will provide much of the framework for such analyses.
Hydrogen safety guidelines, regulatory codes applicable to the distribution of
hydrogen, and safety criteria for liquid hydrogen storage are compiled and presented.

Key words: Explosion, fire, fuel, gasoline, hydrogen, methane, safety.

[*Editor's Note:* Only Table 1 is reprinted here.]

Is Hydrogen Safe?

Table 1. Properties of Hydrogen*, Methane and Gasoline**.

Property	Hydrogen	Methane	Gasoline
Molecular weight	2.016 [1]	16.043 [5]	~107.0 [7]
Triple point pressure, atm	0.0695 [1]	0.1159 [5]	---
Triple point temperature, K	13.803 [1]	90.680 [5]	180 to 220[1] [7]
Normal boiling point (NBP) temperature, K	20.268 [1]	111.632 [1]	310 to 478 [7,8]
Critical pressure, atm	12.759 [1]	45.387 [5]	24.5 to 27 [7]
Critical temperature, K	32.976 [1]	190.56 [5]	540 to 569 [7]
Density at critical point, g/cm^3	0.0314 [1]	0.1604 [5]	0.23 [9]
Density of liquid at triple point, g/cm^3	0.0770 [1]	0.4516 [5]	---
Density of solid at triple point, g/cm^3	0.0865 [1]	0.4872 [1]	---
Density of vapor at triple point, g/m^3	125.597 [1]	251.53 [5]	---
Density of liquid at NBP, g/cm^3	0.0708 [1]	0.4226 [1]	~.70[2] [7]
Density of vapor at NBP, g/cm^3	0.00134 [1]	0.00182 [1]	~0.0045[2] [9]
Density of gas at NTP, g/m^3	83.764 [1]	651.19 [1]	~4400 [9]
Density ratio: NBP liquid-to-NTP gas	845 [1]	649 [1]	156[3] [7,9]
Heat of fusion, J/g	58.23 [1]	58.47 [1]	161 [12]
Heat of vaporization, J/g	445.59 [1]	509.88 [1]	309 [7,9]
Heat of sublimation, J/g	507.39 [1]	602.44 [1]	---
Heat of combustion (low), kJ/g	119.93 [2]	50.02 [2]	44.5 [2,4,9]
Heat of combustion (high), kJ/g	141.86 [2]	55.53 [2]	48 [2,4,9]
Specific heat (C_p) of NTP gas, J/g-K	14.89 [1]	2.22 [1]	1.62[2] [9]
Specific heat (C_p) of NBP liquid, J/g-K	9.69 [1]	3.50 [1]	2.20[2] [9]
Specific heat ratio (C_p/C_v) of NTP gas	1.383 [1]	1.308 [1]	1.05[2] [7]
Specific heat ratio (C_p/C_v) of NBP liquid	1.688 [1]	1.676 [1]	---
Viscosity of NTP gas, g/cm-s	0.0000875 [1]	0.000110 [1]	0.000052 [9]
Viscosity of NBP liquid, g/cm-s	0.000133 [1]	0.001130 [1]	0.002 [9]
Thermal conductivity of NTP gas, mW/cm-K	1.897 [1]	0.330 [1]	0.112 [9]
Thermal conductivity of NBP liquid, mW/cm-K	1.00 [1]	1.86 [1]	1.31 [9]
Surface tension of NBP liquid, N/m	0.00193 [1]	0.01294 [1]	0.0122 [9]
Dielectric constant of NTP gas	1.00026 [3]	1.00079 [1]	1.0035[5] [10]
Dielectric constant of NBP liquid	1.233 [3]	1.6227 [1]	1.93[4] [9]
Index of refraction of NTP gas	1.00012 [1]	1.0004 [1]	1.0017[5] [10]
Index of refraction of NBP liquid	1.110	1.2739 [1]	1.39[4] [9]
Adiabatic sound velocity in NTP gas, m/s	1294 [1]	448 [1]	154
Adiabatic sound velocity in NBP liquid, m/s	1093 [1]	1331 [1]	1155[6] [12]
Compressibility factor (Z) in NTP gas	1.0006 [1]	1.0243 [1]	1.0069
Compressibility factor (Z) in NBP liquid	0.01712 [1]	0.004145 [1]	0.00643[2]
Gas constant (R), cm^3-atm/g-K	40.7037 [1]	5.11477 [1]	0.77
Isothermal bulk modulus (α) of NBP liquid, MN/m^2	50.13 [3]	456.16 [6]	763[6] [11]
Volume expansivity (β) of NBP liquid, K^{-1}	0.01658 [3]	0.00346 [6]	0.0012[2] [9]

Table 1. Properties of Hydrogen*, Methane and Gasoline** (continued).

Property	Hydrogen	Methane	Gasoline
Limits of flammability in air, vol. %	4.0 to 75.0 [13,14]	5.3 to 15.0 [13,14]	1.0 to 7.6 [8,18]
Limits of detonability in air, vol. %	18.3 to 59.0 [15]	6.3 to 13.5 [19]	1.1 to 3.3 [20]
Stoichiometric composition in air, vol. %	29.53	9.48	1.76
Minimum energy for ignition in air, mJ	0.02 [16]	0.29 [16]	0.24 [16,21]
Autoignition temperature, K	858 [17]	813 [8]	501 to 744 [8,18]
Hot air-jet ignition temperature, K	943 [22]	1493 [22]	1313[7] [22]
Flame temperature in air, K	2318 [23]	2148 [23]	2470 [24]
Percentage of thermal energy radiated from flame to surroundings, %	17 to 25 [25,26]	23 to 33 [25,26]	30 to 42 [25,26]
Burning velocity in NTP air, cm/s	265 to 325 [27,28]	37 to 45 [18,19]	37 to 43 [18,24]
Detonation velocity in NTP air, km/s	1.48 to 2.15 [29,35]	1.39 to 1.64 [19]	1.4 to 1.7[8] [30]
Diffusion coefficient in NTP air, cm^2/s	0.61	0.16	0.05
Diffusion velocity in NTP air, cm/s	$\lesssim 2.00$	$\lesssim 0.51$	$\lesssim 0.17$
Buoyant velocity in NTP air, m/s	1.2 to 9	0.8 to 6	nonbuoyant
Maximum experimental safe gap in NTP air, cm	0.008 [31]	0.12 [32]	0.07 [31]
Quenching gap in NTP air, cm	0.064 [16,33]	0.203 [16,33]	0.2 [33]
Detonation induction distance in NTP air	$L/D \approx 100$ [35,36]	---	---
Limiting oxygen index, vol. %	5.0 [34]	12.1 [34]	11.6[9] [34]
Vaporization rates (steady state) of liquid pools without burning, cm/min	2.5 to 5.0 [26]	0.05 to 0.5 [26]	0.005 to 0.02
Burning rates of spilled liquid pools, cm/min	3.0 to 6.6 [25,26]	0.3 to 1.2 [25,26]	0.2 to 0.9 [25,26]
Flash point, K	gaseous	gaseous	~230 [8]
Toxicity	nontoxic [37] (asphyxiant)	nontoxic [38] (asphyxiant)	slight [39] (asphyxiant)
Energy[10] of explosion, (g TNT)/(g fuel)	~24	~11	~10
Energy[10] of explosion, (g TNT)/(cm^3 NBP liquid fuel)	1.71	4.56	7.04
Energy[10] of explosion, (kg TNT)/(m^3 NTP gaseous fuel)	2.02	7.03	44.22

NBP = Normal boiling point.

NTP = 1 atm and 20 C (293.15 K).

* Thermophysical properties listed are those of <u>para</u>hydrogen.

** Property values are the arithmetic average of normal heptane and octane in those cases where "gasoline" values could not be found (unless otherwise noted).

[1] Freezing temperatures for gasoline @ 1 atm.

[2] @ 1 atm and 15.5 C.

[3] Density ratio @ 1 atm and 15.5 C.

[4] @ 1 atm and 20 C.

[5] @ 1 atm and 100 C.

[6] @ 1 atm and 25 C.

[7] Based on the properties of butane.

[8] Based on the properties of n-pentane and benzene.

[9] **Average** value for a mixture of methane, ethane, propane, benzene, butane and higher hydrocarbons.

[10] Theoretical explosive yields.

13

Reprinted from pages 1312–1315 of *Intersoc. Energy Convers. Eng. Conf., 7th, Proc.*, American Chemical Society, 1972, 1533pp.

EXPERIENCE IN HANDLING, TRANSPORT AND STORAGE OF LIQUID HYDROGEN—THE RECYCLABLE FUEL*

J. R. Bartlit, F. J. Edeskuty, and K. D. Williamson, Jr.

Los Alamos Scientific Laboratory, University of California, Los Alamos, New Mexico 87544

ABSTRACT

Hydrogen holds promise as an easily-producible, clean substitute for dwindling fossil fuels. In the past where hydrogen has been used on a large scale, it has sometimes proved advantageous to use it in liquid form for ease in transport and storage. This existing cryogenic technology is found adequate to meet the needs of the most likely future applications for liquid hydrogen which are transportation facilities, remote sites not serviced by pipelines, peak-shaving, and superconducting power lines.

Existing liquefaction facilities, dewar and pipe sizes, flow capacities, design criteria and data, and safety are discussed and compared with future needs.

THE ENVIRONMENTAL DESIRABILITY and the eventual economic necessity of replacing fossil fuels with synthetic fuels is obvious. Reserves of oil and natural gas have estimated lifetimes of the order of 50 years and coal, the dirtiest fossil fuel, has reserves to last about 200 years longer. After that all primary energy must come from either renewable or large, presently untapped energy sources such as solar, nuclear, and geothermal. Because energy can be transported in fluid form for about one-tenth the cost of transmitting the energy equivalent in electricity (1) and because conveniently portable fuels probably will always find application in transportation systems, synthetic fluid fuels which can be made with this solar, nuclear, or geothermal energy must be developed. Such a fuel must be made from renewable or widely abundant substances and ideally should produce little or no pollution either at the production site or in end use.

People living in remote, sparsely populated areas such as New Mexico are especially opposed to the widespread concept of locating polluting energy-producing facilities in those areas so that large population centers can have energy which is clean at the use end (2,3). This is not an acceptable solution because it is both unjust and short sighted since remote areas, especially in the Southwest, are often recreation areas and pollution levels are becoming large enough to pollute very large areas.

HYDROGEN has been suggested by many (2, 4,5,6) as best fulfilling the requirements of a synthetic fuel. It is available by the electrolysis or cracking of water (7) and its only combustion product is water when burned in oxygen, or water plus small amounts of nitrogen oxides when burned in air. While local distribution and long range transport can be achieved using hydrogen only in the gaseous state special cases may indicate the desirability of, or requirement for, some large scale usage of hydrogen as liquid (LH_2). This may be true in spite of the extra expenses of liquefaction and liquid storage. These special cases can be expected to arise in the large scale storage for peak-shaving, stand-by operation, or operation of remote energy depots. Somewhat smaller scale usage may be needed for storage in portable containers either in fuel tankers or for use in professionally operated transport vehicles. The Los Alamos Scientific Laboratory has had, in the Rover nuclear rocket program, considerable experience in the handling, storage, and distribution of both small and very large quantities of hydrogen, especially of LH_2. It is the purpose of this paper to point out the current state of the art of LH_2 technology and to show that little or no further technological development is necessary for safe and efficient handling of LH_2 in quantities of interest.

PRESENT STATE-OF-THE-ART - The largest LH_2 production facility ever built had a capacity of 60 tons/day and several liquefaction plants exist which can produce 30 tons/day. Considerably higher capacity plants could be built without any technological development.

Batch Transport - LH_2 highway transport in batches up to 13,000 gallons is commonplace. Special 20,000 gallon highway trans-

---

*Work performed under the auspices of the United States Atomic Energy Commission.

port trailers have been built. Due to the low density of LH$_2$ (specific gravity referred to H$_2$0 = 0.07) there are no weight restrictions in its transport and the maximum quantities for transport are usually limited only by the total volume of the transport trailer. Railroad tankers of 28,000 gallons are in regular use and special tankers of 34,000 gallons have also been built. LH$_2$ carrying barges have been built with 240,000 gallon capacity. Boil-off losses of 1/4% per day are common for the 13,000 gallon road transport while for the above mentioned larger vessels boil-off losses are even less (8).

Fluid Transfer - For continuous transfer of fluid, the U.S. space program has provided considerable experience. Vacuum jacketed transfer lines for LH$_2$ have been built with inside diameters of from two inches to 20 inches and in lengths up to a few thousand feet. Operating pressures for some of these lines are as high as 2000 psi. Reciprocating pumps for flows of a few gallons per minute at pressures to several thousand psi are available while a wide range of centrifugal and axial flow pumps have been developed for higher flows at lower pressure. At the Nuclear Rocket Development Station in Nevada, LH$_2$ has been transferred at rates up to 35,000 gal/min which is equivalent to 3.7 million SCFM of gas when vaporized. In well chilled lines pressure drops are readily calculated by standard equations such as the Fanning equation. While smaller units are commonplace, LH$_2$ has been converted (9) to ambient temperature gas at the rate of 5000 liquid gal/min in a heat exchanger utilizing water as the heating medium (see Fig. 1). Liquid and gas have been vented and flared at rates from 0 to 30,000 gal/min in flare stacks. Successful operation of these lines has required detailed transient as well as steady state analyses of the thermal stresses for both thin and thick wall portions of the lines. These detailed studies as well as the development of computational techniques for calculating chilldown times and chilldown liquid requirements (10,11) have required a knowledge of two phase flow (12,13) and heat transfer as well as the necessary fluid properties and mechanical properties of the materials of construction.

Liquid Hydrogen Storage Vessels - At the Nuclear Rocket Development Station in Nevada, vacuum-jacketed, perlite-insulated LH$_2$ storage vessels have been built with capacities of up to a half million gallons (see Fig. 2). Boil-off from such dewars is of the order of 0.05%/day or, in other words, the holding time is greater than five years. LH$_2$ dewars as large as 0.9 million gallons have been built having boil-offs of 0.03%/day. The cost of dewars in these sizes is roughly $2.00 per gallon of capacity. It is felt that no further development would be required to build dewars of this type up to four or

Fig. 1. - One-quarter of H$_2$0-LH$_2$ heat exchanger under construction

Fig. 2. - Two 50,000 and two 500,000 gallon LH$_2$ dewars at the Nuclear Rocket Development Station

five million gallons. For non-vacuum jacketed storage, vessels up to tens of millions of gallons should be capable of being fabricated by current techniques.

Successful and safe operation of these LH$_2$ storage dewars has also required much study. Investigation of heat and mass transfer within the dewar has allowed development of techniques of computation of pressurization gas requirements (14), pressure build up rate (15), and thermal stresses. Measurements have been made of thermal gradients both within the hydrogen space and across the insulation space. Quantities of LH$_2$ required for chilldown of these dewars have been both computed and measured (16). Also developed are operational procedures which preclude accumulation of dangerous impurities such as air.

In many respects the technology of handling LH_2 is more completely developed than that of liquefied natural gas or LN_2.

The safety record during 10 years of the above operations involving 200 men and 28 million gallons of LH_2 shows no injuries or loss of property involving LH_2.

<u>APPLICABILITY TO ENERGY NEEDS</u> - As a basis for discussion, consider that in the future the energy requirements of New York City were supplied by hydrogen. What role might LH_2 play in this scheme?

The estimated energy demand for the 7.6 million people in New York City in 1985 for household and commercial, industrial, transportation, and electrical uses has been estimated to be 2.38×10^{15} Btu (2). To size the problem, if through the use of fuel cells, hydrogen-powered mass transit, etc. this energy could all be furnished by hydrogen, a flow of 2×10^{10} SCF/day of hydrogen would be required assuming equal usage throughout the year. To liquefy this amount with perfect efficiency would require over 800 of today's largest liquefaction plants and 6,130 MW of power. By comparing this with the total electrical demand of the City of 12,260 MW, it can be seen it is impractical to liquefy any but a small portion of the total hydrogen flow. Perhaps an amount used for peak-shaving or the amount to supply transportation needs through mass transit could be liquefied. To liquefy hydrogen for transportation needs would require 256 liquefaction plants and ideally 1880 MW. Transportation needs of the City for one hour could be supplied out of one LH_2 storage dewar of 2.3 million gallon capacity, which is well within the existing state-of-the-art.

The problem of piping 2×10^{10} SCF/day equivalent of hydrogen into the City can be met by the equivalent of one high pressure (1000 psi) gas line of seven feet diameter or an insulated pipeline of four and one-fourth feet diameter carrying liquid. This undoubtedly means that where hydrogen can be piped, it will be piped as a gas.

It thus appears because of costs and quantities involved that the use of LH_2 in cities will be limited to specialized applications such as the storage of fuel for transportation both at terminals and aboard vehicles. Another specialized use which should develop in the future is superconducting electrical transmission lines. Ever increasing needs to move very large blocks of power long distances as well as pressures to put all utility lines underground has focused attention on superconductors. Superconductivity can, of course, only be maintained at extremely low temperatures. The upper limit for this has been continually increased through research until today limited superconducting materials exist at LH_2 temperatures

(20 K). Superconducting power lines of the future, therefore, might deliver LH_2 to transportation centers through multiple use transmission lines at essentially no added cost.

SAFETY - In assessing the safety problems of large scale use of LH_2 as a fuel, one must consider both the characteristics of hydrogen as well as those of cryogenic fluids.

Some properties of hydrogen make it more dangerous than natural gas (assumed methane). These are hydrogen's shorter quenching distance (0.06 cm vs 0.25), lower ignition energy (0.02 millijoules vs 0.3), and lower flame emissivity, which makes the flame almost invisible unless something is added to give it color.

Other properties of hydrogen and methane are equivalent from the standpoint of safety. Both are non-toxic; hydrogen has a higher diffusivity (0.634 cm^2/sec vs 0.20) which makes it leak faster, but also dissipate faster; although hydrogen has a wider explosive range in air (4-75%) compared to methane (5-15%), the lower explosive limit is the one which must be avoided and this is essentially the same. Finally there are two properties of hydrogen which present less hazard than natural gas. These are hydrogen's higher ignition temperature (1085°F vs 1000°F) and the fact that it produces no toxic combustion products like CO (17).

In addition to the previously discussed nuclear rocket experience using liquid, there exists a long history of gaseous hydrogen usage. Various synthetic fuel gases designated "producer" gas, "manufactured" gas, "oil" gas, "water" gas, "town" gas, etc. commonly containing from 15 to 85% hydrogen were formerly distributed to residences in most cities with relative safety. The German town of Basilea currently distributes 80% hydrogen for domestic use.

Additional safety considerations are required for hydrogen in liquid form because of the extremely low temperature (-425°F) and increased concentration. In spite of it being the lightest of all liquids (0.6 lb/ gallon) this concentration is equivalent to ambient temperature gas at approximately 25,000 psi. The low temperature promotes increased heat influx through insulation which with the very low latent heat of vaporization (approximately 200 Btu per lb) can cause rapid evaporation. The low temperature has an effect on solid material properties such as strengths, thermal conductivity, specific heats, thermal expansion, etc. Proper design which includes proper material selection can alleviate problems arising from changes in solid properties. The low temperature also allows the possibility of air condensing in the LH_2 and the preclusion of this possibility usually depends upon the use of acceptable operating procedures. Proper system design will determine the necessary placing and sizing of

pressure relief valves as well as provide
the necessary system entrances for purging
and sampling.

CONCLUSION - Hydrogen is attractive as
a clean substitute for dwindling fossil
fuels. Past experience indicates that for
many specialized uses of hydrogen, LH_2 is
the most convenient form for transport and
storage. For the most likely applications
of liquid in a future hydrogen economy,
existing and proven technology is equal to
the task.

REFERENCES

1. D. P. Gregory, D. Y. C. Ng and
G. M. Long, "The Electrochemistry of Cleaner
Environments." by J. O'M. Bockris, Ed., New
York, Plenum Press, 1972.

2. W. E. Winsche, T. V. Sheehan and
K. C. Hoffman, "Hydrogen-A Clean Fuel for
Urban Areas." Paper 719006 presented at
Intersociety Energy Conversion Engineering
Conference, Boston, August 1971.

3. J. Wei, "Energy the Ultimate Raw
Material." Chemtech, p. 142, March 1972.

4. C. Marchetti, "Hydrogen, Master-
Key to the Energy Market." euro-spectra,
Vol. X, No. 4, p. 117, December 1971.

5. L. O. Williams, "A Plan for Elimination of Pollution." Design News, p. 74,
January 5, 1970.

6. R. G. Murray and R. J. Schoeppel,
"A Reliable Solution to the Environmental
Problem: The Hydrogen Engine." Paper 700608
presented at SAE/AIAA/ASME Reliability and
Maintainability Conference, Detroit, July
1970.

7. C. Marchetti, euro-spectra, Vol. IX,
No. 2, p. 46, June 1970.

8. F. J. Edeskuty and K. D. Williamson,
Jr., "Storage and Handling of Cryogens." to
be published in "Advances in Cryogenic
Engineering" K. D. Timmerhaus, Ed., Vol. 17
New York, Plenum Press, 1972.

9. J. R. Bartlit and K. D. Williamson,
Jr., "Further Experimental Study of H_2O-LH_2
Heat Exchangers." "Advances in Cryogenic
Engineering" Vol. 11, p. 561, New York,
Plenum Press, 1966.

10. J. C. Bronson, et al., "Problems in
Cool-Down of Cryogenic Systems." "Advances
in Cryogenic Engineering" Vol. 7, p. 198,
New York, Plenum Press, 1962.

11. D. H. Liebenberg, J. K. Novak and
F. J. Edeskuty, "Cooldown of Cryogenic
Transfer Systems." Paper No. 67-475 present-
ed at AIAA 3rd Propulsion Joint Specialist
Conference, Washington, D. C., July 1967.

12. J. D. Rogers, "Two-Phase Friction
Factor for Parahydrogen Between One Atmosphere and the Critical Pressure." "Advances
in Cryogenic Engineering", Vol. 9, p. 311
New York, Plenum Press, 1964.

13. J. R. Bartlit, "The Integrated
Pressure Drop in Two-Phase Nonadiabatic Flow
of Hydrogen." Los Alamos Scientific Laboratory
Report LA-3177-MS, Los Alamos, N. M., 1964.

14. F. J. Edeskuty, J. R. Bartlit and
A. L. Trego, "Pressurant Gas Usage in a
Large, High Pressure Liquid Hydrogen Dewar."
Presented at XIIIth International Congress
of Refrigeration, Washington, D. C., Aug. 27
through Sept. 3, 1971.

15. D. H. Liebenberg and F. J. Edeskuty,
"Pressurization Analysis of a Large-Scale
Liquid-Hydrogen Dewar." "International
Advances in Cryogenic Engineering" Vol. 10,
p. 284, New York, Plenum Press, 1965.

16. D. H. Liebenberg, R. W. Stokes
and F. J. Edeskuty, "Chilldown and Storage
Losses of Large Liquid Hydrogen Storage
Dewars." "Advances in Cryogenic Engineering"
Vol. 11, p. 554, New York, Plenum Press, 1966.

17. F. J. Edeskuty, Roy Reider and
K. D. Williamson, Jr., "Safety" in "Cryo-
genic Fundamentals", G. G. Haselden, Ed.,
London, Academic Press, 1971.

14

Reprinted from *Cryogenics* **13**:693–698 (1973)

Hydrogen powered automobiles must use liquid hydrogen

L. O. Williams

A problem of particular interest and importance in the evaluation of a future hydrogen—oxygen economy is hydrogen's use as a fuel for highway vehicles.

If, as many authors suggest, hydrogen is the only fuel that will simultaneously keep the environment clean and conserve natural resources, it must eventually be applied to the private vehicles that are a major pollution offender. Although a number of automobiles have been successfully operated on hydrogen (and this represents no large problem even with unmodified engines), the question remains of how to carry the fuel. This problem is primarily in the technology of the fuel tank — that is how to carry the hydrogen on board the vehicle safely, economically, and with a minimum of other penalties. It is unfortunate that this problem has attracted a number of suggested solutions that are less than optimum (for example, high-pressure gas and metallic hydrides) and that detract from the credibility of entire concept.

If, as this author contends, the use of cryogenic liquid hydrogen is the sole, real candidate, its use will have a broad effect on industry. While requiring no basic research, it will necessitate key engineering developments in both the very large cryogenic production facilities and the small vehicle storage vessels and ancillary systems and hardware.

The term 'hydrogen—oxygen economy' or simply 'hydrogen economy' has been used to describe a system in which energy produced at a few locations in massive quantities is used to decompose water into its constituents, hydrogen and oxygen. The hydrogen produced would be used as an all-purpose low-pollution fuel, and the oxygen for water and solid waste treatment. References 1 to 41 discuss many of the attributes of this economy and a number of alternative sources for the prime energy to be used in decomposing water. Whatever term is used to describe the system or whatever energy source is used to power it, the hydrogen—oxygen economy will have a profoundly beneficial effect on the spacecraft earth. Most important are the advantages to be gained from the use of hydrogen — a fuel with a short-term environmental recycle that produces no noxious trace impurities and does not add carbon dioxide to the atmosphere with its unknown long-term effects; a fuel that can be used for essentially all fuel purposes from heating homes to generating electricity to powering transportation; a fuel that is economical to transport and store; and a fuel that is an abudant natural resource.

The oxygen produced as a byproduct can of course be applied in all its current uses. In addition, the very large quantity available and its low cost can make oxygen-enhanced sewage treatment, solid waste processing, and even oxygenation of natural bodies of water economically possible.

The ultimate elimination of the use of fossil hydrocarbon resources as common fuels will make these resources available for other more valuable petrochemical products such as plastics, drugs, and the like.

Barring unpredictable and spectacular breakthoughs in the energy generation technology, the desirability of this economy is beyond question because a fuel burned in air is the most efficient, compact method of generating power in modest quantities. The questions now are when and what are the problems in implementation. In any consideration of implementation, the question of 'How can hydrogen be used as a fuel for vehicular transportation?' arises. The intent of this article is to throw some light on this one specific problem.

The automobile

The work done so far indicates that the conventional internal combustion engine can, without modification, be operated utilizing hydrogen as the fuel.[18,33,34,44] To cite a particular example, a 140 in³ pushrod engine with many hours of service was equipped with a hydrogen air mixer and operated for about an hour using hydrogen as the sole fuel. Although no numerical data were obtained during this operation, the engine operated quite smoothly without any fuel-related problems and demonstrated that hydrogen presents no particular problem in the engine.

Well developed methane and propane carburettors are now widely used on automobile-type engines. Although their flow control orifices are not suitable for use with hydrogen because of the different volumetric ratio required, redesign could be done simply. The ancillary pressure regulators, valves, controls, etc, are also available for these carburettors and they too would require only a modest amount of redesign or resizing to handle hydrogen.

We are left then with only one real problem in vehicle design requiring significant engineering effort — the technology of the fuel tank.

The criteria for the fuel tank are

1 Reasonable volume and weight

2 A refuel time of no more than 10 minutes

The author is with Martin Marietta Aerospace, Box 179, Denver, Colorado 80201, USA. Received 7 September 1973.

173

3   Adequate safety.

Four fuel tank options have been discussed for small vehicles

1   Metallic hydrides [29]

2   High-pressure gas [34]

3   Chemical fuels synthesized from hydrogen

4   Cryogenic liquid.[1,7,21,40]

These will be evaluated in turn.

As a method of evaluating the various options for the hydrogen fuel tank, a reference automobile with the following characteristics was selected

1   Weight, 180 kg

2   Range, 500 km

3   Fuel capacity, 80 (55.3 kg) gasoline

4   Mileage, 6.25 km l^{-1}.

These characteristics establish a baseline amount of fuel required to operate an automobile of specific performance, weight, and range. Ordinary gasoline is of variable composition and has a heat of combustion of 10 to 11 kcal^{-1}. The reference gasoline was assumed to be pure iso-octane with a heat of combustion of 10.98 kcal^{-1}. The fuel capacity of the reference automobile is then equivalent to 6.072×10^5 kcal energy. Since hydrogen has a heat of combustion of 28.9 kcal^{-1}, to provide the required energy a charge of 21.0 kg of hydrogen would be required. This amount was used to generate the weight and volume of the hydrogen fuel tank options.

Metallic hydrides

Metallic hydrides have attracted interest as carriers of hydrogen because they can be stored at ambient temperature and at ordinary pressures. This would hopefully lead to a simple storage tank manufactured in a manner much like today's gasoline tank. Unfortunately this simplicity is an illusion and factors such as methods of extracting or releasing the hydrogen, safety, weight, volume, etc will make application of the hydrides difficult.

For evaluation of alternate sources of hydrogen, the data in Table 1 was prepared. This table considers the properties of the hydrides suggested as hydrogen storage media. It does not include any consideration of the weights and volumes of the tankage, special equipment, or reactants necessary to extract the hydrogen from the hydride.

The table is complete for all elements through titanium with the exception of the rare gases. As can be seen, the weight percentage of hydrogen decreases as the atomic weight increases. Elements with a higher atomic weight than titanium form hydrides. An alloy of titanium and iron has been suggested [30] as well as palladium and lanthanum nickel. However, these higher weight elements form hydrides with such a low hydrogen content that their weight is prohibitive. Because the complex hydrides of the light metals have a relatively high hydrogen content, they are worth evaluation.

Hydrazine and iso-octane are included with the complex light metal hydrides because they are chemically complex.

Here they are considered as sources of hydrogen only. The use of iso-octane type materials as a hydrogen source has been experimentally investigated at Boston University and the results indicated in the table are for this type of application rather than for direct use as a fuel.

The heavier elements that adsorb and release hydrogen easily have attracted much interest because of their potential ease of use. Palladium and lanthanum nickel are included as representative of this type of material.

To evaluate these materials, a number of specific performance gates were selected. The criteria for these gates were arbitrarily established with the hope that they would at least be reasonable. These gates are

1   Is the hydride a potential fuel?

2   Is there a solid residue? (Recycling a solid residue would require a return logistics network as extensive as the supply network, which would result in a large cost penalty in use of the material)

3   The potential fuel to supply the required energy should not weigh more than four times the weight of gasoline (221.2 kg)

4   The potential fuel should not occupy more than four times the volume of gasoline (320 l)

5   There should not be any unusual safety problems

6   The chemical elements used as carriers must be sufficiently abundant to support the conversion of a significant portion of the economy to their use.

The hydrides are compared and their ability to pass or fail the performance gates is shown in Table 1.

The only fuels that survive this evaluation process are hydrogen and the hydrocarbons we wish to replace.

By ignoring toxicity, ammonia and hydrazine might be considered. Since both of these materials give a warning of their presence by a strong odour, the casual danger of poisoning is low. However, in the case of accidents in which the fuel tank is ruptured, the hazard potential can be very great. One deep breath of either would have a high probability of being fatal. Ammonia, along with hydrogen, presents a potential for extensive frostbite if the tank is ruptured and fuel splashes on the passengers. Hydrazine shares with gasoline great persistence. Because it has a boiling point higher than water (114°C), any spilled material will be present in the area of the accident for long periods. This subjects persons involved in the accident to a much longer exposure to the toxicity and fire hazards.

While the toxicity of these fuels may give pause in their consideration, the most telling problem is the manufacturing methods and the cost penalty their use incurs. Currently the only synthetic route to ammonia production is the direct reaction of hydrogen and nitrogen. The most economic method of producing hydrazine is from ammonia. Thus the cost of producing some arbitrary amount of energy from each of these fuels can be expressed as

C = cost of energy as hydrogen
L = cost of liquefaction of hydrogen
S = cost of ammonia synthesis
$L2$= cost of ammonia liquefaction
N = cost of low-oxygen nitrogen
H = cost of hydrazine synthesis.

Atomic No	Name	Formula	Percent hydrogen	Weight required to provide 21.0 kg hydrogen	Specific gravity	Volume required to provide 21.0 kg hydrogen	Safety problems	Ignition temperature	Performance Gates (+ = Pass, − = Fail) 1	2	3	4	5	6	Overall pass?
	Simple hydrides														
1	Hydrogen	H_2	100	21.8	0.07	300	Cold −252°C, fire	585°C	+	+	+	+	+	+	+
3	Lithium Hydride	LiH	12.68	166.0	0.82	202	Caustic, extremely hot fire on combustion	High	+	−	+	+	+	+	−
4	Beryllium Hydride	BeH_2	18.28	115.0	0.6?	191	Extremely hot fire, extreme toxicity	Low	+	−	+	+	−	−	−
5	Diborane	B_2H_6	21.86	96.0	0.417	214	Toxic	Very low	+	−	+	+	−	−	−
6	Methane	CH_4	25.13	83.5	0.415	201	Cold −175°C, fire	575°C	+	+	+	+	+	+	+
7	Ammonia	NH_3	17.76	118.0	0.817	145	Toxic 100 ppm max, cold −30°C	High	+	+	+	+	−	+	−
8	Water	H_2O	11.19	188.0	1.000	188	—		−	−	+	+	+	+	−
9	Hydrogen Fluoride	HF	5.04	416.0	1.005	414	—		−	−	−	−	−	−	−
11	Sodium Hydride	NaH	4.20	500.0	0.92	543	Caustic, extremely hot fire on combustion	High	+	−	−	−	+	+	−
12	Magnesium Hydride	MgH_2	7.66	275.0	1.45	190	Extremely hot fire on combustion	Very low	+	−	−	+	−	+	−
13	Aluminium Hydride	AlH_3	10.08	208.0	1.3?	160?	Extremely hot fire on combustion	Very low	+	−	+	+	−	+	−
14	Silane	SiH_4	12.55	167.0	0.68	247	Toxic 0.1 ppm	−100°C	+	−	+	+	−	+	−
15	Phosphene	PH_3	8.89	236.0	0.746	316	Toxic 0.01 ppm, extremely hot fire on combustion	+20°C	+	−	−	+	−	+	−
16	Hydrogen Sulphide	H_2S	5.92	354.0	1.347	263	Toxic 20 ppm toxic combustion products	Low	+	+	−	+	−	+	−
17	Hydrogen Chloride	HCl	2.77	758.0	1.256	606	—		−	−	−	−	−	+	−
19	Potassium Hydride	KH	2.51	837.0	1.47	569	Caustic, extremely hot fire on combustion	High	+	−	−	−	+	+	−
20	Calcium Hydride	CaH_2	4.79	438.0	1.9	230	Extremely hot fire on combustion	Medium	+	−	−	+	+	+	−
21	Scandium Hydride	ScH_3	6.30	333.0	2.0?	170?	Toxic, extremely hot fire on combustion	?	+	−	−	+	+	−	−
22	Titanium Hydride	TiH_2	4.40	520.0	3.9	133	Extremely hot fire on combustion	High	+	−	−	+	+	+	−
	Complex hydrides														
	Lithium Borohydride	$LiBH_4$	18.51	114.0	0.666	171	Mild toxicity	High	+	−	+	+	+	−	−
	Aluminium Borohydride	$Al(BH_4)_3$	16.91	124.0	0.545	228	Mild toxicity	Very low	+	−	+	+	−	−	−
	Magnesium Borohydride	$Mg(BH_4)_2$	14.93	141.0	1.3?	108?	Mild toxicity	Very low	+	−	+	+	−	−	−
	Beryllium Borohydride	$Be(BH_4)_2$	20.84	101.0	0.9?	113	Extreme toxicity	Low	+	−	+	+	−	−	−
	Lithium Aluminium Hydride	$LiAlH_4$	10.62	197.0	0.917	216		Low	+	−	+	+	−	+	−
	Hydrazine	N_2H_4	12.58	167.0	1.011	163	Toxic 10 ppm	300°C	+	+	+	+	−	+	−
	Iso-octane	C_8H_{18}	15.88	132.0	0.6918	191	Mild toxicity 500 ppm	270°C	+	+	+	+	+	+	+
	Hydrogen absorbers														
	Palladium Hydride	Pd_2H	0.471	4458	10.78	413		Very low	+	+	−	−	−	−	−
	Lanthanum Nickel	$LaNi_5H_6$	1.38	1522	7.56 (Calc)	201		Very low	+	+	−	+	−	−	−

The cost of these fuels can then be calculated as

Cost of liquid hydrogen $= C + L$

Cost of liquid ammonia $= C + L2 + S + N$

Cost of hydrazine $= C + S + N + H$

For ammonia to be cost-competitive with hydrogen, L must be greater than $S + L2 + N$, and hydrazine will be even more expensive because of its added cost of synthesis from ammonia. Keeping these costs less than the cost of liquefaction of hydrogen, L, will be difficult and liquid hydrogen will probably be the least costly way to supply portable energy.

High pressure gas

High pressure gas must be examined because it can be adapted for use with the least development effort. As a result of this low development effort, most of the hydrogen-fuelled automobiles built so far have taken this fuel tank technology approach.

The evaluation of high-pressure cylinders for storage can be approached from a standpoint similar to that used with the hydrides and the same performance gates can be used. The hydrogen fuel charge reference amount is 21.0 kg. Containment of 21.0 kg of hydrogen in a 320 l volume requires a pressure of 800 atm (1 atm = 101 kN m^{-2}) (8.29 x 106.0 kg m^{-2}) calculated from simple ideal gas laws. The actual pressure would be nearly twice 800 atm because of high-pressure deviations from ideal behaviour.

To contain this 800 atm pressure in steel with a tensile strength of 10^8 kg m^{-2}, a sphere would require a wall thickness of 7 cm and would weigh 1 550 kg. Stronger materials of higher cost are available, for example, cryo-formed 304L stainless or maraging steel, titanium, or composite materials containing glass, boron, or carbon filaments. These materials, however, would only result in a weight reduction of a factor of five or so, which is insufficient to make these tanks meet the weight criteria. Producing tanks of a more easily packaged cylindrical shape would increase the weight.

Even if the weight of these tanks could be reduced to an acceptably low level, the most telling difficulty is the hazard associated with a sudden collision of sufficient magnitude to rupture the high-pressure tanks. Even at the commonly used pressures of only 2×10^6 kg m^{-2}, these tanks would rupture as if they contained high explosives, causing great blast damage and greatly increasing the danger of fire. At the calculated pressure level of 8.29×10^6 kg m^{-2}, these tanks would be extremely dangerous as well as heavy and bulky.

Chemical fuels synthesized from hydrogen

The use of various chemical and physical methods for storage of hydrogen has been discussed. The large-scale implementation of a hydrogen economy would also make possible the synthesis of a number of fuels that could be used directly in the engine. Although most of these have already been evaluated as hydrogen sources, some examination of their characteristics when used directly as fuels is appropriate. Table 2 compares these synthetic fuels on a simple weight and volume basis with the reference iso-octane.

Table 2.   Direct fuels necessary to provide 6.072 x 10^5 kcal

	Volume	Weight, kg	Volume ratio	Weight ratio	Heat of combustion, kcal g^{-1}
Iso-octane	80	55.3	1.00	1.00	10.98
Ammonia	167	136.7	2.09	2.47	4.44
Hydrazine	151	152.5	1.89	2.76	3.98
Hydrogen	300	21.0	3.75	0.38	28.90
Methane	122	50.7	1.53	0.92	11.97
Methanol	158	124.9	2.26	1.97	4.86

Table 2 shows that, compared to gasoline, all the synthetic fuels require significantly more volume for storing a unit amount of energy. While hydrogen is by far the least desirable from the volume standpoint, it is by far the best on a weight basis.

All the manufactured fuels should be more expensive than the hydrogen from which they will be synthesized. Ammonia and hydrazine, as previously mentioned, present real problems from their toxicity, and the carbon-containing fuels will present the same environmental problems we are fighting with the current fossil fuels. The advantages offered by these fuels is only one of convenience. If liquid hydrogen can be made convenient, this advantage will no longer be significant.

Cryogenic liquid hydrogen

Compared to the other methods of storing hydrogen, cryogenic storage shows the following advantages

1 Lowest cost per unit energy

2 Lowest weight per unit energy

3 Simple supply logistics

4 Normal refuel time required

5 Probably as low an implementation cost as any technology except for high-pressure gas

6 No unsurmountable safety problems.

The disadvantages are

1 Loss of fuel when vehicle is not in use

2 Large tank size

3 Cryogenic liquid safety engineering problems.

Implementing the cryogenic hydrogen economy

The foregoing discussion presented a strong argument that liquid hydrogen is the only practical form of hydrogen for vehicular transport as an alternative to the hydrocarbon fuels. The use of liquid hydrogen will produce an enormous growth in the cryogenic industry. To realize this growth, several areas of technology will have to be advanced and some new hardware developed. These areas can be outlined as follows.

Production of liquid hydrogen. Whatever the source of the hydrogen, the bulk transport of commercial hydrogen across the continent will be by pipeline. Two options exist for the liquefaction of this pipeline-supplied gas.

Each vehicle refuelling station could have a small liquefier and storage tank to supply its own needs. This option would require the development of small highly reliable liquefiers.

Regional liquefaction plants with large-capacity units and large storage facilities from which liquid hydrogen can be distributed by truck are also possible. The regional approach would require liquefaction and storage facilities significantly larger than any currently in use. For the US alone, the total magnitude of this supply would be on the order of 10^{13} litres per year. If this were divided evenly between 100 regional liquefaction centres, the production rate of each would be 10^{11} litres per year, or 3×10^8 litres per day.

Liquid hydrogen transfer. The transfer of the liquid hydrogen at various points in the supply system will present some specific development problems. For the regional liquefaction option, transfer from large stationary storage containers to over-the-road delivery trucks and from the trucks to the local retail service point will be required. Current hardware and trucks are probably suitable and usable. For either the regional or local liquefaction option, much development is required to provide the simple, fail-safe, foolproof hardware necessary to refuel the private automobiles. It may be desirable and feasible to develop totally automatic refuelling hardware and systems.[42]

Liquid hydrogen storage. The storage of liquid hydrogen on the vehicle will require the development of highly reliable low-heat-leak, low-cost storage vessels. These vessels will be similar in design to current small liquid hydrogen storage vessels — double-walled, vacuum-jacketed, etc — but will be modified to fit the envelope of the automobile and to meet the necessary safety requirements for mobile use. Some work has already been performed in this area in adapting automobiles to run on liquid natural gas.[42]

Warming and cold gas or liquid. Heat exchangers will be required to warm the cold gas or liquid to ambient temperature. If gas is withdrawn, it will be necessary to provide tank pressurization by heating the liquid. If liquid is withdrawn, heat or helium can be used to provide the pressure. The liquid withdrawal mode enables the cooling capacity of the hydrogen to be used to provide air conditioning for the vehicle. The obvious source of heat for heating the hydrogen to ambient temperature is the engine exhaust gas. Utilizing this gas will require care to prevent frost build-up or similar problems from the condensation of water.

Carburettor design. Gaseous carburettors are available for methane and propane. However, as they exist, their volumetric mixture ratio is unsuitable for use with hydrogen. The necessary redesign is, however, extremely simple in that it only requires a change in the area ratio of the gas and air passages.

Engine changes. Changes to the engine are to all extents and purposes zero. As long ago as 1948 [44] it was demonstrated that a carburettor engine free from carbon deposits will run without problems at compression ratios as high as 10:1 utilizing hydrogen as the fuel.

Safety devices. Safety devices will be required to prevent hazards from the use of hydrogen fuel. In a well-designed system that is used as often as every other day, venting will not be required. While most vehicles are used at least this often, some will not be, and every vehicle is occasionally allowed to stand unused for several days. Then it will be necessary to have a safe venting system. Because of the high dispersal characteristics of hydrogen due to its buoyancy and rapid diffusion, it may be safe to simply vent it slowly into the ambient air as is done with gasoline. Unfortunately, extensive safety testing would be required to demonstrate this to everyone's satisfaction. Alternatively, positive safe venting could be obtained by one of the following techniques

1 Catalytic ignition

2 Electric ignition

3 Resonance tube ignition.

These all would require the design of a small flameless burner and stack for product removal.

Catalytic ignition will require development. Currently, catalysts can be made from platinum and palladium. These will work well but will be quite expensive. Raney nickel is much less costly but is very easily poisoned. A new catalyst material based on lathanum cobalt oxide $(LaCoO_3)$ [45] and similar compounds may perform this function well and be quite inexpensive.

Electric ignition would be quite inexpensive but would deplete the battery at a time when no recharging would be available. Over a period of time the battery could become depleted causing failure of the ignition system. An intriguing possibility consists of venting through a small hydrogen—air fuel cell to provide a small amount of electric power.[47] This power could be used to charge the primary battery.

Although a resonance tube device might be quite reliable and inexpensive, it requires high-velocity gas to function.[46]

The leak problem. Good design practice, quality control, and periodic leak checks should prevent leaks in the auto delivery system producing hazards. With protection from leaks and solution of venting problems, the vehicle will be safe in all normal usage environments.

Fuel tank rupture. The final area of consideration would be collisions of sufficient magnitude to rupture the fuel tank. By the very nature of its design, the liquid hydrogen fuel tank will be more difficult to rupture than a conventional gasoline tank and when ruptured the hydrogen would dissipate more rapidly than gasoline. The primary safety consideration is therefore to locate the tank in such a position as to minimize its involvement in the statistically most common, frontal collision [48] and to protect the passenger compartment from splashing of liquid by lightweight splash screens etc.

Conclusions

Implementation of the hydrogen-based economy is desirable because it will end the open-loop combustion of fossil fuels with its attendant air pollution and will preserve the fossil fuel resources for use as hydrocarbons in basic chemical applications.

While much of the hydrogen will be handled in the gaseous form, the mobile sector of the economy and bulk storage will require the use of cryogenic liquid hydrogen. This will produce an enormous growth of the cryogenic industry.

This growth will foster engineering developments but will not require any basic research breakthroughs. In the mobile sector, the use of liquid hydrogen in internal combustion engines will allow a smooth transition to hydrogen—air fuel cells when they become economically feasible.

References

1970

1   Williams, L. O. *Design News* 25 No 1 (1970) 74

1971

2   Escher, W. J. D. No Intake, No Exhaust (Escher Technology Associates, Box 189, St Johns, Michigan 48879, 1971)
3   Gregory, D. P., Ng, D. Y. C., Long, G. M. The Electrochemistry of Cleaner Environments, [J. O'M Bockris (ed)] (Plenum Press, NY 1971)
4   Jones, L. W. *Science* 174 (1971) 367
5   Marchetti, C. *Euro Spectra* 10 No 9 (1971) 117
6   Mayo, A. M. *AAS Newsletter* 10 No 3 (1971) 10
7   Williams, L. O. Proceedings of the Institute of Environmental Sciences (Environmental Awareness Section, 1971)
8   Winsche, W. E., Hoffman, K. C., Sheehan, T. V. Society of Automotive Engineers Paper 719006 (1971)

1972

9   Armagnac, A. P. *Popular Science* 202 No 1 (1972) 64
10   Bacon, F., Fry, T. *New Scientist* 55 No 808 (1972) 285
11   Bockris, J. O'M *Science* 176 (1972) 1323
12   Chopey, N. P. *Chemical Engineering* 79 No 29 (1972) 98
13   *Business Week* 2247 (1972) 98
14   Gregory, D. P. *American Gas Association* 54 (1972) 4
15   Lessing, L. *Fortune* 86 No 5 (1972) 138
16   McCurdy, P. P. (ed) *Chemical & Engineering News* 50 (1972) No 26, 14; No 27, 16; No 28, 27
17   Trotter, R. J. *Science News* 102 (1972) 46
18   Underwood, P., Dieges, P. Society of Automotive Engineers Paper 719046 (1972)
19   Westerman, A. B. (ed) *Battelle Research Outlook* 4 No 1 (1972) 28
20   Wilks, W. (ed) *Technology Forecasts* 4 No 5 (1972) 7
21   Williams, L. O. *Astronautics and Aeronautics* 10 No 2 (1972) 42
22   Winsche, W. E., Hoffman, K. C., Salzano, F. J. National Science Foundation Paper BNL—16694 (1972)
23   The US Atomic Energy Commission, T1D26136 'Hydrogen and Other Synthetic Fuels' Sept 1972

All of the following references were published in the *Proceedings of the 7th Intersociety Energy Conversion Engineering Conference,* Hilton Inn, San Diego, California, 25—29 September

1972

24   Edeskuty, F. J., Williamson, K. D. Los Alamos Scientific Laboratories
25   Hausz, W., Leeth, G., Meyer, C. General Electric Co TEMPO
26   Tanner, E. C. Princetown University; Huse, R. A. Public Service Electric and Gas Co, Newark, New Jersey
27   Gregory, D. P., Wurm, J. Institute of Gas Technology, Chicago, Illinois
28   Martin, F. A. Union Carbide Corporation
29   Wiswall, R. H. Jr, Reilly, J. J. Brookhaven National Laboratory
30   Weil, K. H. Stevens Institute of Technology
31   Jones, L. W. University of Michigan
32   Winsche, W. E., Hoffman, K. C., Slazano, F. J. Brookhaven National Laboratories
33   Murray, R. G., Schoeppel, R. J. Oklahoma State University, and Gray, C. L. Environmental Protection Agency
34   Swain, M. R., Adt, R. R. University of Miami
35   Sorensen, H. International Materials Corporation
36   Escher, W. J. D. Escher Technology Associates

1973

37   Gregory, D. P. *Scientific American* 228 No 1 (1973) 14
38   Marchetti, C. *Chemical Economy and Engineering Review* 5 No 1 (1973) 7
39   Mayo, A. M., Convers, C. C., Burth, P. O. Society of Petroleum Engineers of American Institute of Mining, Metallurgical and Petroleum Engineers Inc SPE paper 4202 (1973)
40   Williams, L. O. *Advances in Cryogenic Engineering* 18 (1973)
41   Zener, C. *Physics Today* 26 No 1 (1973) 48
42   Ginsburg, I., Wagner, T. O. *Automotive Engineering* 80 No 12 (1973) 35
43   Jennings, F. A., Studhalter. The American Society of Mechanical Engineers, Paper 71—PVP—62 (1971)
44   King, R. O., Wallace, W. A., Mahapatra, B. *Canadian Journal of Research* 26 Sec F, No 7 (1948) 264
45   Libby, W. F. *Science* 171 (1971) 499
46   Sinha, R. A Theoretical Analysis of Resonance Tube, the Singer Company, Kearfott Division, Little Falls, New Jersey, Report KD72—82 (1972)
47   Mayo, A. M. Pure Power Inc, 2848 W Kingsley Rd, Garland, Texas 75040, Private Communications
48   Miller, P. M. *Scientific American* 228 No 2 (1973) 78

15

THE SORPTION OF GASES BY IRON

By Arthur F. Benton and T. A. White[1]

[Contribution from the Cobb Chemical Laboratory, University of Virginia, No. 90]

Received December 12, 1931   Published May 7, 1932

Introduction

As a result of numerous researches in recent years in the field of adsorption of gases by catalytically active solids, three processes, apart from compound formation, have become clearly recognized—physical adsorption, activated adsorption and solution. The characteristics and interrelationships of the two types of *ad*sorption are now reasonably clear.[2]

In the attempt to distinguish adsorption from solution, it has been customary to assume that the former is always rapid, while the latter is ordinarily slow. However, since it has been shown that activated ad-

---

[1] Du Pont Fellow in Chemistry.

[2] (a) Benton, This Journal, **45**, 887, 900 (1923); (b) Benton and White, *ibid.*, **52**, 2325 (1930); (c) Taylor, *ibid.*, **53**, 578 (1931); (d) Taylor and Williamson, *ibid.*, **53**, 2168 (1931); (e) Taylor and McKinney, *ibid.*, **53**, 3604 (1931); (f) Benton and White, *ibid.*, **54**, 1373 (1932).

sorption may be a slow process,[2b-f,3] it is clear that rate alone is not a certain criterion. Nevertheless, in the systems copper–hydrogen and copper–carbon monoxide, it was shown[2f] that by suitable procedures involving both equilibria and rates, activated adsorption and solution could be sharply differentiated, and the contributions of each to the total sorption under any conditions could be separately estimated. In the work here reported, the same methods have been applied in a study of the sorption of nitrogen, hydrogen and carbon monoxide by reduced iron.

No extensive investigations of sorption by iron have previously been reported. Sieverts[4] found the solubility of hydrogen in iron wire at 400° was very small, but increased with rising temperature; nitrogen dissolved in gamma but not in alpha iron. The "adsorptions" at lower temperatures have been examined by Taylor and Burns[5] and by Nikitin.[6]

Experimental Details

Since many of the experiments were continued over long periods, it was considered desirable to insure against leaks by substituting mercury seals for stopcocks. The apparatus has already been described.[2b] The sorption under any given conditions of pressure, temperature and time was determined as the difference between the volume of gas admitted to the bulb containing the iron, and the volume remaining in the free space. The latter was obtained by experiments with helium in the manner previously described.

The sorbent (Iron I, 54.6 g.) was prepared from an 8–14 mesh sample of fused ferroferric oxide, containing 0.15% of alumina, for which we are indebted to Dr. P. H. Emmett of the Fixed Nitrogen Research Laboratory, Washington. The oxide was reduced *in situ* in a current of hydrogen for five days at 375° and for fourteen days at 425–450°. Evacuation between runs was carried out at 500° by means of a Töpler pump.

Occasional check runs made at intervals throughout the work revealed no trend in the activity or capacity of the sample. Hence the data obtained for different gases, and for the same gas under different conditions, are directly comparable.

All volumes are given in cc. at 0°, 760 mm., and all pressures in mm. of mercury at 0°.

Results with Nitrogen

The isothermal adsorption of nitrogen at temperatures from −191.5 to 0° is shown in Fig. 1, Curves 1, 2, 3, 4. In this and later figures open circles indicate points obtained from lower pressures, that is, after a fresh addition of gas to the bulb; black circles (and other solid figures) are used for points obtained by the reverse process.

The rate of adsorption at −183, −78.5 and 0° was apparently instantaneous. At −191.5° equilibrium was attained in a few minutes. The fact that the equilibrium was actually reached is shown by the close agreement of points obtained from higher and from lower pressures.

[3] Benton and Elgin, THIS JOURNAL, **48**, 3027 (1926); **49**, 2426 (1927).

[4] Sieverts, *Z. physik. Chem.*, **60**, 129 (1907).

[5] Taylor and Burns, THIS JOURNAL, **43**, 1273 (1921).

[6] Nikitin, *J. Russ. Phys.-Chem. Soc.*, **58**, 1081 (1926); *Z. anorg. allgem. Chem.*, **154**, 130 (1926).

The observed effects of temperature and pressure, together with the practically instantaneous rates, leave no doubt that the sole process occurring here is a surface adsorption of the physical type. Thus, the isotherms show no tendency to reach a saturation limit, and at the higher temperatures the adsorption is very nearly proportional to the pressure.

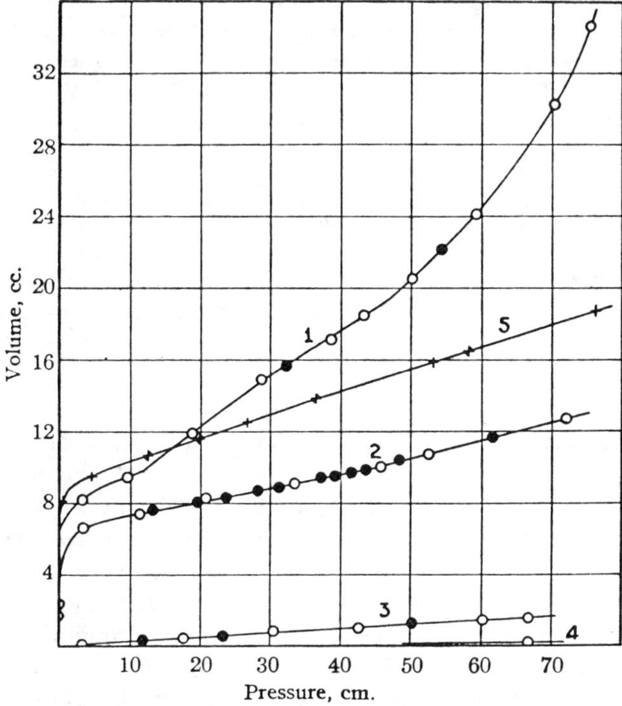

Fig. 1.—Isotherms for nitrogen and carbon monoxide: Curves 1, 2, 3 and 4 are for nitrogen at −191.5, −183, −78.5, and 0°, respectively. Curve 5 is for carbon monoxide at −183°.

Further, at a given pressure the values decrease uniformly with increasing temperature, and the heat of adsorption is relatively small, as shown by the values of Q, in cal. per mole, given in Table I.

TABLE I

HEAT OF PHYSICAL ADSORPTION OF NITROGEN BY IRON I

Vol. adsorbed	p at −191.5°	p at −183°	Q
8.0	22	190	3700
10.0	117	457	2340
12.0	195	641	2050

These heats cover the same range previously found for nitrogen on copper,[2f] and exceed the latent heat of vaporization of nitrogen (1380 cal.) by a factor of only 1.5 to 3.

Higher Temperatures.—The data obtained for the sorption of nitrogen at higher temperatures are given in Table II. These results were obtained by introducing a definite amount (5.91 cc.) of gas into the system, and successively determining the apparent equilibria at the temperatures shown.

TABLE II

SORPTION OF NITROGEN AT HIGHER TEMPERATURES

Temp., °C.	Pressure, mm.	Sorption, cc.
321	418.0	0.43
0	195.1	.42
344	425.1	.54
360	435.6	.56
408	469.6	.54
0	195.1	.42

The sorption at these higher temperatures is considerably greater than at 0°. Since here the physical adsorption must be negligibly small, it is evident that these values represent a second type of sorption. The rate of this process at 0° is very slow, so that little, if any, of the gas taken up at high temperatures is evolved on cooling to 0°.

The data obtained are incapable of furnishing a definite decision as to the nature of the high-temperature sorption. On the other hand, since Sieverts' experiments[4] indicate that gaseous nitrogen at ordinary pressures is insoluble in alpha iron, the evidence points to activated adsorption as responsible for the observed phenomena. This view is rendered still more probable by the fact that at the higher temperatures, iron similar to that here employed is known to activate nitrogen for ammonia synthesis,[7] and there is now available a large body of evidence to support the hypothesis that activated adsorption is an essential step in contact catalysis.

Results with Carbon Monoxide.—The isotherm for carbon monoxide at −183° is plotted in Fig. 1, Curve 5. Here as with nitrogen the rate was instantaneous. Throughout the whole range of pressure the ratio of the adsorption of carbon monoxide to that of nitrogen at this temperature is 1.44 ± 0.02. These facts are convincing evidence that in this case also the process is solely one of physical adsorption.

Measurements were also made with carbon monoxide at −78.5 and 0°, but in neither case was equilibrium reached, even after one hundred hours. Nevertheless, the observed rates clearly show the existence of two different processes. A few typical data on rates are given in Table III; the values shown for zero time are the volumes taken up at the last previous point.

At −78.5° a large fraction of the sorption occurred within three minutes (0.05 hour), after which the rate decreased sharply. At 0° relatively little sorption took place during the first three minutes, but after this time

[7] Almquist and Crittenden, *Ind. Eng. Chem.*, **18**, 1307 (1926).

TABLE III

RATE OF SORPTION OF CARBON MONOXIDE

	−78.5°			0°	
Time, hrs.	P, mm.	V, cc.	Time, hrs.	P, mm.	V, cc.
0.0	0.0	0.95	0.0	2.0	1.66
.05	59.0	2.84	.05	191.4	2.40
12.1	29.0	4.00	.30	162.1	3.22
15.5	27.3	4.07	1.01	138.7	3.88
19.8	25.5	4.13	3.50	113.4	4.59
21.5	25.0	4.15	5.00	106.0	4.80
			13.0	88.0	5.30
0.05	562.2	5.83	35.6	70.0	5.82
.80	555.2	6.10	74.8	57.7	6.16
14.7	530.3	7.07			
27.4	521.8	7.40	0.05	503.7	6.48
38.6	515.9	7.63	1.22	490.7	6.84
50.7	511.4	7.80	5.9	469.8	7.43
62.9	508.9	7.90	21.8	438.0	8.32
87.2	503.4	8.11	50.2	413.2	9.02

the rate was more rapid than at the lower temperature, and the total sorptions reached after a given time were greater.

These results are precisely what would be expected if the total sorption at these temperatures is a combination of physical adsorption and a second process. Since other surfaces give *physical* adsorptions of carbon monoxide and nitrogen in this range of temperature, which are in the ratio[2f] of about 1.5, it may be calculated that here the adsorption of this type, at 500 mm. pressure, should amount to about 2 cc. at −78.5°, and about 0.2 cc. at 0°. Thus the first, rapid part of the observed sorption, which is much greater at −78.5° than at 0°, can be largely accounted for as physical adsorption. The nature of the subsequent process, whether activated adsorption, carbonyl formation or solution, or a combination of these, has not been ascertained.

Results with Hydrogen

Low Temperatures.—The adsorptions of hydrogen at −195 and −183° are plotted in Fig. 2, Curves 1 and 2. The striking discontinuities in these curves have previously been considered in detail, in connection with a general discussion of stepwise adsorption.[8] Neglecting the discontinuities, one observes that the general shape of the isotherms is typical of physical adsorption. There is little or no adsorption at "zero" pressure, and the values continue to increase with increasing pressure without approaching a saturation limit. At both temperatures the rate was apparently instantaneous, and values obtained from the side of higher pressure are in excellent agreement with those approached from lower pressures. The adsorptions decrease with rising temperature, and the heat of adsorp-

[8] Benton and White, THIS JOURNAL, 53, 3301 (1931).

tion calculated from the isotherms by the Clapeyron equation is about 1600 cal. per mole.   These observations show that physical adsorption is the sole process occurring.

Curve 3, Fig. 2, represents an isotherm at −78.5°.   These results must be considered in the light of the following observations.   The first two portions of gas admitted to the iron apparently came to equilibrium immediately at pressures of 50 and 258 mm., respectively, but the adsorption in each case was less than 0.05 cc.   The next portion gave an immediate adsorption of 0.30 cc. at 715 mm.   In each of these cases the pressure remained constant for the twenty to thirty minutes of observation.   Part

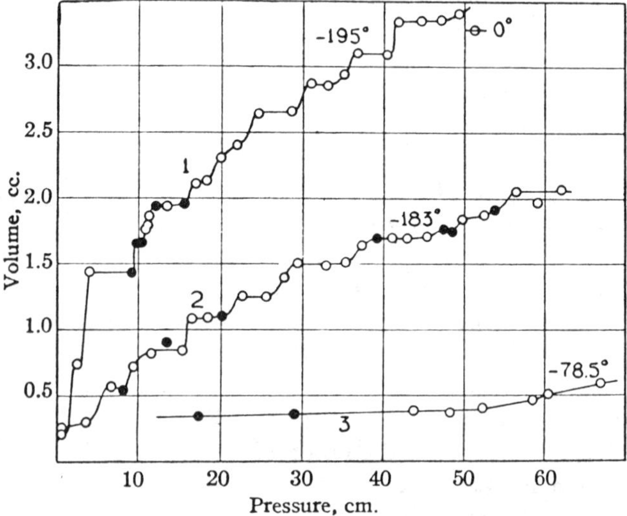

Fig. 2.—Isotherms for hydrogen.

of the gas was then withdrawn from the bulb, with the result that the adsorption at once decreased to 0.05 cc. at 405 mm.   However, this volume did not remain constant, but gradually increased until an apparently constant adsorption of 0.28 cc. was reached after sixteen hours.   From this point on, the values shown in Curve 3 were successively obtained.   At each of these points equilibrium was established very rapidly;   even with times of observation as long as twenty-two hours no change in adsorption occurred.   A repetition of this experiment gave practically identical results, except that here the slow adsorption (0.24 cc. at 86 mm.) took place at the first point, where the system was kept under observation for thirty-nine hours.

The behavior noted is readily accounted for as follows.   At −78.5° there is a small, rapid and readily reversible physical adsorption, which exhibits steps similar to those at lower temperatures, but here the first step occurs at a pressure of about 600 mm. At one point in the course of following this isotherm, where sufficient time was allowed, a very slow activated adsorption of the order of 0.3 cc. occurred. The isotherm subsequently obtained represents the physical adsorption *plus* the 0.3 cc. adsorbed in the activated form. It is remarkable, however, that the activated adsorption, which reached 0.3 cc. in sixteen hours, showed no further increase in three days.

Higher Temperatures.—At 0° there was an immediate adsorption of 0.2–0.3 cc., followed by an extremely slow process which did not come to equilibrium in 35 days. The value reached at the end of this time, 3.28 cc. at 513 mm., is indicated in Fig. 2. At this point the temperature was quickly raised to 110°. The result was a rapid *decrease* of the sorption to 2.06 cc. at 768 mm., but this decrease was followed by a gradual *increase* over a period of three days to 2.47 cc. at 748 mm., without any sign of an approach to final equilibrium.

It may be concluded from these observations that in the neighborhood of 110° the sorption consists of two distinct processes, one of which is rapid and decreases in extent with increasing temperature, while the second is very slow. This conclusion is confirmed by isothermal measurements conducted wholly at 110°, which showed a rapid initial sorption followed

TABLE IV

RATE OF SORPTION OF HYDROGEN AT 110 AND 210°

Time, hrs.	110° P, mm.	V, cc.	Time, hrs.	210° P, mm.	V, cc.
0.0	...	0.0	0.0	...	0.0
.05	22.5	.23	.05	56.5	.38
.12	16.0	.35	.60	48.5	.51
.30	5.0	.58	2.7	45.0	.57
1.72	3.0	.62	6.1	43.0	.60
			25.0	40.5	.64
0.05	113.0	0.69			
.30	102.0	.92	0.05	539.3	1.02
.88	99.0	.98	2.0	522.1	1.30
1.63	97.0	1.02	5.0	516.1	1.40
4.38	93.9	1.08	21.0	505.9	1.57
8.13	92.0	1.12	30.5	503.0	1.62
21.7	88.0	1.20	45	500.5	1.66
			54	499.6	1.67
0.1	315.0	1.42	69	496.8	1.72
1.0	310.0	1.53	78	496.1	1.73
3.0	307.1	1.59	96	493.7	1.77
7.3	303.4	1.66	117	492.4	1.79
24.7	297.7	1.78			
0.05	607.7	1.91			
1.05	605.2	1.96			
4.6	602.2	2.02			
24.3	596.4	2.14			

185

by a very slow process. At 210° also there was a rapid process, which involved smaller quantities of gas than at 110°, and again a slow process ensued. The sorption at these temperatures as a function of time is given in Table IV.

At the conclusion of the measurements at 110°, when a sorption of 2.14 cc. had been reached, the system was rapidly cooled to −78.5°. At this temperature the sorption became constant almost immediately at 2.38 cc. under a pressure of 305.6 mm. It is evident that the gas taken up at 110° was not desorbed on cooling; on the contrary, an increase of about 0.2 cc. occurred, which is undoubtedly physical adsorption and is of the same order as the physical adsorption on the bare surface at −78.5°.

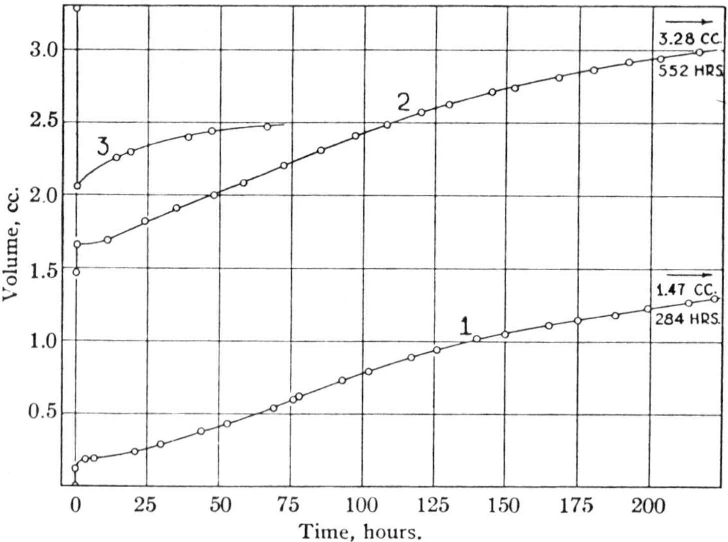

Fig. 3.—Rate of sorption of hydrogen at 0°: Curve 3, after heating to 110°.

The sorption at 0° as a function of time is shown in Fig. 3, Curves 1 and 2. The range of pressure involved here was as follows: for Curve 1, 120.5 to 72.7 mm; for Curve 2, 570.5 to 512.8 mm. In each case, after a rapid initial adsorption of about 0.2 cc., the rate sharply decreased almost to zero, then gradually increased, and finally fell off again. The change of sorption with time after heating to 110° is shown in Curve 3, which brings out clearly the rapid initial decrease of the sorption, followed by a slow increase.

In interpreting these results it is to be noted that the *physical* adsorption, which had already become very small at −78.5°, must be entirely negligible at 0° and above. The two processes which have been distinguished at 110° may provisionally be regarded as activated adsorption and solution.

On this basis the activated adsorption is very slow at 0°, except for the first 0.2–0.3 cc., but rapidly increases in rate with increasing temperature. The final equilibrium, on the other hand, corresponds to a decreasing activated adsorption with increasing temperature. If on a plot of the volumes taken up at 110 and 210° against the time, the slow process at these temperatures is extrapolated back to zero time, a rough estimate may be obtained of the fractions of the total sorption which are due to each individual process. In this manner it was found that (at 500 to 600 mm.) the activated adsorption amounted to about 1.4 cc. at 110° and 1.0 cc. at 210°; the slow process which has been ascribed to solution amounted to about 0.7 cc. at each temperature, but the time of observation was much longer at 210°, so that equilibrium was doubtless more closely approached. Because of the considerable uncertainty in these values, and because the slow process had certainly not reached equilibrium, the isobars for the different types of sorption will not be given here.

If the above view of the situation is correct, it must be concluded that the process of activated adsorption is *autocatalytic*. Thus at 0°, after about 0.2 cc. had been rapidly adsorbed, an induction period of eight to ten hours occurred, during which there was little further sorption. Subsequently the rate increased, passed through a maximum, and then decreased. This behavior is especially clearly shown by Curve 1 of Fig. 3. At 110°, in spite of the rapid rate of this process, there was some slight evidence of the same phenomenon. It is conceivable that only the few tenths of a cc. which was rapidly taken up at 0° (and slowly at −78.5°) represents activated adsorption, perhaps on the edges of the crystals, and that the autocatalytic process involves the formation of a new solid *phase* of iron hydride. Further data will be necessary before this possibility can be profitably discussed in detail. It may be noted, however, that on this view the rapid decrease in sorption on heating from 0 to 110° would indicate that the hydride is unstable at the latter temperature, even at a pressure of 770 mm.; consequently the rapid process at lower pressures observed in the experiments at 110 and 210° could not be ascribed to the formation of this definite hydride.

Discussion

The results for iron are similar in the main to those previously reported for nickel and copper. At low temperatures all three metals exhibit an apparently instantaneous physical adsorption, which decreases rapidly and uniformly with increasing temperature, corresponding to heats of adsorption which are only about two to four times the latent heat of vaporization of the several gases.

At some point activated adsorption first makes its appearance, and rapidly increases in rate at higher temperatures. On both copper and iron the rate of this process is considerably greater for carbon monoxide than for hydrogen, and enormously greater than for nitrogen. Comparison of the different metals shows that the rates of activated adsorption of hydrogen were about equal for the particular samples of nickel and copper employed, but were much less in the case of iron. The rate with carbon monoxide was much less for iron than for copper. (The system nickel–carbon monoxide has not been studied from this point of view.) Doubtless the smaller rates with iron are due in part to the fact that the sample of iron was prepared from fused oxide by reduction at a high temperature, while the other two metals were obtained by cautious ignition of their nitrates and reduction at much lower temperatures.

In many of the cases studied, a third process, provisionally ascribed to solution, has been definitely distinguished. Such a process occurs in the system, iron–hydrogen. With nitrogen and carbon monoxide in contact with iron, this third process is not apparent, but it must be noted that in these latter cases the sorptions have been less extensively examined.

Multimolecular Layers at Low Temperatures.—Inspection of Curves 1 and 2 of Fig. 1 reveals the fact that definite "breaks" occur in each case at an adsorption of about 10 cc. This is apparently the volume of nitrogen necessary to form a complete unimolecular layer of molecules over the surface. The sharpness of the breaks is no doubt diminished by the incipient formation of a second layer before the first is entirely completed. At $-191.5°$ there is a second, less definite break at a volume of 19–20 cc., which presumably signifies the completion of the second layer of molecules. From this point on, the adsorption appears to increase in a regular manner, gradually approaching infinity as the pressure increases toward the vapor pressure of liquid nitrogen at this temperature (about 110 cm.).

On this basis the physical adsorptions at the higher temperatures correspond to a very incomplete covering of the surface with a single layer of molecules. It may be of interest to point out also that the activated adsorption of the various gases never amounts to as much as 10 cc. From this fact the conclusion may be drawn that activated adsorption does not exceed a unimolecular layer, although the latter is closely approached in the case of carbon monoxide at 0°.

Summary

The isothermal sorptions of nitrogen, carbon monoxide and hydrogen by reduced iron have been measured at pressures up to one atmosphere and over a range of temperatures down to 78°K. In many cases also the change in sorption with change in temperature has been examined, with a constant volume of gas in the system.

As in the case of copper previously reported, it is found that in general sorption involves three definitely distinguishable processes, (a) *ad*sorption of the physical type, (b) *ad*sorption of the activated type and (c) a third process which is probably solution. The contribution of the individual processes to the total observed sorption has been estimated.

The distinguishing characteristics of each process have been further elucidated, particularly in respect to rates, equilibria, heats and thickness of adsorbed layers.

With all three gases at $-183°$ and below, the sorption consists solely of physical adsorption. In each case activated adsorption occurs at higher temperatures. The rate of this process is greatest for carbon monoxide and least for nitrogen. With hydrogen at $110°$ and above, the third process, ascribed to solution comes into prominence.

UNIVERSITY, VIRGINIA

16

Reprinted from pages 819–838 of *Hydrogen Energy, Part B*, ed. T. N. Veziroglu, Proc. Hydrogen Economy Miami Energy (THEME) Conf., 1974, Plenum Press, New York, 1975, 1369pp.

HYDROGEN FOR THE SUBSONIC TRANSPORT

Peter F. Korycinski and Daniel B. Snow

NASA LANGLEY RESEARCH CENTER

Hampton, Virginia

ABSTRACT

Subsonic air transports can be designed to use effectively the high energy content of hydrogen. The advantages of hydrogen over other fuels increase as the size and range increase. Although the change from conventional jet fuel to liquid hydrogen is a major departure in airplane design, a substantial base of hydrogen technology exists in both the space and aeronautics programs which can be applied to support the engineering design and development of subsonic hydrogen airplanes.

INTRODUCTION

Much has been said and much more will be said here and elsewhere on the future role of hydrogen in our national economy. Hydrogen has a special attractiveness for aerospace applications; that is, for aeronautics programs and for space programs because of the superior energy content.

Hydrogen technology in aerospace covers the velocity spectrum from subsonic to orbital speeds. The time frame is equally broad. It spans from the mid-1950s into the future. In the 1950s, two projects were started. The National Advisory Committee for Aeronautics (NACA)* modified an existing military jet airplane, the B57 Canberra, and successfully demonstrated the use of liquid hydrogen

---

*Predecessor to the National Aeronautics and Space Administration (NASA).

as a fuel in cruise flight at subsonic speeds. The NACA project
was terminated in 1957 [1]. In this same time period, the United
States Air Force contracted with Lockheed for supersonic, high-
altitude airplanes. The project designated CL-400 was also terminated in 1957 after having progressed to the institution of development of a new hydrogen-fueled jet engine for supersonic flight and
the initiation of the production of the prototype airplane [2].
Interest in hydrogen-fueled subsonic and supersonic airplanes lay
dormant until the current awareness that the high rate of depletion
of natural hydrocarbon fuels may constrain the future of air transportation and that the circumstances require serious consideration
of hydrogen and other fuels.

Although the space program was the beneficiary of much of the
early hydrogen technology, the space program greatly intensified
the effort to advance and develop hydrogen technology. In the
aeronautics program, the technology continued to exist in the form
of long-range research on the air-breathing hypersonic cruise airplane. The proponents of hypersonic flight found in liquid hydrogen not only an ideal fuel for propulsion but also found in the
fuel an ideal heat sink to cool engines and the structure of the
airplane [3]. The steady progress made in hydrogen technology in
the space program led to the use of hydrogen in the upper stages
of the Saturn launch vehicle and in the Centaur space vehicle and
hydrogen will be used in the Space Shuttle. Equally important is
the continually advancing hydrogen technology for hypersonic
cruise airplanes. This is a fortunate happenstance for the pace-
setting hypersonic technology provides a substantial aeronautics-
oriented base for the use of liquid hydrogen as a fuel for supersonic and subsonic airplanes.

Recently published papers on hydrogen-fueled supersonic
transports [4,5,6] note the potential of airplanes designed to
operate at supersonic speeds. Even after the sleek, efficient
shapes of supersonic transports are compromised aerodynamically to
provide the additional volume required to carry liquid hydrogen,
the high energy content of hydrogen pays off in substantial improvements in the load the airplane can carry or the distance it
can fly. Perhaps a better way to say it is that hydrogen advanced
supersonic technology circumvents the range limitation of the jet-
fueled airplanes and opens up the possibility of Trans-Pacific
operations.

The subsonic airplane, the subject of this paper, will continue to be important to air transportation. As the major carrier of the bulk of the world passengers and high-value cargo over
long distances, it will also be a major consumer of fuel. Even
though natural and synthetic jet fuels may be economically available for a long time, liquid hydrogen is an alternate fuel and it

is important that subsonic airplane technology be ready to use
hydrogen when the need arises.

This paper will review air travel and fuel requirements,
alternate fuels, the impact of alternate fuels on airplane design,
hydrogen technology and airplane design, the technology needs, program plans and the outlook for the future.

AIR TRAVEL AND FUEL REQUIREMENTS

Most of us arrived here by air. Air travel has become rather
commonplace throughout the world. It is safe, reliable, fast, and
relatively inexpensive. Fig. 1 shows the growth of air travel [7].
We already are on the steeply ascending portion of the curve. In
1973, the equivalent of just about every man, woman, and child in
this country made a trip on an airline. In view of our current
energy difficulties, there may be a slight jag in the curve and
maybe the slopes will not be as steep as before, once the rising
cost of fuel forces the price of fares up. Nevertheless, we have
no reason to doubt that air transportation will continue to be
the preferred mode of travel.

Until recently the air transportation industry has not really
been sensitive to fuel conservation. Mr. Emile Van Lennep, Secre-
tary General of the Organization for Economic Cooperation and
Development, analyzed the fuel question in relation to air transportation in his Albert Plesman Memorial Lecture [8] which he
presented in Delft in October 1973. He presented comparisons between airplanes and trains, buses, and cars. One of his more
interesting statistics is that a standard passenger "consumes"
four times his own weight (of 154 pounds) of fuel on a round-trip
Trans-Atlantic crossing at a 50 percent load factor. This figure
is roughly equal to average annual per capita consumption of oil
in the non-OECD countries. He observes that "no other form of
transport quite compares with this Promethean exercise in burning
irreplaceable fuel for the sake of speed." Perhaps the first
question Albert Plesman, founder of KLM, one of the world's great
international airlines, would have asked were he alive would be:
"What other forms of transportation are available which by spanning
oceans and continents safely, comfortably and at so modest a cost
can so greatly expand the scope and range of human experience within a life time?" Speed is the only commodity which, relatively
speaking, buys time. The traveling public has always accepted
this fact and has paid the bill for energy consumed.

Much is being done by industry to conserve fuel. We may have
had to accept some compromises in our travel arrangements from home
to here and return because of fewer flights. Not noticeable to

passengers is the fact that the airplanes are being flown at
slightly slower speeds. In the air and on the ground, precautions
are being taken to minimize delays and unnecessary use of fuel.
The air transportation system is basically a high-technology, high-
efficiency system in which a relatively small number of units do a
very big job for a relatively small expenditure of energy. By the
end of 1973, the United States Air Carrier Fleet consisted of 2,246
airplanes [9]. Table 1 shows that all air travel accounts for only
3.2 percent of the total energy used in the United States.

However, the anticipated growth of air travel increases the
gravity of the fuel situation. Fig. 2 shows the growth of the
transportation section of our economy. Rather striking is the
much more rapid growth of the energy requirement of air transportation. A somber view of such forecast is increased competition
among the elements in the transportation sector for a diminishing
supply of petroleum.

ALTERNATE FUELS

Jet fuels currently used by the military and by the civil air
lines are precisely specified products which have evolved over the

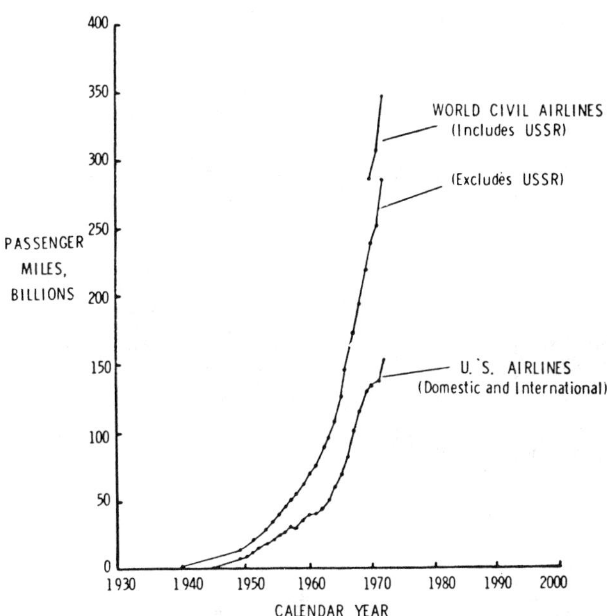

Fig. 1. Growth of air travel.

TABLE 1

Where America's Energy Goes

	Percentage of All Energy Used in U. S.
RUNNING INDUSTRY	37.3%
POWERING TRANSPORTATION	24.8%
Driving cars	13.2%
Driving trucks and buses	5.5%
Flying planes	3.2%
Driving farm and other off-road vehicles	1.2%
Fueling ships and boats	1.0%
Fueling trains	0.7%
HEATING HOMES AND OFFICES	17.9%
PROVIDING RAW MATERIALS FOR CHEMICALS, PLASTICS	5.5%
HEATING WATER FOR HOMES AND OFFICES	4.0%
AIR-CONDITIONING HOMES AND OFFICES	2.5%
REFRIGERATING FOOD	2.2%
LIGHTING HOMES AND OFFICES	1.5%
COOKING FOOD	1.3%
OTHER USES	3.0%

Source: Office of Science and Technology; Chase Manhattan Bank

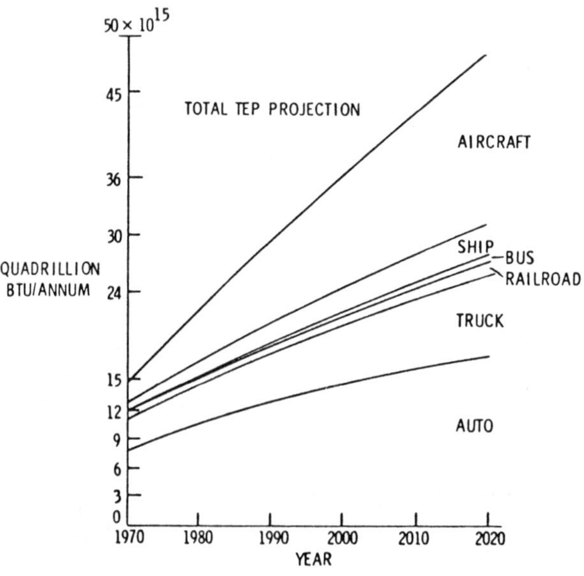

Fig. 2. Projected transportation energy consumption.

years as a result of research to increase their safe handling
characteristics on the ground and in the air, to increase engine
life, and to reduce undesirable exhaust emissions. Airplane design philosophy is almost completely jet-fuel oriented. Similarly,
fuel supply and logistic systems are geared to jet fuel. Any consideration of alternate fuels must maintain an awareness of new
sets of conditions, different for each fuel, which must be insti-
tuted from the initial source of energy through engine exhaust
emissions into the atmosphere. The capital required to build new
industries becomes reasonable only when the alternative is economic castration.

The list of candidate alternate fuels reduces to essentially
three suitable for use in aviation. These are hydrogen, methane,
and synthetic jet fuel. The synthetic jet fuel should closely
approximate jet fuel derived from natural crude petroleum. The
substitute sources of synthetic crude petroleum are oil shale,
coal, and tar sands. The process for making synthetic jet fuel
from these sources will require huge new industrial complexes.
At present, synthetic fuels are produced in laboratories and in
small pilot plants. Limited quantities are available for research purposes.

Methane is a well-documented fuel. Early studies of the
supersonic transport by the NASA Lewis Research Center [10] included methane as an alternate fuel. The 15 percent increase in
energy content of methane at about one-half the weight of jet
fuel indicated a potential for increased range, which even considering the increase in design complexity introduced by a cryogenic fuel appeared sufficiently attractive to warrant further
consideration. Some information on methane in the subsonic airplane will be presented later.

At the time of the Lewis studies, availability and cost of
methane were not major considerations. Now, it is a different
story, and natural gas, the prime source of methane, is rapidly
becoming depleted. Although methane can be derived from many
sources, the quantities of gas required for air transportation
will require complex and costly facilities.

Hydrogen has great potential for air transportation. As
previously mentioned, it is being used in the space program and
will be used for the Space Shuttle. It is essential for the
hypersonic cruise vehicles of the future. It has already received some study for advanced supersonic transports. Cruise
flight of a subsonic airplane on a hydrogen-fueled jet engine was
demonstrated in 1957.

But more must be done to provide a sound engineering base
for the complete air transportation system of which hydrogen, the

fuel, is but one of several key elements. Success in this endeavor
can lead to future air carrier fleets which are essentially independent of hydrocarbon fuels.

AIRPLANE DESIGNS FOR ALTERNATE FUELS

In the 1990-1995 time period aeronautical technology will
have advanced sufficiently so that many new elements will be available to the designer. To satisfy the demand for energy efficient
air transports with useful life of 50,000 hours, such new developments as the supercritical wing, composites for the structure,
active controls, advanced avionics, variable cycle jet engines, to
name a few, will be used whatever fuel is selected. However, some
fuels will impact airplane design much more than others. Synthetic
fuels, for example, should closely approximate current jet fuels
and their impact on design is not expected to be significant.

Methane and hydrogen are cryogenic liquids and their use in
airplane design introduces a host of new systems engineering problems. Cryogenic liquids also compound the difficulty in meeting
the requirement of a useful life of 50,000 hours, especially in
the dynamic environment of airplanes in the air and on the ground.

Of the two cryogenic liquids, hydrogen will be more challeng-
ing to the designer because its very low boiling point of -423°F
and large volume place a premium on excellence on all aspects of
design. Moreover, the fuel volume is large. The same total on-
board energy requires about 4 times as much volume of liquid
hydrogen as of jet fuel. This volume requirement becomes an important consideration in the design of new airplane configurations.

Before Congress passed the law authorizing the construction
of the Alaska Pipe Line, a serious look was taken at the use of
very large airplanes to transport crude petroleum from the Alaskan
North Slope to the United States. These large airplanes would
have operated as shuttles pretty much like the Berlin Airlift
Operation. They may continue to be an option for pipe lines serious limitations in polar regions.

NASA Langley, with the assistance of technical support con-
tractor, LTV Aerospace Corporation, Hampton Technical Center,
initiated a study to examine the potential of a cargo airplane
having a take-off gross weight of 1.5 million pounds [11]. An airplane of this size is twice as large in weight as the Boeing 747.
The design objective was to carry a cargo of 265,000 pounds over a
range of 5,000 nautical miles at a Mach number of 0.85.

Following conventional design philosophies, the jet-fuel airplane designs tended to be long and slender. Changing to hydrogen,

the proportions, Fig. 3, became conventional in form. The change
also resulted in a substantial reduction in weight of the airplane--
from a take-off gross weight of 1.5 million pounds down to 0.915
million pounds--a reduction of 39 percent. Factoring in advanced
technology brought about a further reduction--down to 753,000
pounds, almost a 50 percent reduction in take-off gross weight
from the initial jet-fueled aluminum airplane design. Admittedly,
these were rather preliminary design studies, but the magnitude of
the potential immediately highlighted hydrogen as a driving design
factor for large subsonic airplanes.

About this same time, some advocates for the American supersonic transport began to take a hard look at hydrogen. One of their
major concerns was the unattractive prospect of trying to sell an
airplane which requires a fuel whose supply may be depleted before
the end of the useful life of the airplane.

Other aircraft designers also took a look at hydrogen. Its
potential for conventional and VTOL fighters, for STOL airplanes,
and for helicopters was studied. The answers were different.
For such tightly packaged machines as fighters, the volume requirement of hydrogen for the same mission capabilities required major
configuration changes which would seriously impair performance.
STOL airplanes and helicopters do not carry much fuel and both can
operate on hydrogen as well as jet fuel. However, the use of hydrogen is not advantageous.

A Working Sumposium on Liquid-Hydrogen-Fueled Aircraft was
held in May 1973 at the NASA Langley Research Center to assess the
current state of hydrogen technology as it applies to airplane
design. On rather short notice, the aerospace companies were invited to express their views on hydrogen-fueled airplanes.

Three companies, Boeing, General Dynamics-Convair, and Lockheed, responded with presentations on specific subsonic airplane
designs. Boeing took their 747 airplane and modified the design
by incorporating upper fuselage lobes, Fig. 4, to accommodate the
additional fuel volume required. Lockheed used their 1011 design
and added wing-tip tanks, Fig. 5. The 10.4-foot-diameter wing-tip
tanks do not look unreasonable. Keep in mind that weight is not
the problem with hydrogen that it would be with jet fuel. The
weight of the tank, plus insulation, plus liquid hydrogen is about
the same as that for an equal volume of balsa, the wood used in
model airplane construction. General Dynamics-Convair chose to
use their Advanced Transport Technology design, Fig. 6, and was
able to accommodate the hydrogen fuel within the fuselage below
the passenger compartments.

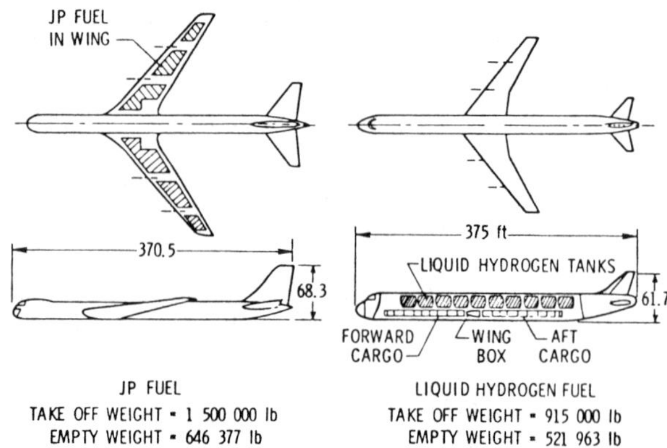

Fig. 3.   Subsonic cargo airplanes designed to carry payloads of 265,000 pounds for a range of 5,070 nautical miles.

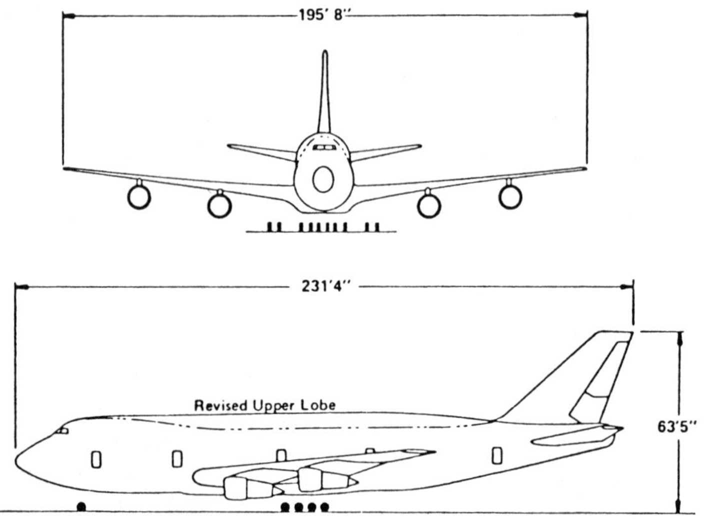

Fig. 4.   Boeing 747 modified to carry all liquid hydrogen fuel in the expanded upper lobe.

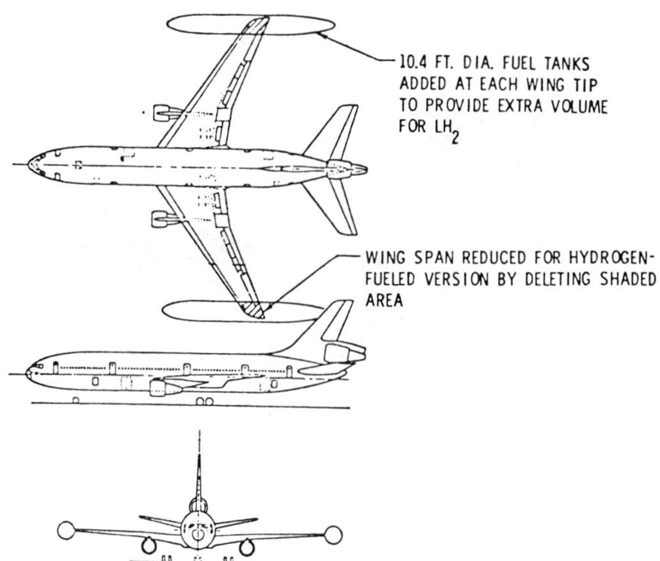

Fig. 5.   Lockheed L1011 modified to carry all liquid hydrogen fuel
in wing-tip tanks.

In Table 2, the major mission capabilities and airplane
weights are summarized for both jet-fueled and hydrogen-fueled
subsonic airplanes. Although the size and range vary, as does to
a lesser extent the cruise Mach number, there is consistant agreement in the reduction in take-off gross weight. In these studies,
most of the reduction is due to reduction in fuel weight.

The consistency of results from the earlier study at Langley
and from the quick-look studies by three major aerospace companies
spurred further effort. In-house, a number of configurations were
selected to expand the range of possibilities. Examples of these
configurations are shown in Figs. 7, 8 and 9.

All are variations on the basic B-747 design. The configuration of Fig. 7 has liquid hydrogen tanks in the fuselage above the
passengers. Of the three designs, this design has the lowest take-
off gross weight, lowest operating weight empty, and the lowest
fuel weight. These weight reductions result in a lighter airplane
to carry the same payload. Lower thrust engines and lower drag
translate into a significant reduction in energy consumed per
passenger mile.

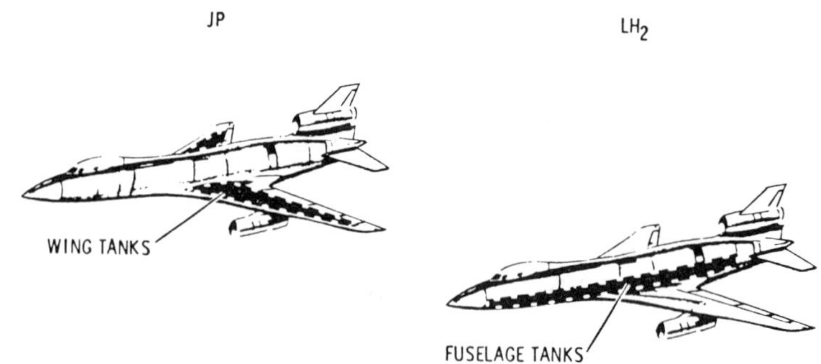

Fig. 6. General dynamics: Convair advanced transport technology design. All liquid hydrogen fuel carried in fuselage tanks.

TABLE 2

JP-Fueled Subsonic Airplanes and LH$_2$

	BOEING (1)		LOCKHEED (2)		CONVAIR (3)	
PAYLOAD, LBS. RANGE, n. mi. MACH NO.	123,000 5,000 0.86		56,000 3,400 0.82		40,000 3,000 0.80	
	JP	LH$_2$	JP	LH$_2$	JP	LH$_2$
TOGW, LBS.	775,000	574,000	430,000	318,000	285,740	201,000
FUEL, LBS.	268,000	90,500	137,000	46,650	88,775	26,500
WING SPAN, FT.	195	195	155	140 plus tip tanks	139	116.5

DERIVATIVES OF (1) 747, (2) L1011 AND (3) ATT DESIGNS.

An example of a configuration in which liquid hydrogen is carried in wing-tip tanks is shown in Fig. 8. In this case, separation is provided between the passengers and the fuel. Inspection and servicing of the fuel tanks is less difficult. The size of the fuselage is reduced because it does not contain the fuel.

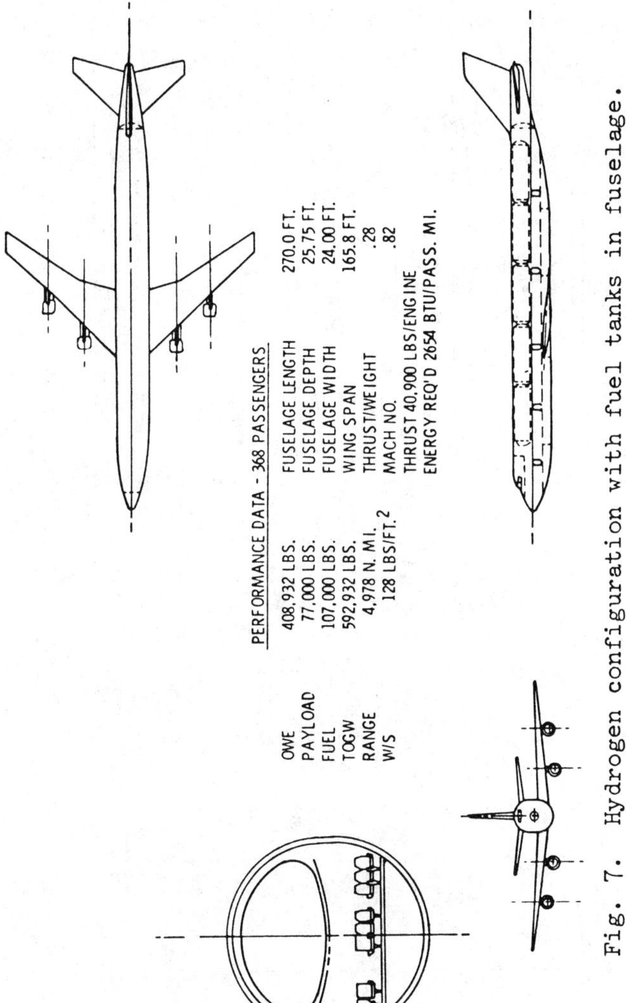

Fig. 7. Hydrogen configuration with fuel tanks in fuselage.

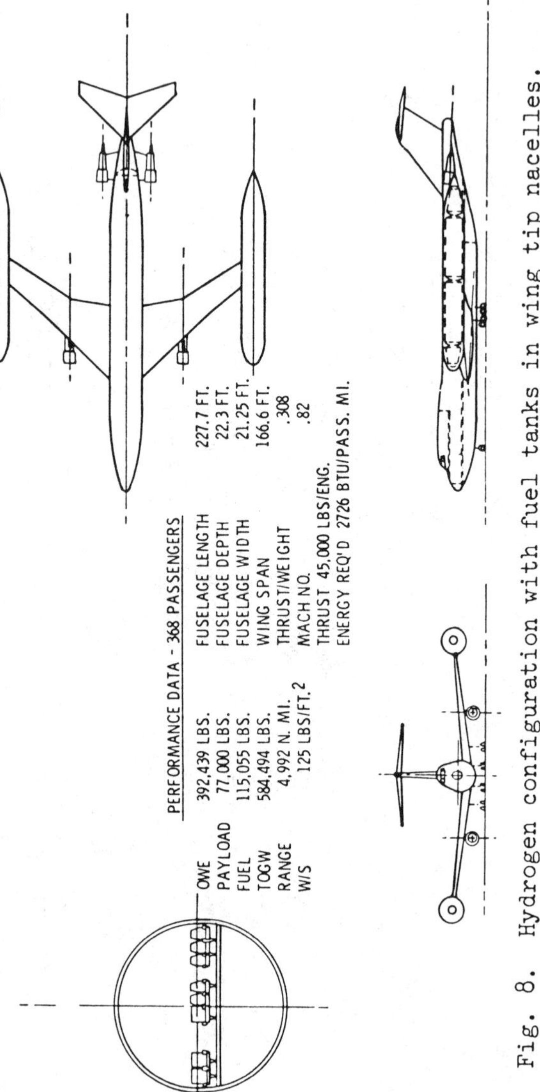

Fig. 8. Hydrogen configuration with fuel tanks in wing tip nacelles.

A methane-fueled airplane design is illustrated in Fig. 9.
The fuel is carried within the fuselage below the passenger compartment. The fuselage is smaller than the hydrogen airplane
(Fig. 7). Of the three configurations, this methane airplane is
the heaviest, requires the largest jet engines, consumes more
energy than the hydrogen airplane with fuel in the fuselage.

In October 1973, the NASA Langley research Center published
a Request for Proposals for an engineering study of the long-
range subsonic hydrogen airplane for possible introduction to
service in the 1990-1995 time period. The study will also provide an assessment of the problem and technology requirements
peculiar to hydrogen-fueled subsonic aircraft and identification
of action to be taken in order to accelerate development of the
required aircraft technology. In December 1973, Lockheed was
selected to make the study which will be completed in about 1
year.

HYDROGEN TECHNOLOGY AND AIRPLANE DESIGN

It may appear that there is already on-the-shelf a substantial body of technology applicable to the design of hydrogen-
fueled airplanes. In reality, advances in cryogenic engineering,
and specifically in aerospace hydrogen technology, are but be-
ginnings of a whole body of new technology which must be available
to the designers of hydrogen transport airplanes.

The change from military airplanes and space vehicles to commercial air transports is neither simple nor small. Major considerations in the design of transports are safety, economy, long
life, and compatibility with the air traffic control system.

The space program made hydrogen routinely acceptable to flight
and ground crews. To be routinely acceptable to the air traveler,
much more has to be done to win confidence of the public in the
safety of hydrogen airplanes. Whereas aviation fuels have a well-
documented history of hazards to passengers and to people and
property on the ground, practically no history exists for hydrogen.
Some may recall or may have read about the tragic loss of the German Dirigible, the Hindenburg, during mooring operations at the
Lakehurst Naval Air Station, New Jersey, in May 1937. But as spectacular as that accident was, it has little relation to liquid
hydrogen-fueled airplanes.

Even though the use of hydrogen is small, large quantities
are manufactured, transported by truck and by rail, stored in
large tanks above ground and transferred between truck, storage
tank, and space vehicle. Hydrogen truck trailers have been

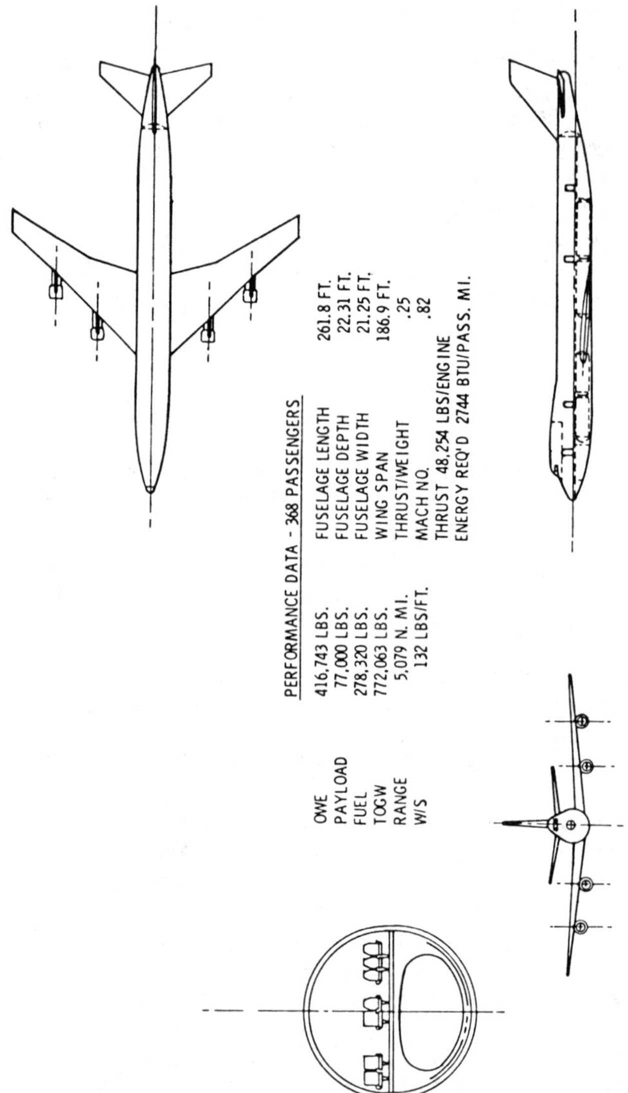

Fig. 9. Methane configuration with fuel tanks in fuselage.

involved in a few highway accidents, but to date only one death is
recorded. Hydrogen-fueled space launch vehicles have been struck
by lightning without damage. That the safety record is excellent
is a tribute to the men who design and operate hydrogen systems.

In a hydrogen airplane, one safety consideration is the location of passengers. Because of hydrogen fuel volume requirements,
as noted earlier in this paper, passengers may be seated above
hydrogen tanks, below, and in between. However, there are alternative arrangements. Passenger-fuel separation can be achieved by
locating fuel in nacelle pods. Another possibility is putting
people in nacellelike pods and carrying fuel within the fuselage.
Analyses will have to be made to determine the limits of accepta-
bility of locations of passenger compartments.

Economy is a consideration which appears throughout airplane
design. With respect to the vehicle, one important element is the
fuel. The rising price of jet fuel not only narrows the gap between jet fuel and liquid hydrogen but the cost of fuel becomes
an increasingly important element in the direct operating cost of
the airplane. Apart from dollar economy, fuel and energy conservation have become important considerations in the design of
any new airplane. An equally important consideration is the development costs of hydrogen fuel and tank systems. More on this
subject later.

To the air carrier, the introduction of new airplanes involves
much preparation to effectively and productively integrate the airplane into service. When the new airplanes use a vastly different
fuel, the carrier's problems increase manifold.

A major factor in the economics of hydrogen airplanes compared to jet-fueled airplanes is that the hydrogen airplanes
usually will be bigger and will carry greater payloads because
their designs tend to optimize at the larger size.

Airplanes for commercial air transportation are designed for
a useful life of 50,000 hours. Many components are cycled several
thousands of times. Major extensions in space hydrogen technology are required before the airplane designer can design, say,
fuel tanks, with confidence that they will survive the required
number of cycles in the environments encountered in flight and on
the ground.

Compatibility of hydrogen-fueled air transports with the air
traffic control system should be comparable with jet-fueled airplanes. Hydrogen airplanes may fly at higher altitudes and thus
alleviate traffic at lower flight levels. Operations at air terminals may be more favorable for the hydrogen airplanes. If the

designs capitalize on the lighter wing loadings, the approach speeds, take-off and landing runs may be less than for jet-fueled airplanes.

Airplane design is so firmly based on jet fuel that no significant effort has yet been made to conceive and develop hydrogen fuel designs. The designs studied so far are basically jet-fuel designs modified to accommodate hydrogen. Unlike jet-fuel airplanes which are performance- and weight-oriented, hydrogen airplanes must be volume oriented to fully capitalize on the properties of hydrogen. Lesser performance, in speed, for example, may be acceptable.

With respect to jet engines, the situation is similar. In the mid-1950s Pratt and Whitney modified a J-65 jet engine to use gaseous hydrogen (Fig. 10) and built the 304 hydrogen expander cycle engine [12]. The 304 engine, Fig. 11, was designed to propel the Lockheed CL-400 airplane at speeds up to Mach 2.5. More recent studies by Pratt and Whitney and by General Electric of hydrogen-fueled engines for the NASA Advanced Supersonic Technology program indicate that the savings in weight in changing from jet fuel to hydrogen are small. However, no major effort has yet been made to design from scratch a jet engine for hydrogen.

Exploratory and preliminary studies of hydrogen airplanes identified fuel volume as a major design problem. As shown earlier, subsonic hydrogen airplane designs propose carrying liquid hydrogen within the fuselage and in nacelles below the wing, blended into the wing and at the wint tips. The number of tanks and their locations have a major impact on airplane configuration and on the locations of passengers, baggage, and cargo. Each tank must be designed to achieve high volumetric efficiency within the space available and to minimize fuel loss due to boil-off and leakage. Airplane configurations must be conceived which optimize the blend of fuel volume and aerodynamic performance.

Insulation concepts developed from the space program must be reassessed for suitability for high-cycle, long-life use in flight and ground environments of the airplane.

Fuel systems will be much more complex than for the jet-fuel airplane. These complexities arise not only because of the cryogenic hydrogen, but also because of the need for double-wall vacuum plumbing of fuel pipes, manifolds, valves, pumps, and related hardware. Fortunately, much space technology exists on ground and vehicle systems which should be transferable with appropriate modification to airplane designs.

The above sampling of technology needs represents challenges to the subsonic airplane designer. Although these challenges are of a lesser order of difficulty than those confronting researchers

Fig. 10.   Comparison of Pratt and Whitney J-65 jet engines:
hydrogen - upper photo; jet fuel - lower photo.

Fig. 11.   The Pratt and Whitney model 304 hydrogen expander cycle
jet engine for the Lockheed CL-400 Mach Number 2.5
high altitude reconnaissance airplane.

and designers of supersonic and hypersonic transports, it is a
reasonable place to start. A strong, early response to these
challenges will set in motion a progressive trend which will
assure a bright future for the aerospace industry.

CONCLUDING REMARKS AND OUTLOOK

The potential of liquid hydrogen as a fuel for the next gen-
erations of long-range subsonic transports is continue growth of
low-cost air travel unconstrained by dependence on hydrocarbon
fuels. Although the engineering design problems of the airplane
are formidable, none appear to be insurmountable from a technical
point of view. The economic problems of production, distribution,
and liquefaction of hydrogen are equally difficult. While it may
be unrealistic to expect the aerospace industry to solve the
economic problems, it may well spearhead the introduction of hydrogen and its acceptance by other industries and the public to
our national economy.

REFERENCES

1. Lewis Laboratory Staff, NACA Lewis Flight Propulsion Labora-
    tory: Hydrogen for Turbojet and Ramjet Powered Flight, NACA
    Research Memorandum, RM E57D23.

2. Rich, Ben R., Lockheed CL-400 Liquid Hydrogen Fueled M 2.5
    Reconnaissance Vehicle, Lockheed California Company. Paper
    presented at the Working Symposium on Liquid-Hydrogen-Fueled
    Aircraft, NASA Langley Research Center, Hampton, Virginia,
    May 15-16, 1973.

3. Small, William J., David E. Fetterman, and Tom F. Bonner, Jr.,
    Potential of Hydrogen Fuel for Future Air Transportation Sys-
    tem, The Second Intersociety Conference on Transportation,
    Denver, Colorado, September 24-27, 1973.

4. Brewer, G. D., The Case for Hydrogen-Fueled Transport Air-
    craft, Lockheed California Company. AIAA/SAE 9th Propulsion
    Conference, Las Vegas, Nevada, November 5-7, 1973.

5. Loftin, Laurence K., Jr., Transport Aircraft - 1980 and Beyond.
    The Second Intersociety Conference on Transportation, Denver,
    Colorado, September 24-27, 1973.

6. Escher, William J. D., Prospects for Liquid Hydrogen Fueled
    Commercial Aircraft, Escher Technology Associates, Report
    PR-37, September 1973.

7.   Anon., Aerospace Facts and Figures - 1973/74, Aerospace In-
     dustries Association of America, Inc., Washington, D.C.

8.   Van Lennep, Emile, Aviation Needs and Public Concerns. The
     Seventh Dr. Albert Plesman Memorial Lecture sponsored by the
     International Air Transport Association, Delft, October 23,
     1973.

9.   Anon., Airline Passenger, Freight Traffic Up in 1973. Release
     dated January 2, 1973, Air Transport Association of America,
     Washington, D.C.

10. Whitlow, John B., Jr., and Gerald A. Kraft, Potential of
     Methane-Fueled Supersonic Transports over a Range of Cruise
     Speeds Up to Mach 4, Lewis Research Center, Cleveland, Ohio
     NASA TM X-2281, May 1971.

11. Advanced Transport Technology Staff, A Fuel Conservation
     Study for Transport Aircraft Utilizing Advanced Technology
     and Hydrogen Fuel, LTV Aerospace Corporation, Hampton Tech-
     nical Center, Hampton, Virginia, NASA CR-112204, November 10,
     1972.

12. Anschutz, Richard H., Hydrogen Burning Engine Experience,
     FRDC, Pratt and Whitney. Paper presented at the Working
     Symposium on Liquid-Hydrogen-Fueled Aircraft, NASA Langley
     Research Center, Hampton, Virginia, May 15-16, 1973.

Reprinted from Science 182:1299–1304 (1973)

Methanol: A Versatile Fuel for Immediate Use

Methanol can be made from gas, coal, or wood.
It is stored and used in existing equipment.

T. B. Reed and R. M. Lerner

In the short period of a decade we have developed a healthy concern about our pollution of the environment and an awareness that we will soon face a shortage of convenient forms of energy. Hydrogen has been suggested (1, 2) as a universal, nonpolluting fuel, since it can be produced from water and burns cleanly to water. Although we may some day see a "hydrogen economy," we have yet to find ways of making hydrogen cheaply, of storing and transporting it, of adapting it to the automobile, and of using it safely.

In several recent, comprehensive studies of potential fuels, hydrogen has been compared with other synthetic fuels such as methanol, ethanol, hydrazine, and methane (3–5). In these studies methanol has been described as being superior to hydrogen in many ways, and as providing an especially attractive alternative fuel to gasoline. In this article we discuss the advantages of using methanol as a fuel and suggest ways in which it could be introduced immediately into our fuel economy.

Methanol, CH_3OH, can be thought of as two molecules of hydrogen gas made liquid by one molecule of carbon monoxide. It thus shares many of the virtues of pure hydrogen. Last year in the United States, 3.2×10^9 kilograms (10^9 gallons or 1 percent of present gasoline production) of methanol were manufactured and sold at an average price of $0.06 per kilogram ($0.183 per gallon) (6). It can be made from almost any other fuel—from natural gas, petroleum, coal, oil shale, wood, farm and municipal wastes—so that a methanol economy would be flexible and could draw from many energy sources as conditions change. Methanol is easily stored in conventional fuel tanks and can be shipped in tank cars, tank trucks, and tankers; it can be transported in oil and chemical pipelines. Of most importance is the fact that up to 15 percent of methanol can be added to commercial gasoline in cars now in use without it being necessary to modify the engines. This methanol-gasoline mixture results in improved economy, lower exhaust temperature, lower emissions, and improved performance, compared to the use of gasoline alone. Methanol can also be burned cleanly for most of our other fuel needs, and it is especially suited for use in fuel cells for generating electricity. The sources, distribution, and uses of methanol are shown diagrammatically in Fig. 1.

Properties of Methanol and Related Fuels

Methanol, which is also called methyl alcohol, wood alcohol, or methylated spirits (7), is a colorless, odorless, water-soluble liquid. It freezes at −97.8°C, boils at 64.6°C, and has a density of 0.80. It is miscible with water in all proportions, and spillages are rapidly dispersed. Methanol burns with a clean blue flame and is familiar to most people as the alcohol used for heating food at the table or as the alcohol in Sterno. Mixtures with between 6.7 percent and 36 percent of air are flammable. The autoignition temperature of methanol is 467°C, which is high compared with 222°C for gasoline (5). This may account for the high octane number, 106, of methanol; a typical gasoline has an octane number of 90 to 100 (8).

The energy content of a number of fuels is shown in Table 1. Hydrogen produces the most energy on a weight basis; hence its use in rockets where volume and cost are secondary considerations. Petroleum products such as gasoline have the highest energy of the fuels listed on a volume basis and are second highest on a weight basis; hence gasoline will long be preferred for airplanes where lowest weight is at a premium. However, of all the liquid fuels, methanol produces the second highest amount of energy on a volume basis.

Although methanol is not the cheapest fuel (see Table 2), its properties make it competitive with the other fuels. Ethanol, which has many of the desirable fuel properties of methanol and could be used in most of the applications discussed herein, costs, in this country, about three times more than methanol. In less industrialized countries, however, ethanol may be an attractive fuel because it can be produced from agricultural products by fermentation.

In the manufacture of methanol, the output of the plant can be increased by 50 percent if small amounts of other alcohols can be tolerated in the product (9, 10). Such a mixture is called "methyl-fuel," and it contains more energy than pure methanol because of the presence of ethanol, propanol, and isobutanol. It can be produced in larger quantities at a lower price than pure methanol, and, in general, has superior properties as a fuel. In this article, we consider methanol and methyl-fuel to be synonymous.

Dr. Reed is a member of the solid state division and Dr. Lerner is a member of the communications division at the Lincoln Laboratory, Massachusetts Institute of Technology, Lexington, Massachusetts 02173.

Historical Uses of Alcohol for Fuel

During the last 50 years in the United States, methanol and other alcohols have not competed successfully with the abundant supplies of petroleum. Before this time, however, alcohols were used extensively as fuels. Alcohol, for example, became a popular fuel for lighting in about 1830, when it replaced malodorous fish and whale oils. In about 1880, kerosene replaced alcohol as a lighting fuel because of its sooty flame which gave more light; a clean flame produces no light without special additives. During the middle of the last century, France was partially on a methanol fuel economy. Wood was distilled in the provinces to give alcohol, which was burned in Paris for heating, lighting, and cooking. This was more economical than transporting wood to Paris and then disposing of the ashes.

During World Wars I and II, when gasoline shortages occurred in Germany and France, vehicles of all sorts, including tanks and planes, used wood burners in the rear or in trailers. Wood chips were distilled to make alcohol vapors that included carbon monoxide and hydrogen; these vapors would (barely) drive the vehicle. In 1938, 9000 wood-burning cars were used in Europe. "Power alcohol" (ethanol) was also used by France and Germany to supplement gasoline supplies and stimulate alcohol production for anticipated use in munitions production (11).

In about 1920, manufacturers in the United States began to produce methanol for use as a solvent, for plastic manufacture, and for fuel injection in piston aircraft. In 1972, 3.2×10^9 kg (10^9 gallons) of methanol were produced (6, 7), equal to about 1 percent of the amount of gasoline produced.

Methanol in the Internal Combustion Engine

It has been claimed that hydrogen is an ideal fuel for the internal combustion engine (12). It certainly causes little pollution, but is difficult to store, high in price, and difficult to burn efficiently in the engine without it knocking and backfiring (13). These problems arise because of the very wide flammability limits and the very high flame velocity of hydrogen. Perhaps an engine can be invented which takes advantage of these unusual properties.

Methanol used as an additive or substitute for gasoline could immediately help to solve both energy and pollution problems. We will first discuss its use in place of gasoline and then present new results to show that 5 to 15 percent of methanol added to gasoline could produce disproportionate improvements in the fuel economy, pollution levels, and performance of cars now in use.

A number of studies (14, 15) of methanol and ethanol have been conducted in the last 50 years to test their suitability as substitutes for gasoline in the internal combustion engine. Existing engines can be converted to use pure methanol by decreasing the ratio of air to fuel consumed from about 14 for gasoline to 6 for methanol, by recycling more heat from the exhaust to the carburetor, and by providing for cold starts. The conversion is estimated to cost about $100 per vehicle (16). A municipal vehicle converted in this way has been operating satisfactorily in Santa Clara, California, for the past year. Compared with gasoline, the use of methanol in a standard test engine (without catalytic treament of exhausts) yielded one-twentieth of the amount of unburned fuel, one-tenth of the amount of carbon dioxide, and about the same amount of oxides of nitrogen NO_x as gasoline (3, 17). Table 3 shows that a 1972 Gremlin fueled with methanol almost met the 1976 federal standards for emissions and had five times lower emissions than a similar car operated with gasoline (3, 4, 18). In these studies the reduced emissions were attributed to methanol being able to burn without misfire at an air-to-fuel ratio 25 percent higher than gasoline; exhaust temperatures with methanol were 100°C cooler; more spark retard was possible with methanol because of its higher flame speed (18). It was suggested that greater performance and economy could be expected in an engine designed specifically for methanol, and that such design should encompass higher compression ratios and a fuel injection system. In another study, on a one-cylinder research engine, it was found that 10 to 20 percent leaner mixtures could be tolerated with methanol than with gasoline. The amounts of unburned hydrocarbons, CO, and NO_x produced were lower with methanol than with gasoline,

Table 1. The energy content of some fuels, shown on the basis of weight and volume.

Fuel	Formula	Heat of combustion (low)*	
		Kjoule/ g	Kjoule/ cm³
Liquids			
Hydrogen.	H_2	124.7†	8.7
Methanol	CH_3OH	20.1	15.9
Gasoline	C_8H_{18}	44.3	30.9
Solids			
Hydrides	VH_2‡	4.7	28.4
Coal	$C_{29}H_{42}$	32.2	41.8
Wood	$C_{32}H_{46}O_{22}$	17.5	14.2
Gases			
Hydrogen	H_2	124.7	0.0010
Methane	CH_4	61.1	0.0044

* Combustion to CO_2 and H_2O (gas). † Conversion factors are: 0.948 kjoule = 1 Btu; 2.10 kjoule/g = 10^3 Btu/lb; 0.27 kjoule/cm³ = 10^3 Btu/gallon; 33.4×10^4 kjoule/cm³ = 10^6 Btu/ft³. ‡ Vanadium hydride is given as an example.

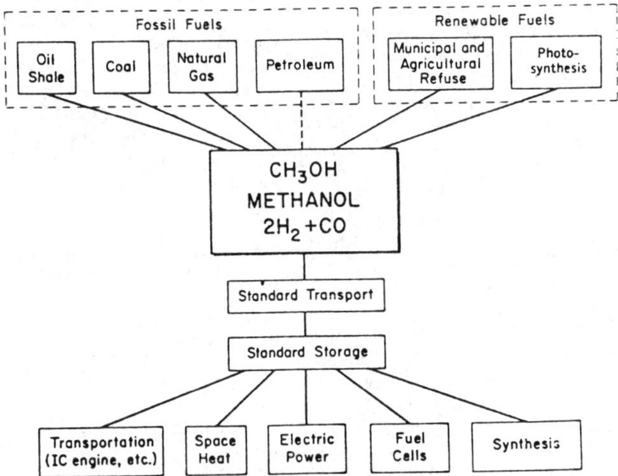

Fig. 1. Sources, transport, and possible applications of methanol.

while the amounts of aldehydes produced were higher (*19*).

From these results it seeems clear that if gasoline becomes scarce or too expensive, we can design cars that will operate on pure methanol and cause less pollution. Specific fuel consumption will certainly be higher on a weight or volume basis (see Table 1), necessitating a larger fuel tank; but specific energy consumption (energy per kilometer) will certainly be lower because higher compression ratios and simpler pollution controls can be used.

The principal drawback to the immediate use of pure methanol as a gasoline substitute is that not enough is available. We have recently tested the possibility of adding 5 to 30 percent of methanol to gasoline (*20*). A number of unmodified private cars (year models 1966 to 1972) were tested and operated over a fixed course with varying concentrations of methanol. It was found that (i) fuel economy increased by 5 to 13 percent; (ii) CO emissions decreased by 14 to 72 percent; (iii) exhaust temperatures decreased by 1 to 9 percent; and (iv) acceleration increased up to 7 percent. The results obtained on a 1969 Toyota (1900 cm^3 engine, 85 brake horsepower, 8 : 1 compression ratio) are shown in Fig. 2. This car has now been driven about 8000 kilometers fueled alternately with gasoline and with mixtures of gasoline and 10 to 30 percent methanol. There have been no mechanical problems. A most striking feature observed with two unmodified cars was the elimination of knock and "Diesel operation" (continued operation after the ignition is

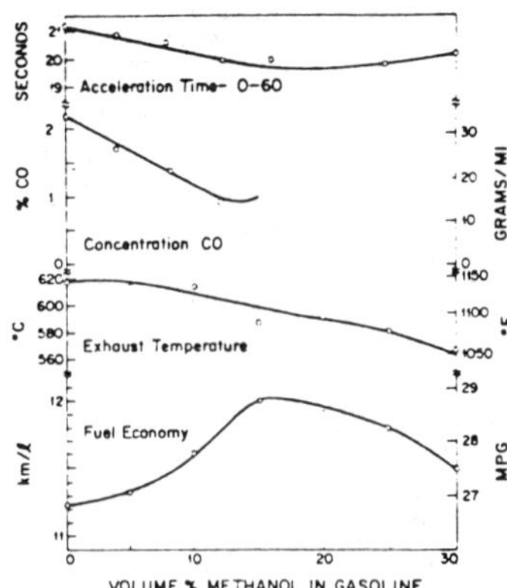

Fig. 2. Performance of a 1969 Toyota Corona with methanol-gasoline mixtures. *MPG*, miles per gallon.

turned off) when gasoline containing 5 percent methanol was used.

An octane rating of 106 for pure methanol hardly seems sufficient to explain these improvements. The methanol is said to have a "blending octane value" (BOV) of 130 (*9, 10*) defined by

$$BOV = [O_b - O_g(1 - x)]/x$$

where O_b and O_g are the octane numbers of the blend and the gasoline (*8*), and x is the volume fraction of methanol in the gasoline. From this, 10 percent of methanol added to gasoline with an octane rating of 90 would be expected to yield a fuel with an octane rating of 94, equivalent to the addition of 0.13 gram of tetraethyl lead per liter of gasoline (*9*). Ethanol has a BOV of 110 to 160, depending on the octane of the gasoline (*15*).

To account for the disproportionate effects of methanol on the octane value and other properties of gasoline, we propose the following mechanism (*20*). When methanol is synthesized from CO and 2H$_2$ at 50 to 300 atmospheres and 200°C, −90.8 kilojoules of heat must be removed (see Eq. 1). In the compression stroke of the internal combustion engine, methanol can dissociate at very low temperatures and reabsorb this energy, cooling the charge and quenching premature combustion. For example, at 10 atmospheres methanol is 18, 85, and 99.7 percent dissociated at 100°, 200°, and 300°C, respectively [calculated from data in (*7*)]. The CO and H$_2$ formed on dissociation increase the flame velocity of the charge, giving more complete and efficient combustion.

Other Aspects of Methanol Utilization

Although methanol is suggested here principally as a fuel for automobiles, it could also be used advantageously in most other fuel applications if it becomes sufficiently plentiful. It is a safe, clean fuel for home heating and can also be burned in power plants to generate electricity without polluting the atmosphere. In a recent set of pilot-plant and full-scale power boiler demonstrations, methyl-fuel was tested against No. 5 fuel oil and natural gas (*9*). In the tests with methyl-fuel it was observed that (i) no particulates were released from the stack; (ii) the amount of NO$_x$ in flue gases was less than the amount emitted from natural gas and much less than that emitted from the oil; (iii) the CO concentration was less than that from oil and gas; (iv) no sulfur compounds were emitted; (v) the amounts of aldehydes, acids, and unburned hydrocarbons produced were negligible; and (vi) soot deposits in the furnace from previous oil firing were burned off with methyl-fuel, thereby allowing higher heat-transfer rates and higher efficiency.

Methanol is one of the few known fuels suited to power generation by fuel cells (*21*). In principle, the fuel cell can convert chemical energy to electricity with much higher efficiencies than heat engines such as turbines. Although methanol is not as simple to use in a fuel cell as hydrogen, it can be stored and shipped more easily. Recently a fuel cell has been developed that gives more than 30,000 hours of continuous operation on methanol and air. It uses tungsten carbide and charcoal as electrodes and sulfuric acid as electrolyte (*22*).

The costs of storage and shipment of methanol and other fuels are shown in Table 4. The storage of methanol mixed with gasoline may present cer-

---

Table 2. Production cost of the energy contained in some fuels. The costs are for large plant capacities, assuming that 15 percent of the plant cost is spent annually on profit, interest, depreciation, and maintenance [data from (*3*, p. 12)].

Fuel	Source	Cost ($/10⁶ kjoule)
Gasoline	Crude oil*	1.00
Methanol	Natural gas†	1.49
	Coal‡	1.40
	Lignite§	1.18
Methane gas	Wellhead	0.14–0.37
	LNG imported‖	0.76–0.95
	Coal	0.76–0.95
Hydrogen gas	Natural gas	0.92
	Coal	1.25
Liquid hydrogen		2.37

* Gasoline produced at $0.118 per gallon. † Natural gas at $0.040 per 10³ ft³. ‡ Coal at $7 per ton or $0.25 per 10⁶ kjoule. § Lignite at $2 per ton or $0.14 per 10⁶ kjoule. ‖ LNG, liquefied natural gas.

Table 3. Emissions from a 1972 Gremlin (i) that uses gasoline and (ii) that was modified for use with methanol fuel and equipped with a catalytic converter (*18*). Projected (as of 1973) federal standards (for 1975 to 1976) are included for comparison.

Fuel	Emissions (g/mile)		
	Unburned hydrocarbons	CO	NO$_x$
Gasoline	2.20	32.5	3.2
Methanol	0.32	3.9	0.35
Federal standards	0.41	3.4	0.40

tain problems because of the solubility of methanol and water in gasoline. Only about 0.01 percent of water is soluble in pure gasoline, and therefore excess water from condensation sometimes accumulates in storage tanks and causes corrosion. Gasoline containing 10 percent methanol will dissolve ten times as much water and so can keep the tanks dry; in fact proprietary gas-tank drying agents generally contain methanol. However, water in larger quantities will remove almost ten times its own weight from gasoline containing 10 percent methanol, so that unless a storage tank is first dried out, problems may arise when a mixture of gasoline and methanol is first put in it (*15*).

Although methanol is miscible with gasoline at room temperature, less than 10 percent is soluble in some gasolines at 0°C. However, the volatile constituents added to gasoline in cold weather to aid ignition increase this solubility. Also, the higher alcohols in methyl-fuel increase the solubility of methanol. For instance, 2.4 percent of isobutanol in one gasoline increased the solubility of methanol from 3 to 10 percent at 0°C (*15*).

Methanol, although not highly toxic, can be lethal if ingested. It would therefore be prudent to avoid the names methyl alcohol and wood alcohol in any labeling of methanol containers. Methanol vapors are also poisonous, but no more so than those of many other common substances. For example, the maximum allowable exposure to methanol vapor is 200 parts per million (ppm), while the value for ethyl alcohol is 1000 ppm; for benzene, 10 to 25 ppm; octane, 400 ppm (octane and benzene are typical constituents of gasoline); trichloroethylene, 100 ppm; and carbon tetrachloride, 10 ppm (*23*).

Methanol Manufacture

Methanol can be made from many sources, as shown in Fig. 1. Until about 1925 it was made (along with acetic acid and tars) by the destructive distillation of wood. Since that time, most methanol has been synthesized from CO and H_2 (*7*) according to Eq. 1:

$$CO + 2H_2 \rightarrow CH_3OH(gas);$$
$$\Delta G = -90{,}800 + 229T \text{ (joule/mole)} \quad (1)$$

where ΔG is the free energy change and T is the temperature.

In the original high-pressure process, pressures of 300 atmospheres at 200°C

Table 4. Costs of storage and transportation of methanol, gasoline, and liquid and gaseous hydrogen (*3*).

Fuel	Cost ($/$10^6$ kjoule)	
	Storage	Transport over 100 km
Methanol	3–21	0.027
Gasoline	2–15	0.018
Liquid hydrogen	300–1000	1.55
Gaseous hydrogen	350	0.035

were used in the presence of a zinc-chromium oxide catalyst, and yields of over 60 percent were obtained. In 1968 the Imperial Chemical Industries (ICI) developed a low-pressure process using 50 atmospheres at 250°C and a highly selective, copper-based catalyst. This process produces much purer methanol. A number of processes in which intermediate pressures are used have since been developed (*6, 9*).

The CO and H_2 (synthesis gas) for manufacturing methanol can be obtained by partial oxidation of any carbonaceous fuel with oxygen or water. At present it is obtained almost exclusively from methane by partial oxidation with water. This source of methanol will not long be useful in this country, because there is not now enough natural gas available for our domestic heating needs. However, in the Near East the energy value of the methane flared off at the oil wells would be sufficient to supply much of our energy needs. Plans are now under way to construct methanol plants so that this gas can be converted to methanol at the wellhead and shipped in conventional tankers to this country (*24*).

In a recent study (*25*) it was estimated that methanol could be produced on the Persian Gulf and landed on the East Coast of the United States at a cost of $1 per 10^6 kjoule ($0.061 per gallon; $1.05 per 10^6 British thermal units). The construction of refrigerated tankers for methane is also being considered. Liquefied gas in refrigerated tankers would cost $1.46 per 10^6 kjoule. These estimates, which include the cost of construction of new plants or refrigerated tankers, would vary according to the distance of transportation.

Methane gas is also produced biologically by the decomposition of natural wastes, such as pig and chicken manure and sewage. It has been claimed that such methane can be used for powering automobiles, where it has an operational cost equivalent to a cost of $0.03 per gallon for gasoline (*26*).

It is also claimed that enough fuel could be made from this source to meet all present fuel needs in the United States, and that the use of such a process would reduce by half the problem of sewage and animal-waste disposal (*27*). In experiments with cars and trucks converted to use methane, the U.S. General Services Administration has reported clean, reliable operation (*28*). However, the type of cylinder required to contain compressed gaseous methane severely limits the amount of fuel that can be carried; a six-cylinder sedan has a range of 80 km (50 miles), each cylinder measuring 6.7 m^3 and weighing 100 kg (*29*). Conversion of the organic wastes to methanol rather than methane would make this fuel source much more practical.

For the next few decades, coal is the most attractive candidate for methanol production. Coal has long been used for the production of synthesis gas, according to the endothermic reaction (with ΔH being the heat change):

$$C + H_2O(gas) \rightarrow CO + H_2;$$
$$\Delta H = +131.4 \text{ kjoule} \quad (2)$$

Although synthesis gas contains CO, which is poisonous, it is used for industrial power and for heating homes in many European cities without further conversion. It represents a clean, gasified coal (*30*). Much work is in progress to develop methods to obtain methane and hydrogen from coal for use as pipeline gas. The same technology can be applied to the manufacture of methanol from synthesis gas. It is estimated that the cost of making methanol from coal would be $1.40 per 10^6 kjoule ($0.085 per gallon) for a plant making 20,000 metric tons per day (*3*). If lignite is used as the starting material instead of coal, the resulting ash may contain 0.40 percent uranium, equivalent to commercial uranium ore, as well as other valuable minerals such as molybdenum, vanadium, arsenic, germanium, selenium, cobalt, and zirconium (*3*). Efforts are being made to develop practical methods of gasifying coal in the ground, eliminating the need for strip mining and consequent landscape destruction.

Some day we will run out of fossil fuels. By coupling the manufacture of methanol with the disposal of wastes, we could supplement our fuel supply and thereby prolong the existence of fossil fuels, and simultaneously clean up the landscape. A recent patent (*31*) describes an "oxygen refuse converter" that can dispose of our refuse and at the same time generate useful energy. In a shaft furnace shown in Fig. 3, unseparated trash or sewage sludge is fed into a hopper at the top. Low-cost oxygen (0.2 kg of O_2 per kilogram of refuse) is fed into this furnace near the bottom, creating a 1500°C zone that melts the metals and glasses found in refuse. These melts, drawn off as slag and metal, have 2 percent of the original refuse volume, while all other products are gaseous or water soluble. Carbon, burning in the high-temperature zone, produces CO, which rises through the furnace. The hot CO creates an intermediate-temperature zone where carbohydrates and plastics are broken down to a gas containing, typically, 47 percent CO, 28 percent H_2, 17 percent CO_2, and 5 percent CH_4 by volume. Finally, in the uppermost section the incoming refuse is dried as the gas mixture cools to about 100°C.

This gas mixture stores 8.0 kjoule per gram of refuse (7×10^6 Btu per ton), or 76 percent of the original refuse energy. Because oxygen rather than air is used in burning, the output gas is high in heat content, low in volume, and relatively easy to scrub to remove fly ash and chlorine.

The United States produces about 1.8×10^{11} kg of solid refuse each year. The energy in the gas from this refuse is 1.4×10^{15} kjoule, or 2 percent of the 7.4×10^{16} kjoule (7.0×10^{16} Btu) consumed each year (*32*). If this gas were converted to methanol, it could supply about 8 percent of the fuel for our transportation needs. Although initially developed for refuse, converters of this type could also be used to convert farm waste and the waste from lumbering into more useful forms of energy such as methanol.

Forests, which are one means of capturing solar energy, formed the principal energy source for this country until about 1875. Commercial forests now cover about 23 percent of the land area of the United States, or 2.1×10^{12} m². These forests intercept from the sun about 5.8×10^6 kjoule/m² per year, or a total of 1.2×10^{19} kjoule per year (*33*). If the conversion of solar energy with an efficiency approaching 1 percent could be achieved by improved forest management (*34*), the annual energy harvest might be 1.2×10^{17} kjoule per year, more than our present energy needs of 7.4×10^{16} kjoule per year. The advantage of utilizing forests for the production of methanol is that whole trees can be used, not merely those fractions that make good lumber or pulp. It has been calculated that between 5 and 20 percent of our commercial forests operated as "energy plantations," could supply all of our electrical power (*35*).

Recommendations

The management of energy resources is the management of the lifeblood of our economy, and whatever future energy sources evolve should be those that best blend environmental and technical solutions with economic reality. Therefore we look forward in the next few years to vigorous, public, and possibly polemic debate on the virtues of various fuels.

Recently it has been recommended that "the various energy planning agencies should now begin to outline the mode of implementing hydrogen energy delivery systems in the energy economy" (*2*). On the contrary, we see methanol as a more benign solution to our fuel problems. The use of methanol would produce the least dislocation of our economy and industry, and would solve environmental as well as energy problems. Since it is compatible with gasoline and existing automobiles, it can be introduced gradually as a fuel as production increases. We will not then have to scrap refineries, automotive facilities, or our cars.

The course of these debates should be influenced by the results of research on all of the various fuels that might be used to supply our energy needs. We suggest that there are three main areas for research into the large-scale production of methanol. First, since a plant must operate under different and less exacting constraints in producing methyl-fuel than in producing the present industrial grade of alcohol, the existing methods of methanol production should be reexamined and optimized for the production of methyl-fuel. Second, while methanol is now being produced primarily from natural gas, likely to be augmented by production from coal (*31*), other processes should be developed for utilizing all major sources of carbonaceous materials for methanol synthesis. Third, methanol is now produced primarily by partially oxidizing other fuels to CO and H_2 and by rebuilding these to methanol with some loss of the potential energy value of the original fuel. We should look for methods of direct conversion that do not entail this energy loss.

Finally, we recommend that the various energy planning and regulatory agencies for fuels should strongly consider altering existing regulations to accommodate the introduction of new fuels. For instance, the blending of methanol with gasoline could be encouraged by considering it an environmentally beneficial additive, rather than as a fuel to be taxed. This would certainly make methanol-blend gasolines cheaper than gasoline and encourage production all over the world. We think also that it would be in the national interest to reexamine the tariff now imposed on imported methanol if it is to be used for fuel.

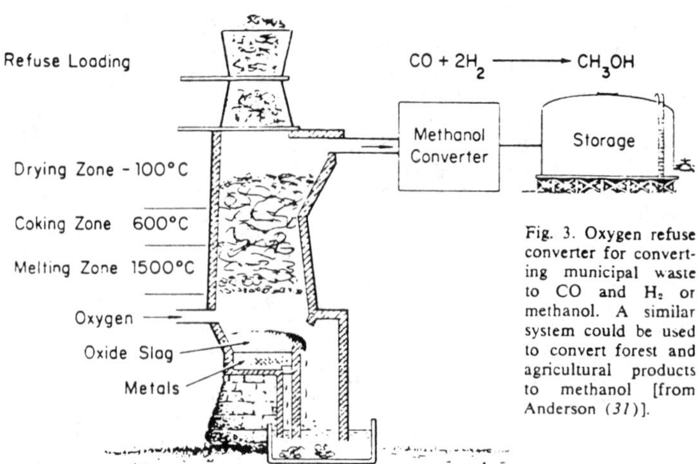

Fig. 3. Oxygen refuse converter for converting municipal waste to CO and H_2 or methanol. A similar system could be used to convert forest and agricultural products to methanol [from Anderson (*31*)].

Summary

We believe that methanol is the most versatile synthetic fuel available and its use could stretch or eventually substitute for, the disappearing reserves of low-cost petroleum resources. Methanol could be used now as a means for marketing economically the natural gas that is otherwise going to waste in remote locations. If methanol were used as an additive to gasoline at a rate of 5 to 15 percent, for use in internal combustion engines, there would be an immediate reduction in atmospheric pollution, there would be less need for lead in fuel, and automobile performance would be improved.

With increasing production of fuel-grade methanol from coal and other sources, we foresee the increasing use of methanol for electrical power plants, for heating, and for other fuel applications. We hope that a practical methanol fuel cell will be commercially available by the time that methanol becomes plentiful for fuel purposes.

Methanol offers a particularly attractive form of solar-energy conservation, since agricultural and forest waste products can be used as the starting material. Indeed, at 1 percent conversion efficiency the forest lands could supply the entire present energy requirements of the United States.

References and Notes

1. D. P. Gregory, D. Y. C. Ng, G. M. Long, in *The Electrochemistry of Cleaner Environments*, J. O'M. Bockris, Ed. (Plenum, New York, 1972); C. Marchetti, *Euro-Spectra* **10** (No. 4), 117 (1971); J. R. Bartlit, F. J. Edeskuty, K. S. Williamson, Jr., in *Proc. 7th Intersoc. Energy Conversion Eng. Conf.* (1972), paper 729205, p. 1312; F. C. Tanner and R. A. Huse, in *ibid.*, paper 729207, p. 1323; D. P. Gregory and J. Wurm, in *ibid.*, paper 729208, p. 1329; F. A. Martin, in *ibid.*, paper 729029, p. 1335; R. H. Wiswall, Jr., and J. J. Reilly, in *ibid.*, paper 729210, p. 1342; see also other papers in this volume [Proceedings of the Intersociety Energy Conversion Engineering Conferences are available from the American Institute of Aeronautics and Astronautics, 1290 Avenue of the Americas, New York, N.Y. 10019; see also references (*12, 13, 17, 22, 30,* and *35*)]; D. P. Gregory, *Sci. Am.* **228** (No. 1), 13 (1973); see also collected papers in *International Symposium and Workshop on The Hydrogen Economy*, S. Linke, Ed. (Cornell Univ. Press, Ithaca, N.Y., in press).
2. W. E. Winsche, K. C. Hoffman, F. J. Salzano, *Science* **180**, 1325 (1973).
3. Synthetic Fuels Panel, "Hydrogen and other synthetic fuels; a summary of the work of the synthetic fuel panel" (U.S. Atomic Energy Commission, Report TID-26136, September 1972).
4. National Academy of Sciences, *Report by the Committee on Motor Vehicle Emissions* (National Academy of Sciences, Washington, D.C., February 1973).
5. D. P. Gregory and R. B. Rosenberg, "Synthetic fuels for transportation and national energy needs" (paper presented at the Society of Automotive Engineers National Meeting Symposium on Energy and the Automobile, Detroit, May 1973).
6. Stanford Research Institute, *Chemical Economics Handbook* (Stanford Research Institute, Menlo Park, Calif.), sections 674.5021-674.5023.
7. R. D. Kirk and D. S. Othmer, Eds., *Encyclopedia of Chemical Technology* (Interscience, New York, 1967), vol. 13, p. 370.
8. The octane number of a fuel represents its rating for antiknock properties, measured as the percentage by volume of isooctane in a reference fuel consisting of a mixture of isooctane and normal heptane and matching in knocking properties the fuel being tested. The higher the octane number, the less likely is the fuel to detonate. The octane number of methane, 106, was determined by the re-research method.
9. D. Garrett and T. O. Wentworth, *Methyl-Fuel—A New Clean Source of Energy* (Division of Fuel Chemistry, American Chemical Society, Washington, D.C., August 1973), paper 9; T. O. Wentworth, *Environ. Sci. Technol.* **7**, 1002 (1973).
10. D. Garrett, personal communication.
11. S. Egloff, *Ind. Eng. Chem.* **30**, 1091 (1938).
12. K. H. Well, in *Proc. 7th Intersoc. Energy Conversion Eng. Conf.* (1972), paper 729212, p. 1355.
13. R. E. Billings and F. E. Lynch, *History of Hydrogen-Fueled Internal Combustion Engines and Performance and Nitric Oxide Control Parameters of the Hydrogen Engine* (Energy Research Corp., Provo, Utah, papers 73001 and 73002, 1973); R. R. Adt, Jr., D. L. Hirschberger, T. Kartage, M. R. Swain, in *Proc. 8th Intersoc. Energy Conversion Eng. Conf.* (1973), paper 739092, p. 194.
14. J. A. Bolt, *A Survey of Alcohol as a Motor Fuel* [Society of Automotive Engineers (2 Pennsylvania Plaza, New York, N.Y.); conference paper SP 254, pp. 1-13, June 1964] (65 references are included); E. S. Starkman, H. K. Newhall, R. D. Sutton, *Comparative Performance of Alcohol and Hydrocarbon Fuels* (Society of Automotive Engineers, conference paper 254, pp. 14-33, New York, June 1964).
15. American Petroleum Institute, *Use of Alcohol in Motor Gasoline—A Review* (American Petroleum Institute, Publ. No. 4082, Washington, D.C., 1971).
16. R. K. Pefley, personal communication.
17. ——, M. A. Sand, M. A. Sweeney, J. D. Kilgroe, in *Proc. 6th Intersoc. Energy Conversion Eng. Conf.* (1971), paper 719008, p. 36.
18. G. Adelman, D. G. Andrews, R. S. Devoto, *Exhaust Emissions from a Methanol-Fueled Automobile* (Society of Automotive Engineers, paper 720693, August 1972).
19. G. D. Ebersole and F. S. Manning, *Engine Performance and Exhaust Emissions: Methanol versus Iso-octane* (Society of Automotive Engineers, paper 720692, August 1972).
20. R. E. Lerner, T. B. Reed, E. D. Hinkley, R. E. Fahey, in preparation.
21. G. Siprios, in *Proc. 20th Annu. Power Sources Conf.* (1966), p. 46.
22. L. Baudendistel, H. Böhm, J. Heffler, C. Louis, F. A. Pohl, in *Proc. 7th Intersoc. Energy Conversion Eng. Conf.* (1972), paper 729004, p. 20
23. N. I. Sax, *Dangerous Properties of Industrial Materials* (Reinhold, New York, ed. 3, 1968); Occupational Safety and Health Association, *Federal Register* **36**, 10504 (1971).
24. J. Schons, personal communication; G. Smith, *New York Times*, 29 September 1973; Anonymous, *Chemical Week*, 19 September 1973.
25. B. Dutkiewicz, "Methanol competitive with LNG on long haul," paper presented at the 52nd Annual Meeting, Natural Gas Producers Association, Dallas, 26-28 March 1973; *Oil Gas J.*, 30 April, p. 166 (1973).
26. *Clean Air Water News* **4**, 38 (1972); B. Grindrod, *Mother Earth News No. 10*, 14 (1971).
27. H. L. Bohn, *Environment* **13**, 4 (1971).
28. "GSA expands duel-fuel vehicle experiment," *Clean Air Water News* **3**, 494 (1971).
29. E. Gross, *Sci. News* **97**, 73 (1970).
30. R. A. Graff, S. I. Dobner, A. M. Squires, in *Proc. 7th Intersoc. Energy Conversion Eng. Conf.* (1972), paper 729172, p. 1153.
31. J. E. Anderson, U.S. Patent 3,729,298 (1973); *New York Times*, 21 April 1973.
32. H. C. Hottel and J. B. Howard, *New Energy Technology* (MIT Press, Cambridge, Mass., 1971), p. 4.
33. W. R. Cherry, *Astronaut. Aeronaut.* **11** (No. 8), 30 (1973).
34. P. Tonge, *Christian Science Monitor*, 5 October 1973, pp. 13-14.
35. G. C. Szego, J. A. Fox, D. R. Eaton, in *Proc. 7th Intersoc. Energy Conversion Eng. Conf.* (1972), paper 729168, p. 1131; G. C. Szego and C. C. Kemp, *Chem. Technol.* **3**, 275 (1973).

18

Reprinted from pp. v–xxi of *Survey of Alcohol Fuel Technology*, Office of Energy Research and Development Policy, National Science Foundation, vol. 1, 1975, 117pp. Courtesy of The Mitre Corporation.

SURVEY OF ALCOHOL FUEL TECHNOLOGY

B. Baratz, R. Ouellette, W. Park, and B. Stokes

EXECUTIVE SUMMARY

This report summarizes a study on alcohol fuel technology sponsored by the National Science Foundation during the months of June through December, 1974. The goals of this study have been to collect a base of information on alcohols as fuel, to synthesize current research on the subject, and to make an effort to tabulate current and ongoing research. The report also lists current researchers, their research topics, and their locations. The second volume of this report is a bibliography covering all aspects of alcohol fuel technology.

The use of alcohol fuel is not new. Alcohols have been used throughout the world as petroleum fuel substitutes, and during war periods in Europe their use was extensive. However, with the large supply and relatively low cost of petroleum since World War II, the use of alcohol fuel has been minimal.

Today, with petroleum shortages and associated rising costs, interest in alcohol fuel is returning. Moreover, with concern about the environment, researchers are reassessing the environmental advantages of burning alcohols.

Only monohydric alcohols with up to four carbon atoms are studied in this report. These include methanol, ethanol, the two monohydric propyl alcohols, and the four monohydric butyl alcohols.

SOURCES OF ALCOHOL

There are many sources of alcohol. Figure S-1 is a flow diagram depicting the various routes to alcohol production; Table S-1 gives details about some of these processes. Today, only a few of these routes are used for alcohol production. Most methanol comes from natural gas and most other alcohols from cracked petroleum feedstocks. Some industrial ethanol is produced from foods such as grain and

216

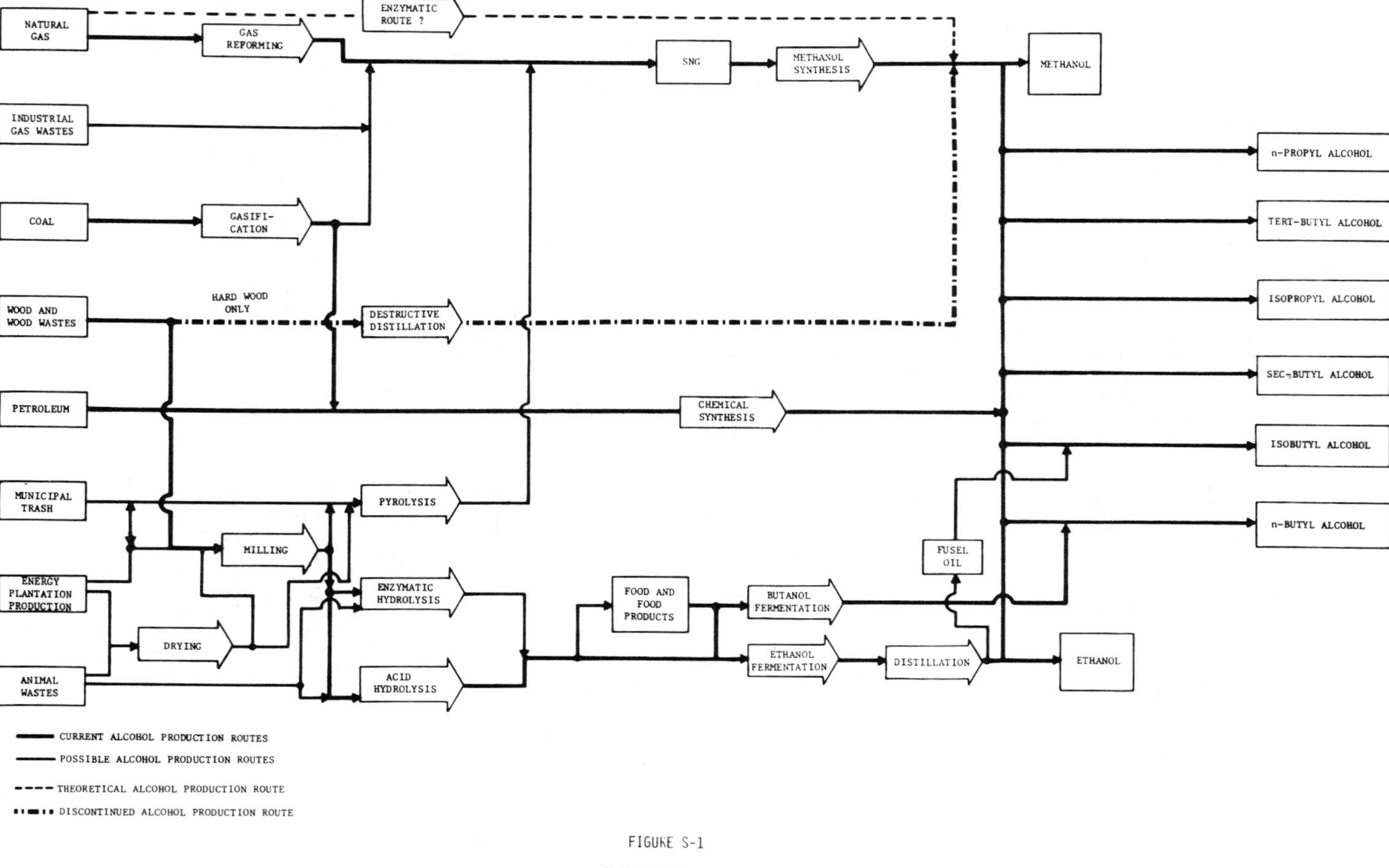

FIGURE S-1

ALCOHOL PRODUCTION

217

TABLE S-I

SUMMARY OF CONVERSION PROCESSES LEADING TO ALCOHOL PRODUCTION

	DESCRIPTION	PROCESS NAMES	TEMPERATURE	PRESSURE	STATE OF DEVELOPMENT	ADVANTAGES	DISADVANTAGES
COAL GASIFICATION	Process of converting coal to synthetic natural gas	Lurgi Koppers Totzek Winkler	-800°C 1650°C 800-1000°C	20-30 Atmospheres Atmospheric 1.5 Atmospheres	Commercial	Large coal resources in United States	High capital costs
METHANOL SYNTHESIS	Process of conversion of SNG to methanol	ICI Lurgi Vulcan Cincinnati	200-300°C 230-250°C	50-100 Atmospheres 50-60 Atmospheres	Commercial		
DESTRUCTIVE DISTILLATION	Original wood alcohol process - cooking of hard woods in the absence of air - resulting distillate fluids contain methanol		~525°C	Atmospheric	All commercial U.S. plants now discontinued	Methanol can be produced on a small-scale basis	Methanol yields are low
CHEMICAL SYNTHESIS	Various chemical processes used commercially today in the production of the majority of ethanol, propyl, and butyl alcohols (see report)		High	High	Commercial		All processes here use petroleum feedstocks which are now in short supply
MILLING	Processes to reduce feedstock size	Black Clawson Ball milling	Normal	Atmospheric	Commercial	Process decreases the hydrolysis reaction times	Can be a significant enzyme factor in the hydrolysis processes
PYROLYSIS	Process of burning with insufficient oxygen - yielding SNG	Union Carbide	500-900°C	Atmospheric	Test Plant	Pilot municipal trash disposal process with energy benefit	Feedstock material must be reasonably dry
ENZYMATIC HYDROLYSIS	Enzymatic decomposition of cellulose to glucose	None	50°C	Atmospheric	Pilot Plant	Utilizes cellulose as a feedstock Inexpensive process to generate glucose	Process times are slow Substrate must be clean and well milled
ACID HYDROLYSIS (WEAK ACID)	Acid hydrolysis of wood wastes to glucose		180-190°C	-20 Atmospheres	Commercial	Disposes of wood waste yielding liquid fuels	Process is expensive and slow
ACID HYDROLYSIS (HIGH PRESSURE)	Acid hydrolysis of cellulose is present in municipal trash to glucose	None	220-330°C	-30 Atmospheres	Pilot Plant Planned	Form of substrate not critical Process times are rapid	Equipment is expensive
ETHANOL FERMENTATION	Enzymatic process of decomposition of hexose sugars to ethanol		20-50°C	Atmospheric	Pilot Plant	Can be done small or large scale	Process times are slow (40 hours)
ETHANOL DISTILLATION	Process to separate ethanol to a 95% concentration	Atmospheric Still Pressure Still	60-80°C 60-80°C	-Atmospheric -2-3 Atmospheres	Commercial	Can be done small or large scale	

molasses through the process of fermentation and distillation, yet this approach is still a minor source. If there is promise for alcohol as an alternative fuel to petroleum, it appears that production must develop from other sources such as coal.

Methanol

A major source of methanol could be coal, which would involve coal gasification into synthetic natural gas (SNG), followed by a methanol synthesis process. Both of these processes are well developed today and descriptions are given in the body of this report. Several studies have been made of the feasibility and associated costs of methanol production by coal gasification. The most recent and conservative estimate puts the privately financed costs of methanol production at 16.5 to 19.5 cents per gallon depending on the coal source and the gasification procedure. As methanol has about one-half the BTU capacity of gasoline we also compare these prices on a cost per 10^6 BTU basis in Table S-II. For public utility financing the costs per 10^6 BTU of methanol from coal would be about equivalent to the cost of gasoline.

A major reason that coal gasification is not being developed today is the uncertainty of U.S. industry about the future price of foreign crude oil. Should coal gasification be encouraged, government incentives or price supports could reduce this risk.

Methanol can also be produced from biomass sources. Prior to the 1930s, all methanol was produced from hard woods by destructive distillation (whence its vernacular name "wood alcohol"). Although this process is no longer commercially used for methanol production, other gasification processes can convert biomass into SNG which, as in the case of coal, can then be converted to methanol.

The most well developed process is pyrolysis (burning with insufficient oxygen). Numerous pyrolysis processes have been developed and some are in test plant stages. Although most pyrolysis studies were

TABLE S-II

FORECAST ALCOHOL PRODUCTION COSTS (NOVEMBER 1975)

	COST/GALLON	COST/10^6 BTU
Methanol via Coal Gasification	24.8¢–29.2¢	$3.89–$4.59
Methanol via Pyrolysis of Municipal Trash	27.5¢	$4.31
Ethanol via Enzymatic Hydrolysis	75¢ (Approx.)	$8.87 (Approx.)
Ethanol via High Pressure Acid Hydrolysis	81¢–$1.11	$9.57–$19.67
Gasoline – Approximate Current Refinery Price	35¢	$2.77

CURRENT MARKET PRICES FOR ALCOHOLS (NOVEMBER 1975)

Methanol	38¢–49¢	$5.99–$7.72
Ethanol	$1.04–$1.14	$12.29–$13.47
n-Propyl Alcohol	$1.57	$15.98
Isopropyl Alcohol	63¢–70¢	$6.93–$7.70
n-Butyl Alcohol	$1.15–$1.49	$10.95–$14.18
Isobutyl Alcohol	$1.07–$1.47	$10.32–$14.19
Sec-Butyl Alcohol	$1.43	$13.72
Tert-Butyl Alcohol	$1.75	$17.34

initially centered on municipal trash gasification, the pyrolysis process could be applicable if biomass feedstock is sufficiently dried and in small enough form. If drying and/or milling is required, there would be additional costs.

The City of Seattle has made extensive studies of various processes for eliminating municipal trash and associated landfill problems. Of the 14 processes that were studied, the pyrolysis of municipal trash and subsequent methanol synthesis was recommended. In the fall of 1974, this conclusion was reconfirmed by an independent consulting firm. Using the estimated cost of municipal trash at $5.00 per ton, the cost of methanol production is given in Table S-II. It was estimated that if all City of Seattle government vehicles were to operate on the methanol from the above process, they would require only 15 percent of the total methanol produced. The City of Sea tle is presently searching for markets for the methanol if the above process is implemented.

Other biomass gasification processes are not as yet well defined. A steam or hydro-gasification process and a catalytic gasification process have been suggested. As both of these processes are forecast to be high temperature and pressure schemes, it appears that costs associated with them could be high.

Although several alcohols can be formed through enzymatic processes, no enzymatic routes yielding methanol are known. Recently, however, some microbiologists in England have discovered that methanol occurs as an intermediate product in aerobic bacterial oxidation of methane to carbon dioxide. If an enzymatic route of methane to methanol were discovered, it would be meaningful to small-scale energy production. As a desirable but undiscovered route, this process has been depicted by dotted lines on Figure S-1.

<u>Ethanol</u>

Ethanol, the well-known drinking alcohol, is produced commercially in two major ways: by the direct hydration of ethylene gas and by the process of fermentation and distillation using various foods such as grain and molasses as feedstocks. Although there is no chemical difference between the ethanol produced from these processes, the alcohol used for beverages must by law be produced in the latter manner. Nevertheless, approximately 80 percent of ethanol production today is from ethylene, showing the very strong current dependence of ethanol production on petroleum.

With extensive food and grain excesses in the U.S. and Canada prior to the 1970s, ethanol fuel was often considered as a way to utilize these surpluses effectively. The conclusions of studies on this idea were uniform: even with no cost for the feedstock, ethanol fuel costs were excessive relative to the costs of petroleum fuel. Although the costs of petroleum are significantly higher today, food costs are also higher and the large grain reserves gone. It is not surprising to see that the current market costs of ethanol are almost three times those of methanol (see Table S-II for current market prices of alcohols as compared with gasoline).

There may be some reduction of these high production costs from recently discovered enzymatic processes that convert cellulose to glucose. Scientists at the Army Natick Laboratories in Natick, Massachusetts, have developed strains of fungus that are relatively active in cellulose decomposition. After promising initial results, development of this enzymatic hydrolysis process is now at the pilot plant stage.

A major advantage of the enzymatic hydrolysis process is that the conversion of cellulose to glucose occurs at atmospheric pressures and close-to-room temperatures. On the other hand, process times are in the range of 24 to 48 hours and, unfortunately, the form of the cellulose substrate is critical. To obtain effective conversion rates, the substrate must be reasonably clean and very well milled.

After enzymatic hydrolysis, the resulting glucose syrup can be filtered off and then converted to ethanol through the fermentation-distillation process. Although the hydrolysis process is relatively inexpensive, the costs of fermentation and distillation are significant. Initial cost estimates of ethanol production from this hydrolysis process are sketchy but place the cost at about 50 cents per gallon.

Acid hydrolysis is another way of converting cellulose to glucose. Today, there is a small amount of ethanol produced using acid hydrolysis of wood wastes. This process acts at atmospheric pressures and has relatively long process times. A recently developed process shortens the acid hydrolysis process time to a matter of seconds through the use of high pressure and very concentrated acid. An advantage of this process is that the substrate physical form is not critical. It can be applied effectively to coarsely milled municipal trash. The new acid hydrolysis process should be at the pilot plant stage soon.

The hydrolysis processes form an important possible route to the production of liquid fuels from biomass. The sources of raw material are extensive, including the large amounts of biomass that are produced from photosynthesis and also the large amounts of municipal and animal wastes. There is considerable interest today in the energy plantation concept, both land and marine based, from which large amounts of plant material can be grown and used for their energy content. Research in this whole area is recommended since renewable energy sources will be of importance in the future.

Associated with ethanol production and use are the legal restrictions established to protect the substantial taxes required for beverage ethanol. These restrictions are discussed in detail in the legal factors section of this report. As the law now reads, small-scale ethanol

production is strongly discouraged by the expensive bonding requirements placed on ethanol producers. Should ethanol fuel use become
large-scale, even with denaturing controls, the risk of illegal beverage ethanol activity will be significantly greater.

<u>Propyl and Butyl Alcohols</u>

Brief descriptions of the chemical processes that yield these
alcohols are given in the body of this report. As noted above, virtually all of these alcohols are derived from cracked petroleum hydrocarbon feedstocks. There is a fermentation process that yields n-butyl
alcohol derived from food products but this is currently a minor source.

ALCOHOL FUEL USES

<u>Methanol-Gasoline Blends</u>

Satisfactory engine operation is possible on current automobiles
that are fueled with up to 15 percent methanol-in-gasoline blends and
requires no carburetion readjustment. However, the performance results
from this blended fuel differ substantially for different models. For
this reason there has been recent and heated debate on the effectiveness
of methanol-gasoline blends as automotive fuel. Some very recent analyses seem to synthesize the research results of the past few years and
apparent discrepancies are gradually being ironed out.

With the addition of methanol to gasoline, automatic carburetor
leaning occurs because of the stoichiometric differences in the two
fuels. For older cars that have rich carburetion, adding the methanol
to gasoline gives better performance, better fuel mileage (in spite
of the BTU loss), and lower exhaust emissions. However, for newer
model cars with leaner carburetion to minimize exhaust emissions, the
effect of adding methanol to gasoline is excessive leaning, a drop in
performance, a loss in mileage (proportional to the BTU loss), and
little change in the emission results.

Some advantages of the methanol-gasoline blend are outlined below:

1) Methanol has octane ratings substantially higher than gasoline. The addition of methanol to gasoline increases the antiknock capacity of the fuel. In recent years, with the demands of reduced exhaust emissions and the phasing out of lead additives in gasoline, engine compression ratios have had to be significantly lowered to prevent engine knock. With the addition of methanol to gasoline, these compression ratios could possibly be increased again to obtain more efficient use of the fuel.

2) If methanol is produced from non-petroleum sources such as coal, the addition of methanol to gasoline would save the corresponding amount of petroleum for more critical uses.

3) As noted above, in older cars, methanol-gasoline blends will give reasonable engine operation without any engine or carburetor readjustment.

A number of problems and possible difficulties inherent in the methanol-gasoline concept need to be resolved.

1) The primary area of difficulty has to do with water contamination. Although alcohol and gasoline are completely miscible, the addition of small quantities of water to methanol-gasoline blends causes fuel phase separation. In the separation the methanol and water appear in solution at the bottom of the container and the gasoline at the top. Present transportation and storage fuel tanks contain significant amounts of water. If methanol and gasoline were blended at the refinery these storage tanks would have to be dried and special effort would be required to keep them that way. Methanol and gasoline could be mixed at the fuel pump. However, care would still be required in the prevention of water contamination within the car fuel system itself.

2) Alcohols are excellent solvents. With an existing automotive fuel system, this solvent action has been shown to cause corrosion in fuel tanks and swelling of seals.

3) The addition of alcohol to gasoline causes a more than linear increase in the Reid Vapor Pressure (RVP) of the fuel. In some cases this increase could result in RVP that surpasses legal RVP limitations for some seasons in some states. Moreover, there may be increased vapor lock problems resulting from these increased vapor pressures.

4) The leaning effects of adding methanol to gasoline may cause performance deficiencies such as automobile stalling, hard starting, and lean surge.

<u>One Hundred Percent Methanol</u>

The present automobile engine can be retrofitted to run successfully on 100 percent methanol. This requires substantially larger fuel jets than found on standard carburetors and also requires some kind of induction heating to give adequate methanol vaporization. This was done very successfully in 1970 with the modification of an American Motors 1970 Gremlin to run on 100 percent methanol. This car was the winner of the 1970 Clean Air Car Race. More recently, Volkswagen has made similar modifications on one of their European models.

Some of the advantages of 100 percent methanol fuel are outlined below:

1) Methanol has a much leaner "lean misfire limit" than does gasoline. Running at mixtures near this limit produces much lower NO_x emissions than gasoline. The Gremlin that won the 1970 Clean Air Car Race was able to meet the original 1975-76 standards on NO_x without the use of exhaust gas recycle. (Hydrocarbons and carbon monoxide emissions were maintained within limits with a catalytic converter.)

2) The results of testing on this vehicle demonstrated that adequate performance characteristics are possible running a methanol powered vehicle at very lean equivalence ratios.

3) As the octane rating of 100 percent methanol is significantly higher than that of gasoline, engine compression ratios could be increased substantially to obtain more efficient fuel use.

Some of the disadvantages of 100 percent methanol automobile fuel are:

1) Methanol and gasoline are not interchangeable without engine retrofitting.

2) There are substantial cold start problems with the 100 percent methanol vehicles tested to date. Methanol has a high heat of vaporization. At cold temperatures, without some special methanol vaporization technique, the engines are almost impossible to start. With the rich mixtures required to warm up the engine, the emissions results are poor.

3) As the BTU capacity of methanol is about one-half that of gasoline, approximately double the quantity of methanol is required to give the same work. This implies a doubling of the fuel tank capacity of the vehicles and also requires a doubling of the capacity of any gasoline transportation network that might be converted to methanol.

4) Aldehyde and unburned alcohol emissions from methanol powered vehicles are substantially higher than from gasoline powered vehicles.

In light of current interest and technology, the idea of 100 percent methanol powered automobiles is more than the reemergence of an old idea. Research into engine design that optimizes the advantages noted above and minimizes some of the problems has not yet been done. More emissions studies are needed. Fleet operation in high air pollution areas appears to be an effective place to use and test 100 percent methanol vehicles. It is interesting to relate this possibility to the methanol production process utilizing municipal trash studied by Seattle and mentioned earlier.

<u>Ethanol-Gasoline Blends</u>

Much of what was said of methanol-gasoline blends is also valid for ethanol-gasoline blends. As discussed in the text, most research into this subject was accomplished in the early 1960s and summarized in earlier studies. With the high costs of ethanol there has not been as much current interest in ethanol-gasoline blends as in methanol-gasoline blends. However, the State of Nebraska is continuing to look at the ethanol-gasoline idea in their "gasahol" program and the details are given in the body of this report.

Other countries have used ethanol-gasoline blends recently or are using them now. Brazil has been producing ethanol from excess sugar as a sugar price support. The percentage of ethanol that appears in the gasoline is proportional to the amount of excess sugar produced that year. In South Africa, the SASOL Corporation markets an alcohol-hydrocarbon blend produced from coal by a Fischer-Tropsch process. Ethanol is the primary alcohol constituent of this automobile fuel. They have had substantial commercial experience with the ethanol-gasoline blend.

Advantages of ethanol-gasoline blend automobile fuel are listed below:

1) Ethanol has an octane rating(s) which is higher than gasoline though not as high as methanol. The octane rating of the fuel is increased in ethanol-gasoline blends.

2) Ethanol-gasoline blends will give reasonable engine operation without any engine or carburetor readjustment.

Disadvantages are as follows:

1) The water contamination problem is the same as with methanol-gasoline blends. The SASOL Corporation apparently has proprietary information concerning techniques they use to minimize this problem.

2), 3), and 4) (same as with methanol-gasoline blends)

5) Ethanol costs today are substantially higher than those of gasoline or methanol.

6) The addition of ethanol to gasoline does not significantly change the emissions from those of 100 percent gasoline.

Methanol for Electric Power Generation

In 1971, a group of electric power companies sponsored a test of methanol fuel on a 50 megawatt boiler generator located in New Orleans. Although the results are somewhat sketchy, the emissions of both NO_x and CO were reported to be lower than when the boiler was fired by either natural gas or fuel oil.

In the Spring of 1974, the General Electric Company performed methanol fuel tests on a gas turbine generator. The emissions from 100 percent methanol in comparison to number two fuel oil were as follows:

	NO_x	CO
100% Methanol	60% Reduction	300% Increase

The substantial reduction in the NO_x emissions is important. Although the increase in the CO emissions is large, as the CO maximum emissions proposed by the EPA are 900 percent greater than those of number two fuel oil, the CO emissions from methanol are still well within this level.

Some retrofitting of both the boiler and gas turbine systems is required before successful methanol operation is possible. Twice the fuel quantity has to be used to achieve the equivalent BTU output of the fuel oil. Whereas pumps used in the fuel oil operations are lubricated with the fuel oil itself, this is impossible with methanol fuel, so externally lubricated pumps would be required.

OTHER ALCOHOL FUEL USES

There has been recent interest among natural gas companies in the idea of conversion of overseas natural gas to methanol and shipping it back to the United States in this form for reconversion to methane and use within the natural gas network here. The development of a dual coal gasification and methanol operation in which the gases are used within the natural gas network and the methanol is stored for reconversion to methane and use during high demand months has been studied. Other more specific uses of methanol and the higher alcohols are given in the main report.

TRANSPORTATION AND STORAGE OF ALCOHOL FUEL

Alcohols can be stored and transported in much the same manner as gasoline. Two major considerations must be kept in mind in the development of an alcohol fuel dispersion network.

First, because of the reduced BTU capacities of alcohols in comparison with gasoline, to achieve the equivalent energy transfer from the alcohol transportation network its capacity must be larger in proportion to the BTU ratio of gasoline to alcohol. For methanol this ratio implies that the transportation network will have to be about twice the size of a gasoline network. Clearly this requirement will add substantial transportation and storage costs.

Second, whereas gasoline and water are not miscible, a strong affinity exists between alcohol and water. Present gasoline storage systems contain substantial amounts of water which collects at the

bottom of the storage vessel. Special care would be necessary to
prevent water in any kind of alcohol transportation and storage network.
Slight fresh water contamination in alcohol used in pure form for fuel
should not be serious. However, if alcohol and gasoline are blended as
for automobile fuel, only slight amounts of water can cause fuel separation. The drying and reblending of separated fuel would most likely be
quite expensive.

ENVIRONMENTAL CONSIDERATIONS

Although the data on emissions resulting from alcohol fuel burning
have been discussed in the fuel use section of this summary, general
environmental questions require investigation. The primary emissions
concerns today are unburned hydrocarbons, carbon monoxide, and nitric
oxides. It is apparent that with the burning of alcohols, some emissions
of aldehydes and unburned alcohol also occur. The widespread environmental effects of these emissions are unknown. Research in this area
would be most useful.

All of the alcohols are toxic, with ethanol being the least toxic.
Methanol vapors are more toxic than gasoline vapors although only
slightly. Methanol is quite poisonous when taken internally. Even
though the toxicology of the various alcohols has been established in
well-controlled laboratory testing, as outlined in the Appendix II, the
severity of the toxic problems from widespread fuel use is not known.
It is very important for public safety purposes that this topic be well
investigated.

A specific environmental problem that must receive immediate
investigation is the impact on marine life of a large-scale methanol
spill. As noted, methanol is quite toxic when taken internally and,
unlike petroleum fuels, is completely miscible in water. It is quite
possible that a methanol spill could be disastrous to marine life.

CONCLUSIONS

In a broad perspective, the development of alcohol fuel has the advantage of having many routes. So many tradeoffs are available that some of these routes turn out to be infeasible, many other routes are possible. One of these tradeoffs toward energy is the route to food production. Others include a possible tradeoff to gasoline production using alcohol as a chemical intermediate and obvious tradeoffs to the chemical industry.

The growing body of research data indicates that alcohols could become a source of energy in the future. However, before widespread use of alcohol fuels, either directly or as a blend, can become a reality, much research must be done in cost-competitive production techniques, transportation, storage, and end uses.

Part IV

KINETIC ENERGY STORAGE: FLYWHEELS

Editor's Comments
on Papers 19 Through 24

Flywheels have been used to smooth man's energy requirements for certain special technical applications longer than any other type of energy storage. The earliest known example is the potter's wheel, which probably was first made to support the clay as it was rotated and later was made much more massive than necessary for support when it was found that heavier wheels gave a nearly constant speed of rotation. The earliest potter's wheels, which were made of stone, date back nearly five millennia and were pushed at irregular intervals by the potter or an assistant; the large mass of the wheel allowed the artist to make more delicate and perfect pots.

As technology developed during the Middle Ages and the Renaissance and man began using complex gears and other mechanical devices, flywheels occasionally were used to moderate the effects of irregular power generation and energy use. During this period, however, even though they were used in some water pumps, pile drivers, and mills, they were mainly a not-too-well understood curiosity in the form of tops and gyroscopes.

With the beginning of the industrial revolution of the nineteenth century, flywheels began to be used as short-term energy-storage devices, usually in conjunction with a steam engine. The ideas of angular momentum and rotational energy were well understood by some of the engineers developing these engines, and flywheels were constructed on the basis of a firm but modest understanding of material characteristics. Nevertheless, an inordinate number of flywheels were destroyed due to stress fatigue during normal operation or occasional excursions beyond their design speed.

The power cycle of a single piston steam engine is very irregular. Power is generated for about 35 percent of the cycle, and the average power output is only about 20 percent of the peak power. The loads for many steam engines, however, were roughly constant. If a speed variation of 1 percent was acceptable, it was necessary to have an energy-storage device capable of storing energy produced in about 50 cycles. Based on these criteria alone, a 200 hp steam engine operating at 60 rpm would require a 7-foot diameter, 8500-lb flywheel. Often in the nineteeth century a large steam engine would be used to power a complete factory, a textile mill, or a large machine shop. Power would be transmitted from floor to floor and from one work area to another by belts and long rotating shafts. Each of these factories had a flywheel attached to the steam engine. Unfortunately few of them were adequately described in the journals of the time.

With few exceptions, internal combustion engines have flywheels that are an integral part of the engine design. Automobile engines have a flywheel that also functions as a clutch plate and as a gear for starting the engine electrically. This flywheel maintains smooth operation of the engine at low rpm. After the vehicle is moving, the flywheel serves no useful energy-storage purpose because the mass of the vehicle itself acts to smooth the power variations from the motor. For the same reason the steam engines on trains do not need flywheels because the momentum of the train serves to smooth the engine's power production. On airplane engines, the propeller serves as a flywheel, in addition to providing propulsion.

Recently flywheels have been proposed for a variety of new and demanding applications ranging from diurnal energy storage on electric utilities to small, highly stressed units for primary automobile propulsion. Technical evaluations of flywheels have shown that they are most effective where high power density is required and low energy density is acceptable. Based on these studies, it appears they may find some application in hybrid automobiles in which a small internal combustion engine or a battery provides the primary power and a small flywheel is added to the vehicle to boost power for acceleration or passing. The flywheel can be charged directly from the primary power source or through regenerative braking. It is unlikely, however, that they can be effective for diurnal energy storage.

Although very few articles on flywheels from the nineteenth century describe design and operating characteristics in detail, two very thorough descriptions appeared in 1892 in the proceedings of the recently formed American Society of Mechanical Engineers. The first, Paper 19, describes the destruction of a large, 116,000-lb, cast-iron rimmed flywheel and the design of its replacement, a wooden-rimmed flywheel that was considered to be safer. The second paper, Paper 20, also addresses the problems of flywheel failure, includes a simple but reasonable analysis of the strains produced in a flywheel during operation, and gives suggestions for the construction of safe flywheels.

Paper 21 looks at several structural forms that can be used for flywheels and compares the cost-effectiveness of these different structures. It also mentions some of the more likely applications of flywheels in modern technology. Paper 22 proposes the use of a flywheel to improve the efficiency of an automobile by maintaining internal combustion engine operation near peak thermal efficiency. Paper 23 is representative of the materials measurements necessary for flywheel design and the analysis of stresses encountered during normal operation.

Paper 24 outlines the design and operating characteristics of the largest flywheel ever constructed, a 223-ton, 2.9-m diameter forged flywheel that is used to deliver 10-s, 170-MVA pulses to a set of coils used for plasma physics experiments.

BIBLIOGRAPHY

Anon. 1975. Economic and Technical Feasibility Study for Energy Storage Flywheels. *Energy Res. Dev. Admn. Rept. ERDA 76–65.* Rockwell International.

Anon. 1979. Project Summary Data Thermal and Mechanical Energy Storage Program. *U.S. Dept. Energy Rept. DOE/ET-0091.*

Chang, G. C., and R. G. Stone, eds. 1975. *1975 Flywheel Tech. Symp. Proc.,* U.S. Energy Res. Dev. Admn. and Lawrence Livermore Lab. Univ. Calif.

Clerk, R. C. 1963. *The Utilization of Flywheel Energy.* SAE Paper 711A, Society of Automotive Engineers Int. Summer Meet. Montreal, Canada.

Dugger, G. L., A. Brandt, J. F. George, L. L. Perini, D. W. Rabenhorst, T. R. Small, and R. O. Weiss. 1972. Heat-Engine/Mechanical-Energy-Storage Hybrid Propulsion Systems for Vehicles. *Johns Hopkins Univ. Appl. Phys. Lab. Rept. CP011.*

Mallon, B. 1979. DOE/STOR Bibliography for Flywheel Energy Systems, 1978. *Lawrence Livermore Rept. UCRL-52794.*

Mallon, B., and R. Kuhn. 1979. DOE/STOR Bibliography for Flywheel Energy Storage Systems, 1977. *Lawrence Livermore Rept. UCRL-52637.*

Post, R. F., and S. F. Post. 1973. Flywheels. *Sci. Am.* **229**:17–23.

19

Reprinted from *Am. Soc. Mech. Eng. Trans.* **13**:618–632 (1892)

A NOVEL FLY-WHEEL.

BY CHAS. H. MANNING, MANCHESTER, N. H.

(Member of the Society.)

On October 15, 1891, the fly-wheel of a large pair of Corliss engines belonging to the Amoskeag Mfg. Co., of Manchester, N. H., exploded from centrifugal force, causing great destruction of property and the loss of three valuable lives. It was to replace this wheel that the one about to be described was built; but before speaking of the new wheel, it may be well to say something of the engine and the old wheel.

The engine was built by the Corliss Steam Engine Co., of Providence, R. I., in 1883, and has two cylinders 36 inches diameter, with 6 feet stroke of piston, and was running at the time of the accident at a little over 61 revolutions per minute. One cylinder was connected with a Bulkley condenser, and the other was working non-condensing. The fly-wheel of cast-iron and "built up," was 30 feet diameter and 110 inches face, with one set of 12 arms, and weighed 116,000 lbs. This wheel was turned for three belts, the two outside ones being 40 inches wide each and running back between the cylinders to a jack-shaft running to Mill No. 5; the central belt, 24 inches wide, leaving the fly-wheel in the opposite direction and connecting to a jack-shaft running to No. 7 Mill.

Figure 166 shows the general situation of all this, although it should be mentioned that the two jack-shafts do not stand in line with each other but diverge about two and three-eighths feet per hundred, and the engine shaft bisected this angle, which necessitated the use of guide pulleys on all the belts.

The engine had been run under precisely the same conditions for several days, including a water-wheel at part-gate attached to No. 7 jack-shaft and furnishing about 200 H.P., and one on

---

* Presented at the San Francisco Meeting, May, 1892, of the American Society of Mechanical Engineers, and forming part of Volume XIII. of the *Transactions*.

No. 5 jack-shaft with gate closed to one-tenth and giving no power.

The indicator cards are on record for every day up to that of the accident, and show for several days previous about 1,950 H.P., and of this probably 1,850 was transmitted to the belts. The belts were heavy, double leather, and the total number of days' work they had done was 879, less than three years, as this

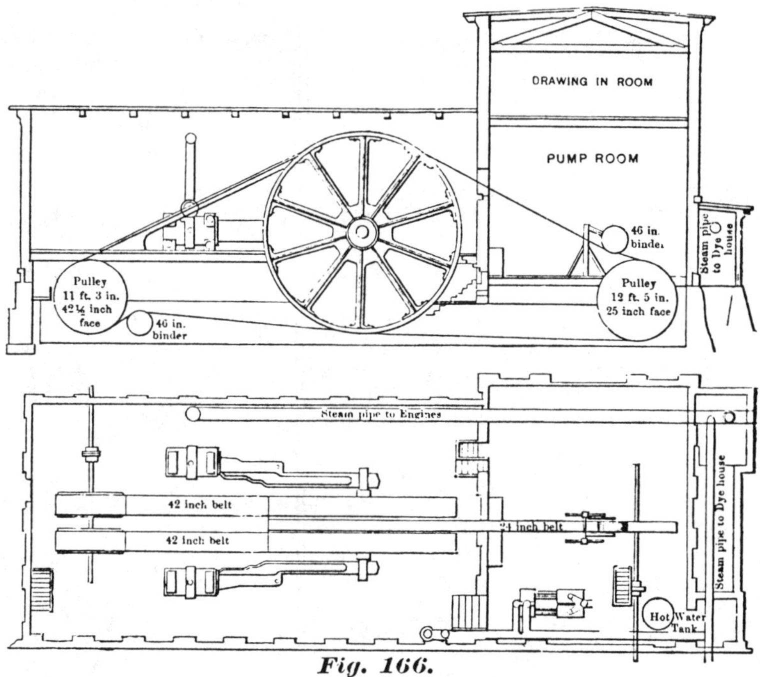

Fig. 166.

engine is only used in times of low or back water, when the water-wheels will not furnish sufficient power.

The belts had been carefully examined within a few days and were in good condition, and though heavily loaded were not overtaxed.

The first unusual occurrence was a loss of speed in No. 5 Mill, doubtless caused by slipping of the twin belts on the jack-pulleys ; the operatives, thinking the engine about to stop from some cause, threw off their work, when speed immediately came up once more and the machines were started ; this caused the belts to slip again and work was more generally thrown off, thus relieving the engine suddenly, the speed increasing sufficiently to attract the attention of the engineers, who immediately closed the throttles, but too late to save the wheel, which went to pieces at a speed of seventy-three or four revolutions per minute. On the No. 7 jack-shaft drive no irregularity of speed was noticed until the acceleration a few seconds previous to the explosion, when many of the looms "knocked off" automatically, but

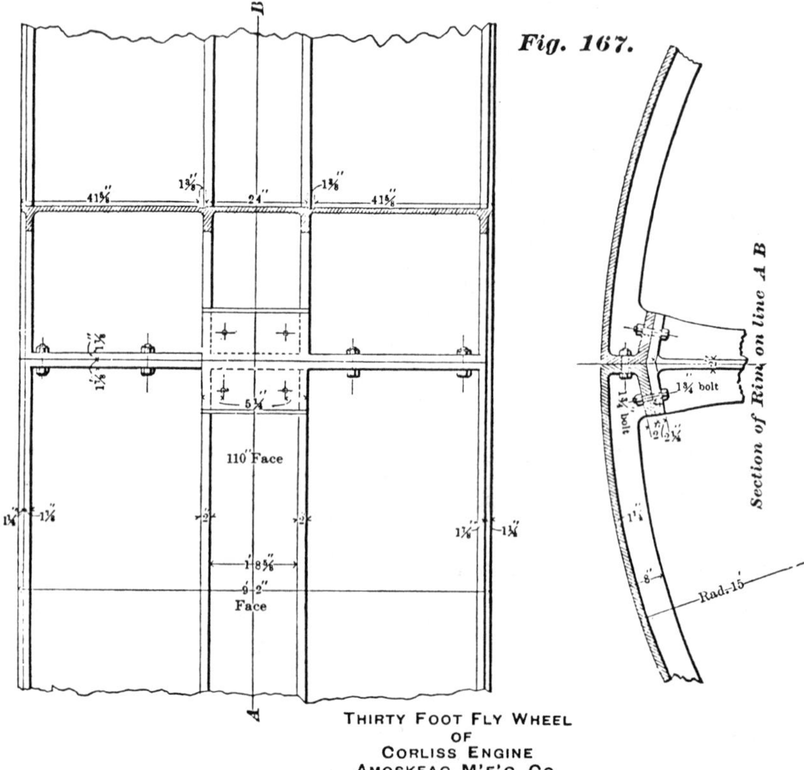

THIRTY FOOT FLY WHEEL
OF
CORLISS ENGINE
AMOSKEAG M'F'G. CO.

one particular loom ran until the speed stopped after the explosion. Experiment demonstrated that this loom would "knock off" before it attained a speed corresponding to 75 revolutions of the engine.

The governor and its belt were intact and in excellent order after the explosion, with the exception of the trip rods, which were broken.

At this speed, had the castings been ordinarily good, the wheel should have been perfectly safe and, as will be seen by reference to Fig. 167, the bolting of the sections together with five $1\frac{3}{4}$-inch iron bolts, made these joints theoretically the weakest places; but in fact there were very few of the bolts broken, some of them even pulling out large washers of cast-iron; but the rim castings, as well as the ends of the arms, were full of flaws caused chiefly by the drawing and shrinking of the metal. Fig. 168 is a fair representation photographed from pieces of the wheel.

Fig. 168.

Specimens of the metal were tested for tensile strength and varied from 15,000 lbs. per square inch in sound pieces to 1,000 lbs. in spongy ones. None of these flaws showed on the surface, and a rigid examination of the parts before they were erected failed to give any cause to suspect their true nature or they never would have been accepted. Experiments were carried on for some time after the accident in the Amoskeag Company's foundry in attempting to duplicate the flaws, but with no success in approaching the badness of these castings.

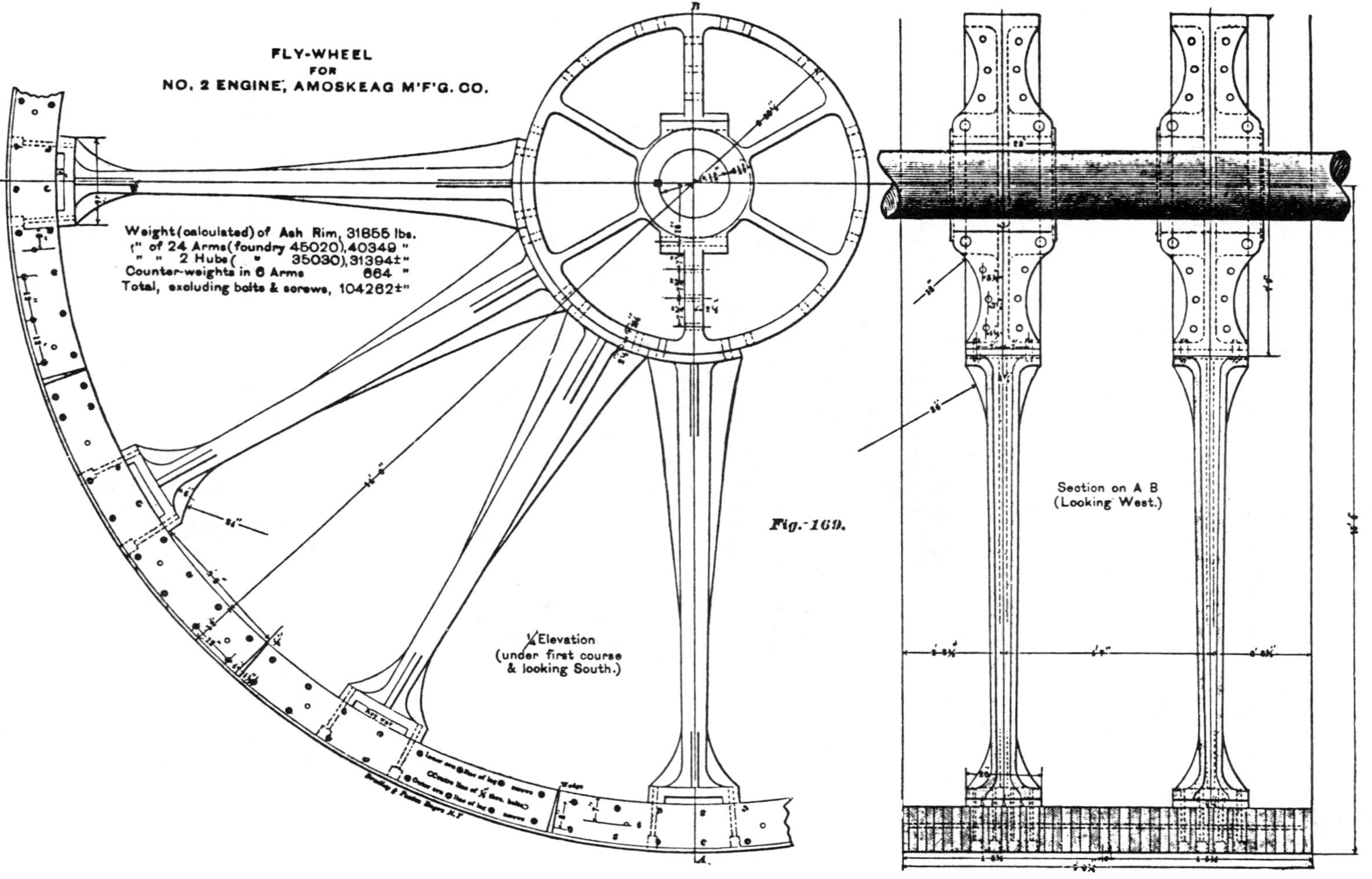

FLY-WHEEL
FOR
NO. 2 ENGINE, AMOSKEAG M'F'G. CO.
Weight (calculated) of Ash Rim, 31855 lbs.
1" of 24 Arms (foundry 45020), 40349 "
" " 2 Hubs (" 35030), 31394± "
Counter-weights in 6 Arms 664 "
Total, excluding bolts & screws, 104262± "
¼ Elevation
(under first course
& looking South.)
Fig. 169.
Section on A B
(Looking West.)

DESIGN OF THE WHEEL.

It has been the practice of this company for some years past
to make all pulleys having a rim speed of 3,000 feet or over with
wood rims on cast-iron spiders, and after due consideration it

Fig. 170.

was determined to replace this fly-wheel with a wooden-rimmed
one on two sets of arms, and the design shown in Fig. 169 was
the result of this decision.

As it was not desirable to remove either of the cranks, the
hubs were made in halves, and in erecting the wheel the design

as shown was departed from to the extent of placing the joints of the two hubs at right angles to each other on the shaft. The bolting together of the half hubs was made ample to hold the halves of the wheel together when running at normal speed of 61 revolutions per minute, independent of any other strength in the wheel. The arms were so placed that the inner end served as a butt-strap over the end of this joint. The arms were so designed that each one would safely carry the belt strain coming on its half the wheel, if the other eleven arms were entirely relieved, and the bolting to the hub is sufficient to carry its portion of the rim if cut adrift from the remainder of the rim. Sharp angles and changes of section in the arms were avoided as much as possible, as will be seen in the arm section, both where it joins the hub and the rim. The circular cross section was selected in preference to the elliptic on account of the greater certainty of uniform distribution of metal.

The counterbalancing of cranks and connecting rods was obtained by placing heavy cast-iron plugs in the hollows at the outer end of the three arms directly opposite each crank, and these plugs were secured in place by 1-inch bolts running through the centre of the arms, and large washers fitted to the inner ends of the arms. Though the total weight in the wheel is not much less than that of the old one, the weight in the rim is only about one-half, but it has shown itself to be ample for a very steady speed.

As for safety, the speeds being the same in both cases, the hoop tension in the rim per unit of cross section would be directly as the weight per cubic unit, and its capacity to stand the strain directly as the tensile strength per square unit, therefore the tensile strengths divided by the weights will give relative values of different materials.

Cast-iron weighing 450 lbs. per cubic foot and with a tensile strength of 1,440,000 lbs. per square foot would give a value of $\frac{1,440,000}{450} = 3,200$, whilst ash, of which the rim was made, weighing 34 lbs. per cubic foot, and with 1,152,000 lbs. tensile strength per square foot, gives a result $\frac{1,152,000}{34} = 33,882$, and $\frac{33,882}{3,200} = 10.58$, or the wood-rimmed pulley is ten times safer than the cast-iron when the castings are good.

This would allow the wood-rimmed pulley to increase its speed to $\sqrt{10.58} = 3.25$ times that of a sound cast-iron one with equal safety.

CONSTRUCTION.

In the foundry work on hub and arms great care was taken to produce perfectly sound castings, and as tough as possible. When poured, the metal was allowed to cool in ladles, as much as safe to avoid "cold shuts," and after pouring, it was thoroughly churned to get rid of all gas and avoid blow-holes, and no casting was uncovered for forty-eight hours after pouring.

The first arm was tested to destruction for cross-breaking strength, being bolted as to the hub and loaded at outer end, and it stood well above calculation. In the machine shop, after the hubs had been bored and turned, the arms were bored by being placed one at a time in a stationary saddle and the tool carried on the face-plate of the lathe. The arms being clamped to the hubs, they were drilled together.

The wreck of the old wheel being cleared from the shaft, the key-way was continued toward each end by a rotary milling tool travelling on guides clamped to the shaft, the tool being driven from a small stationary engine placed for the purpose. One hub was then put on, keyed fast, and its arms put in place and bolted on. The shaft was then revolved and marks scribed on the ends of the arms by which to place the first course of wood.

The wood was Western ash plank, thoroughly kiln-dried, eight courses being of 4-inch plank, seven of 2-inch, and the remainder of 3-inch, all of which was reduced about one-half inch in dressing, so that there are forty-four courses in all, each course or ring made of twelve pieces. The first ring of $3\frac{1}{2}$-inch plank was set by the scribe marks so that the butts came over centre of arms, and with bolt heads let in $2\frac{1}{2}$ inches, with washers under the heads and the nuts on the inner end. The butts were all left three-quarters of an inch open at the outer ends. The next course was applied with the butts breaking joints about 22 inches, and so on with each course, and the thickness of the courses was so arranged as to bring the arm-bolts through the centre of a $3\frac{1}{2}$-inch course. Each course was carefully dressed to exact thickness on a buzz planer, and each piece was thoroughly glued and secured in place with sixteen $3\frac{1}{2}$-inch lag bolts. The holes for these bolts and the countersinks for the heads were bored to template before leaving the wood-working shop; then each piece was clamped in the place it was to go, on the part already erected, the bolts were scribed through, and then the piece being removed, the holes were bored by a gauged auger, set on a flexible shaft and run by power. The piece was then glued, put in position, and the lag bolts screwed in by a socket-wrench set in the same flexible shaft. This shaft was driven with the usual rope drive, and the weight on the tension-pulley was so adjusted that when all the strain was on the lag bolt that it would safely stand, the rope would slip.

This allowed the lag screws to be set up very rapidly and to an even strain whilst the glue was still hot.

The wheel was turned one-twelfth of a revolution each time, and every ring completed before another was commenced. After the face had grown to about two feet in breadth, several round turns of manila rope about one inch in diameter were taken around it, and the ends then led off to the opposite sides around a suitable drum several times, and the ends spliced. This drum was revolved by engine power in either direction with friction clutches, so that the wheel as it grew and increased in weight was very readily handled. At the proper time the second hub and set of arms were put on, and the rim continued over this.

By reference to Fig. 169 it will be seen that between each pair of arms there are three seven-eighths inch bolts extending through the rim from side to side, and the holes for these bolts were bored in each piece before it was put in place, but left small and then rimmed to size after the pieces were all in place. The courses being completed, the next step was to drive well-fitted hard-wood wedges in the 504 spaces between the butts. These were driven on hot glue with a heavy sledge, as were also the hard-wood keys between the outer ends of the arms and the rim in the key-ways shown in the design. The wheel was then belted to the jack-shaft which was driven by a water-wheel, and slide-rests being set up, it was turned in position on the outside, and as far as the arms would permit on the inside, the remainder of which was finished by hand. All but the driving face was thoroughly painted, and into the driving face a finish of raw linseed oil and beeswax was worked under heavy friction with the wheel revolving, this being done to exclude moisture.

When completed the wheel was run up to a speed of 76

revolutions per minute, being a surface speed of nearly 7,200 feet per minute, and at this speed it ran absolutely true, and when stopped failed to show the least hair crack in varnish on inside of rim or where secured to arms. Since its completion, there having been an ample supply of water, it has been used but little, but so far has given entire satisfaction.

The cost of this wheel complete was $7,000 nearly, which is less than that of its sinful predecessor. In this cost is included all patterns and special appliances, and it could be duplicated for much less money. Safety was the greatest consideration, and it is firmly believed that this wheel is as safe and durable as any in existence.

DISCUSSION.

Mr. A. K. Mansfield.—This paper deals with a new wheel, and, incidentally, with the old one which it replaced. I think we will all agree that Mr. Manning has designed the wheel in a thoroughly mechanical manner, and that it will probably stand the test required of it. But we ought to be, I think, concerned chiefly with the old wheel, from the fact that in the Eastern part of the country there have been quite a number of accidents from wheels bursting lately, which makes this a very important subject to engineers. Quite a number of our large wheels have gone to pieces to very nearly the same time; so much so that I fear some users of large wheels which have not gone to pieces will naturally feel great anxiety.

Mr. Manning, on page 620, says: "At this speed, had the castings been ordinarily good, the wheel should have been perfectly safe, and, as will be seen by reference to Fig. 167, the bolting of the sections together with five $1\frac{3}{4}$-inch iron bolts made these joints theoretically the weakest places." I think perhaps it would have been better for him to say, "made these joints very weak places." He shows us that he has designed a wheel to perform the same service as the one that went to pieces, several times stronger than that was theoretically. Yet he seems to consider that the wheel which went to pieces was theoretically strong enough. This seems a little inconsistent; and suspecting it to be so, I have made a few figures which may be interesting.

It is very easy for us to calculate what the centrifugal tension is, in a wheel running at any speed, tending to part the wheel

247

through one of its sections—to part it in two halves. Applying the well-known formula, I found that the centrifugal force tending to part this wheel—that is, to divide it in two halves—was about 360,000 lbs. at 60 revolutions, which speed, I take it, the engine was designed to run at. This is based on the weight of the rim alone ; arms ignored. Adding one-half of the total weight of wheel to this, which is proper, for the force of gravity of this half also tends to divide the wheel in halves, this figure becomes 418,000 lbs. If we assume that the hub where the wheel is bolted on the shaft carries one-third of the strain, which may be proper to assume in the case of this wheel, it being very easy to get large bolts in at the hub (the paper does not show how it was bolted together there)—if we assume, therefore, that at the hub the strain taken by the bolts was one-third the centrifugal force of the rim, in addition to the whole centrifugal force of arms and hub, it leaves one-third of the former force to be accounted for, or to be resisted, at the rim on each side. One third of this strain is 139,000 lbs. There are five bolts. One-fifth of this figure is 27,900 for each bolt. These bolts inside of the thread have a cross section of about $1\frac{3}{4}$ inches, which makes the force per square inch of bolt 15,920 lbs., or, roughly, 16,000 lbs. Now this means that if these bolts were placed in such a position that they directly resist the force tending to separate the wheel, the strain on them would be, roughly, 16,000 lbs. to the square inch, which is a greater strain than we ordinarily allow in practice in such parts of machinery. But the bolts are not placed where they resist the strain directly ; they would have to be placed directly in the rim, which cannot be ; they are placed inside of the rim, some little distance away. In designing wheels I have generally placed these bolts in such place that I consider there is a leverage of. 2 to 1 on the bolts, and I think in the present case it is fair to assume there was a leverage of 2 to 1. If so, they were under a strain of about 32,000 lbs. per square inch. Now, at once you may think the bolts should have given out and not the casting. Mr. Manning states that not many of the bolts gave out, or something to that effect. It is quite sufficient if any of them gave out. An examination of the diagram of the old wheel will show that, considered roughly—bearing in mind that we have no figures in some parts on which to base calculation—it would seem that the remainder of the wheel was designed to conform in strength to the strength of the bolts, or in weakness, I should rather say. But besides that, there were flaws in the casting. I doubt if there are many wheels of that sort made in which there are not flaws. I think if you break up wheels of that kind, or other large castings of that sort, you will nearly always find flaws of the kind mentioned. The fact is, as near as I can judge—I think, perhaps, you will agree with me if you do this figuring over after me—that in this particular case the wheel is a badly designed wheel; the bolts should have been three times as strong, and the wheel casting about the bolts about three times as strong as it was. This is an interesting fact—it seems to me an interesting fact—and it is a manifestly important matter to us, considering the number of wheels which have gone to pieces lately. What caused the other wheels to go to pieces I don't know. This wheel went to pieces under a higher speed than it was intended to run at; it seems to have been too weak to run at the speed at which it was designed to run.

Mr. James McBride.—There is one very remarkable thing in regard to these two wheels. If I understand the paper, the first wheel, the writer stated, weighed 160,000 lbs.; the new wheel weighs only about half that much, and regulates the engine perfectly.

The first wheel was twice as heavy as there was any necessity for. Now is it not possible if the first wheel had been of less weight—half as heavy as it was—and made of good material, it would not have gone to pieces? There seems to be considerable discrepancy between the opinions of the designers of the two wheels. As Mr. Mansfield says, I think it was badly designed. If an engine-builder designs a wheel for an engine and it breaks, and another man comes along and puts one on half as heavy, that does the work and runs the engine steadily, evidently the first man was at fault.

* The breaking of this *heavy weak* cast-iron wheel, and the substitution of a *strong light* wooden wheel emphasizes the fact not generally known, that nearly all very large cast-iron band wheels which serve as balance wheels on steam-engines are much heavier than is really necessary, owing to the great difficulty of constructing such large wheels from light castings. Builders are obliged to make the segments very thick to insure good castings, and to allow sufficient stock for finishing, and are generally sure to

---

*Added after adjournment.

make all parts heavy enough to meet those ends. Hence they get wheels that are too heavy.

Mr. Manning has done a good thing in producing a *light* wheel which is heavy enough for regulation, and yet many times stronger than an all cast-iron wheel.

Mr. Mansfield.—I would like to add a little to what I have said. According to the figuring I made on the wheel, I found the factor of safety might be considered to be about 2, referring to the ultimate strength of the material. A wheel may run twenty or any number of years with perfect safety, provided it is subject to no undue strain, but only the strain which it is designed to bear. We know in this case the wheel received those undue strains. The conclusion I come to would be, naturally, I think, the factor of safety was not great enough—that is, it was not as great as the best judgment of engineers teaches them to apply in such cases.

Mr. T. J. Borden.—Twenty-three years since I put in a wheel 30 feet in diameter, 9 feet 2 inches face, from the same parties who made the Amoskeag wheel, and a year earlier another mill in the same city put in one of the same diameter, with only two inches less face, from the same makers. There are about fifty wheels in the city of Fall River, ranging from 24 feet to 30 feet in diameter, and most of them from 7 feet to 9 feet 2 inches face. Most of them are of the same general construction and proportion as the Amoskeag wheel, but are usually run at about ten per cent. less surface velocity. The average period of service of the entire number is probably about fifteen years.

I have known of no failure to stand the service to which they have been subjected. I have some doubts as to the giving way of the Amoskeag wheel having been due to centrifugal force. Other conditions existed in that case which may have been the prime cause of the accident. Some reference has been made to other wheels going to pieces recently. I know of one quite recently which, from facts that have come to my knowledge, I have no doubt gave way from the weakness of the shaft, it having been in use but a very short time and had shown weakness from the beginning of its operation.

The wheel should not depend solely upon the bolts in the flanges of the rim segments to hold it together. Each arm of the wheel should be able to sustain itself and the section of the rim attached to it independent of any other arm. The end of each arm in the Amoskeag wheel, and all other wheels from the same makers, is attached to the centre of a segment of the rim, and the tensile strength of each of those arms and the four bolts securing the rim segment to it was sufficient to afford a large margin of safety. Hence the wheel should have been able to withstand the centrifugal force regardless of the bolts in the end flanges of the rim segments. If dependence was to be placed upon the segments being bolted sufficiently to withstand the centrifugal force regardless of the arms, there should undoubtedly have been a greater number of bolts at each joint, or a stronger form of flanges used.

I am inclined to think that if that wheel had been run with half the bolts taken out of the flanges of the segments, the wheel would have held together, from the fact that each segment would have been held by the arm to which it was attached.

Mr. F. H. Laforge.—I would like to ask the speaker if he knows of any wheels running with as wide a face as the wheel in question had when it broke. I have not read the paper carefully to say just what the width of the face is, but my impression is that this was an extraordinarily wide-faced wheel.

Mr. Borden.—The two 30-foot wheels to which I previously alluded as having been in service twenty-three and twenty-four years were of the same diameter as the Amoskeag wheel; one was 108 inches face and the other 110 inches face. I think the Amoskeag wheel was 110 inches face.

Mr. George W. Dickie.—There is one question which has not come up in this discussion, nor in the reading of the paper, which is quite a prominent factor; that is, that this wheel burst under a pressure that it was not designed for. Wheels, like boilers, are built to stand a certain pressure; here was a wheel built to run 61 revolutions, and it was run at 73. There is no explanation given of why that engine was running at 73 revolutions when it was designed to run at 61. The pressure on the wheel would increase as the square of the revolutions, and consequently it was subjected to a far greater strain than it was designed for. It seems to me that there was something else than the wheel at fault, and that the wheel here branded as the sinful predecessor of the one which took its place was more sinned against than sinning. The governor of the engine had evidently been out of place or inoperative. In condemning structures of that kind, all the causes should be taken into account, and if the structure was subjected to a strain that it was never designed for, why, condemnation would not be proper.

Mr. Wilfred Lewis.—It looks to me as though the broken wheel was very badly designed. The bolts to unite the sections are put through the projecting flanges without any supporting ribs, and although the tension in the rim was not excessive, the eccentric loading might strain the flanges up to the breaking point very easily. At 7,200 feet a minute the centrifugal tension would be about 1,400 lbs. to the square inch, and the application of that load on flanges without supporting ribs seems to be a very good reason why the wheel should have broken as it did.

*Mr. C. H. Manning.**—In closing this argument, I have very little to say in way of criticism of the remarks that have been made. I differ entirely from Mr. Mansfield in doubling the strain on the bolts on account of any leverage on them, as it could only be true if the cast-iron was perfectly rigid in one direction and very pliable in the other, and any increase of strain due to this cause is very slight, and I see no reason to change the language of my paper, especially as Mr. Mansfield practically agrees with my statement, that, theoretically, the bolts were weaker than good castings.

If Mr. Mansfield will examine the wreck of the wheel, he will be convinced that the bad casting did the mischief.

According to Mr. Mansfield's figures, the bolts would have been taxed beyond their elastic limit every time the wheel went to speed; consequently would have stretched, which they did not do.

There was no inconsistency in making the wheel stronger than the old one was theoretically, as, from the nature of the material used, I could not well do otherwise, as the strength is a function of speed and $\dfrac{\text{tensile strength}}{\text{weight}}$, and making the rim thicker or thinner, as long as rigidity is retained, would not affect its strength. I have probably broken up as many large castings as most men with thirty odd years spent amongst machinery, and I never saw such flaws or so many of them; and a distinguished ex-president of our Society, with still more years of experience behind him, said he had never seen their like, and only wondered that the wheel ever held together to be turned.

---

* Author's closure.

Mr. McBride was mistaken as to the figures in the paper: the old wheel weighed 116,000 lbs., and the new one 104,000, but the latter having two sets of arms, against one in the old, made a much greater difference in rim weights, they being about as 74,000 to 32,000 in round numbers. Mr. McBride is undoubtedly right that most builders of fly-wheel pulleys, in order to get sufficient face, get too much weight.

Mr. Borden's statement that " other conditions existed in that case which may have been the prime cause," doubtless means that the jack-pulley may have given out first ; but that is simply impossible, as several tons of the fly-wheel wreck were under the wreck of the jack. The other wheel, which went to pieces on account of a weak shaft, at Willimantic, was of about the same size, and also had but one set of arms, which I consider a very grave fault, as by applying the weight to the shaft through two sets of arms it is brought much nearer the bearing, and the bevelling of the shaft avoided. By reference to Fig. 167, Mr. Borden will see that he is mistaken about the arm bolting to the middle of the segment, as the segments abut on the arm.

If Mr. Dickie has ever had any experience with Corliss engines he must know that when the governor is set for a heavy load the engine will overrun at a light load. The governor was not only in excellent working order, but it was aided by a Gale regulator in addition, but yet it failed to perform its office. To be absolutely reliable, a cut-off governor should fail to engage the valve at all when the engine is over speed, but with single eccentric and any considerable range of cut-off this is impossible with the Corliss engine.

20

Reprinted from pages 251–260 of *Am. Soc. Mech. Eng. Trans.* **14**:251–265 (1892)

STRAINS IN THE RIMS OF FLY-BAND WHEELS PRODUCED BY CENTRIFUGAL FORCE.

BY JAMES B. STANWOOD, CINCINNATI, O.

(Member of the Society.)

THE strains developed in fly-wheels and pulleys are of an extremely complex nature, so that the numerous rules and formulæ employed for proportioning them are almost entirely empirical in character. Experience and judgment have dictated largely the thickness of the rim, while the eye has played an important part in determining the shape and taper of the arms.

In text-books on machine design are found analyses of the strains in pulley arms due either to the pull of belt, the inertia of the rim, or the centrifugal pull on the arms due to weight of rim, etc. These text-books also state that the strength of the rim is not affected by its thickness as regards centrifugal force ; for with an increase of thickness there goes an increase of weight with a corresponding increase of centrifugal force and area of cross-section to resist it. From this reasoning wheels of good metal, perfectly true and in good running balance, are only limited in speed by a periphery velocity, roughly assumed at about a mile per minute, or 88 feet per second ; a speed giving a very low strain per square inch of rim—to wit, about 800 lbs. per square inch of rim cross-section.

Yet in view of these statements we find wheels bursting at low speed, others running safely at high speed.

In saw-mill practice, where belts travel at high speed, bursting pulleys have given so much trouble, that heavier rims and arms for this service have become common when cast-iron wheels are used. Sometimes even solid disc wheels are employed.

In electric lighting plants many receiving and tightening

---

* Presented at the New York meeting, November, 1892, of the American Society of Mechanical Engineers, and forming part of Volume XIV. of the Transactions.

pulleys have failed, causing serious or casual disaster, according to circumstances.

Within a few years very large fly-band wheels with wide thin rims have taken the place of fly-wheels with square rims and have also been operated at very high speed. The writer knows of one case where the periphery velocity on a 17′–9″ wheel is over 7,500 feet per minute.

In band saw-mills the blade of the saw is now operated suc-

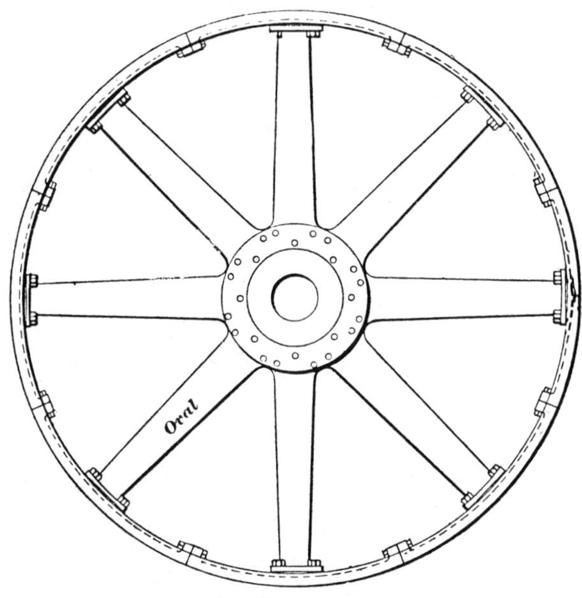

Fig. 107.

cessfully over wheels 8 and 9 feet in diameter, at a periphery velocity of 9,000 to 10,000 feet per minute. These wheels are of cast iron throughout, of heavy thickness, with a large number of arms. Who would dare to operate an ordinary 9-foot pulley at that speed?

In shingle machines and chipping machines where cast-iron discs from 2 to 5 feet in diameter are employed, with knives inserted radially, the speed is frequently 10,000 to 11,000 feet per minute at the periphery.

In view of the disasters so common of late to large band fly-wheels running at slower speed than these, it seems probable that the strains set up in wheels by centrifugal force have not been fully considered.

Last spring the writer was one of three experts, called in to decide upon the cause of a serious fly-wheel accident. From the testimony submitted by a well-known builder of large.wheels, he was led to investigate a special strain developed by centrifugal force in band-wheel rims. This strain is due to the fact that all materials have elasticity, and stretch when under strain. The wheel in question was 22 feet in diameter, 50 inches face, and

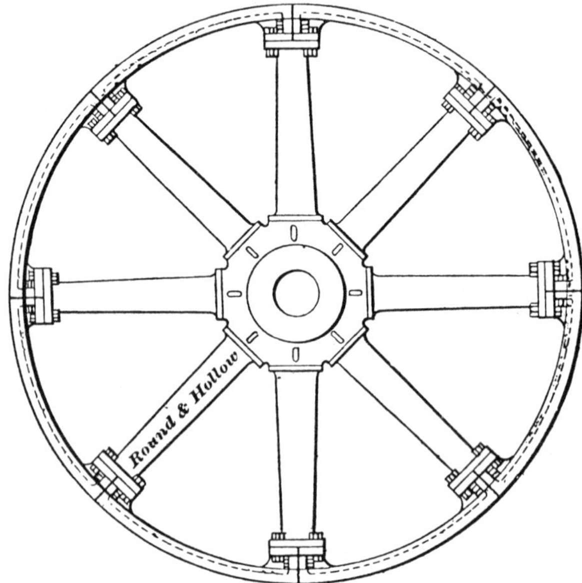

Fig. 108.

had a periphery speed of 5,000 feet per minute. It was built in sections, as so many large wheels usually are (see Fig 107). The segments were secured to arms in such a manner that the segmental joints lay half way between the arms. This construction was criticised by the aforesaid builder, who indicated by a sketch (see Fig. 108) a better method. This formed the starting point for the investigation, which the writer wishes to submit for discussion.

A thin annular ring (Fig. 109) 1 inch wide and t inches thick, revolving about a central axis A, is subjected to a simple tensile strain, similar in every respect to that found in a boiler shell, its amount per square inch T being $\dfrac{V^2}{10}$, when V equals velocity of rim, in feet, per second. This can be shown as follows :

Let r = radius of ring in inches.

$R =$ "   "   " feet $= \dfrac{r}{12}$

$V =$ velocity "   "   " per second.

The tensile strain T in pounds per square inch of rim section,

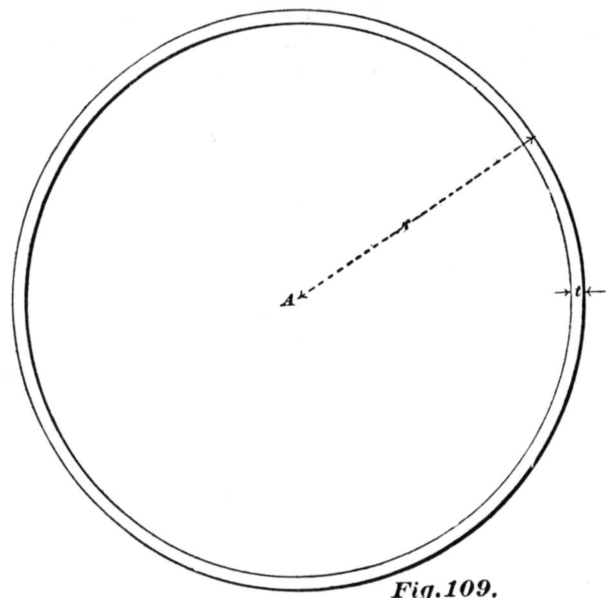

Fig. 109.

due to bursting pressure per square inch p with a ring thickness t in inches, is :

$$T = \frac{pr}{t} \quad . \quad . \quad . \quad . \quad . \quad . \quad . \quad (1)$$

Each pound weight of ring exerts a bursting pressure due to centrifugal force $=$

$$\frac{V^2}{32.2\,R} \quad . \quad . \quad . \quad . \quad . \quad (2)$$

or for each inch in length of ring it will be

$$\frac{V^2}{32.2} \times \frac{12}{r} \times .261\, t = p \quad . \quad . \quad . \quad . \quad . \quad (3)$$

substituting in (1) we have:

$$T = \frac{\dfrac{V^2}{32.2} \times \dfrac{12}{r} \times .261\, t \times r}{t} = \frac{V^2}{10} \text{ nearly} \quad . \quad . \quad . \quad (4)$$

Under this strain the ring expands. If $\dot{C}$ be taken as the modulus of elasticity, this expansion will amount to $\dfrac{.314\ V^2 d}{C}$ in the circumference of the ring.

If arms are inserted in the ring, and are supposed at first to have no weight and to be rigid, with no elasticity, their effect

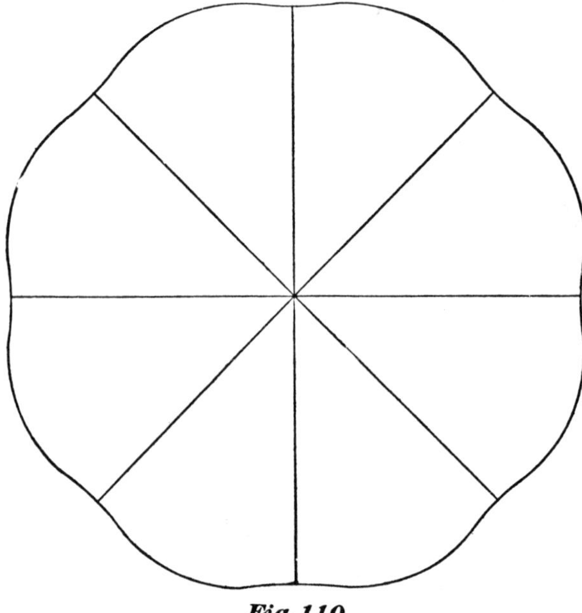

Fig. 110.

will be to pull the expanded ring into its original diameter at the arms. It will then have the form diagramatically exaggerated, as shown by Fig. 110.

Under these conditions what will be the strain upon the ring? The ring between each arm resembles now a uniformly loaded rectangular beam fixed at both ends; it is also under a tensile strain of $\dfrac{V^2}{10}$ pounds.

The greatest strain in pounds per square inch F to which any fibre is subjected, can be expressed generally thus:

$$F = \frac{p l^2}{2 t^2} + \frac{V^2}{10} \quad \cdots \quad \cdots \quad (5)$$

where l represents in inches the distance from center to center of arms, p the uniform load in pounds per square inch on the length l, and $t =$ thickness of rim in inches. The width is assumed to be unity.

This formula can be expanded by substituting for p its value $\frac{V^2}{32.2} \times \frac{24}{d} \times .261\, t$ (from 3) where $d =$ diameter of ring in inches $= 2r$, and for l its value $\frac{3.14\, d}{N}$ (6) where $N =$ number of arms.

Then

$$F = \frac{\dfrac{3.14^2 d}{N^2} \times \dfrac{V^2}{32.2} \times \dfrac{24}{d} \times .261\, t}{2\, t^2} + \frac{V^2}{10} \quad \cdots \quad (6)$$

$$= \frac{.95\, d\, V^2}{N^2 t} + \frac{V^2}{10} \quad \cdots \quad \cdots \quad (7)$$

$$\text{or } F = V^2 \left(\frac{.95\, d}{N^2 t} + \frac{1}{10} \right) \quad \cdots \quad \cdots \quad (8)$$

$$\text{or } t = \frac{.95\, d}{N^2 \left(\dfrac{F}{V^2} - \dfrac{1}{10} \right)} \quad \cdots \quad \cdots \quad (9)$$

But the ends of the arms do not remain rigid; the arms themselves expand longitudinally under the influence of their own centrifugal force, and they are also elongated by the strain outward imposed upon them by the expanding ring. In fact a compromise is effected; the arm stretches out to the ring, the ring yields in towards the arm, and the bending action in the ring depends upon this undeterminate amount. It appears, therefore, that the strain F cannot equal the value as derived by equation (8). By a careful comparison of rim thicknesses, as determined by equation (9), with good practice t is found to be too large   On the contrary, $\frac{t}{2}$ gives consistent results, and reasonably so, for roughly the arms can be assumed to expand one-half of the amount that the ring expands diametrically. When the arms are few in number, and of large cross section, the ring will be strained transversely to a greater degree than with the same number of lighter arms. To illustrate the necessary rim-thicknesses for various rim velocities, pulley diameters, number of arms, etc., the following table is submitted, based upon the formula

$$t = \frac{.475\,d}{N^2\left(\dfrac{F}{V^2} - \dfrac{1}{10}\right)} \quad \ldots \ldots \ldots \quad (10)$$

This is half the value given by (9), and the value of F is taken at 6,000 lbs. per square inch.

THICKNESS OF RIMS IN SOLID WHEELS.

Diameter of pulley in inches.	Velocity of rim in feet per second.	Velocity of rim in feet per minute.	No of arms.	Thickness in inches.
24	50	3,000	6	$\frac{2}{10}$
24	88	5,280	6	$1\frac{5}{32}$
48	88	5,280	6	$1\frac{5}{16}$
108	184	11,040	16	$2\frac{1}{2}$*
108	184	11,040	36	$\frac{1}{2}$*

If the limit of rim velocity for all wheels be assumed to be 88 feet per second, equal to 1 mile per minute, $F = 6,000$ lbs.; the formula becomes

$$t = \frac{.475d}{.67N^2} = 0.7\,\frac{d}{N^2}. \quad \ldots \ldots \ldots \quad (11)$$

When wheels are made in halves or in sections, the bending strain may be such as to make t greater than that given above. Thus, when the joint comes half way between the arms, the bending action is similar to a beam supported simply at the ends, uniformly loaded, and t is 50% greater. Then the formula becomes

$$t = \frac{.712d}{N^2\left(\dfrac{F}{V^2} - \dfrac{1}{10}\right)} \quad \ldots \ldots \ldots \quad (12)$$

---

* These are the calculated thicknesses for a band saw-wheel 9 feet in diameter with 390 revolutions per minute. The actual thicknesses as made were found to be 2¼ inches with 16 arms, and ¾ inches with 36—1⅜ inches steel arms.

or for a fixed maximum rim velocity of 88 feet per second, and $F = 6,000$ lbs.,

$$t = \frac{1.05d}{N^2} \quad . \quad . \quad . \quad . \quad . \quad . \quad . \quad . \quad (13)$$

It is a fact that wheels do spring out at the joint when cast in halves or sections; for the writer has frequently tested them when they have shown this state of affairs. For this reason the construction for segmental wheels, shown by Fig. 108, is preferable to that shown by Fig. 107.

Wheels in halves, if *very thin rims* are to be employed, should

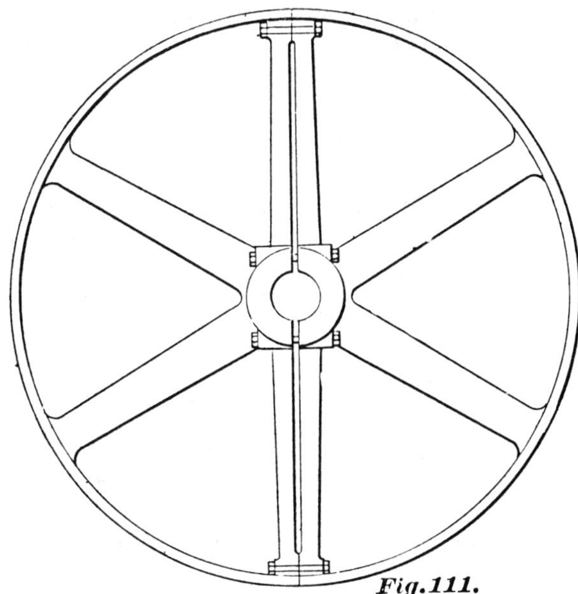

have double arms along the line of separation (Fig. 111). Compare the thickness of the rim of a wheel made in segments or in halves, the joint half-way between the arms (Fig. 107), with the thickness of the rim as constructed in Fig. 108. For example, take a 20-foot wheel at 88 feet rim velocity per second, with 10 arms. When constructed according to Fig. 107, t must be $2\frac{1}{2}$ inches; according to Fig. 108, $1\frac{11}{16}$ inches.

Attention should be given to the proportions of large receiving and tightening pulleys. The thickness of rim for a 48-inch wheel (shown in table) with a rim velocity of 88 feet per second, is $\frac{15}{16}$ inch. This should be carefully noted. Many wrecks have been caused by the failure of receiving or tightening pulleys whose rims have been too thin.

A word of caution should be given to engine builders and engineers. Wheels may be specified to be made too light in weight; for with a given diameter and weight there is a minimum safe-weight dependent upon the principles already set forth. Fly-wheels calculated for a given coefficient of steadiness are frequently lighter than the minimum safe-weight. This is true especially of large wheels.

A rough guide to the minimum weight of wheels can be deduced from our formulæ. The arms, hub, lugs, etc., usually form from one-quarter to one-third the entire weight of the wheel. If b represent the face of a wheel in inches, the weight of the rim (considered as a simple annular ring) will be

$$w = .82 \; dtb \text{ lbs.} \qquad \qquad (14)$$

If the limit of speed is 88 feet per second, then for solid wheels

$$t = 0.7 \; \frac{d}{N^2} \qquad \qquad (11)$$

For sectional wheels (joint between arms)

$$t = \frac{1.05d}{N^2} \qquad \qquad (13)$$

Substituting in (14) we have : weight of rim for solid wheels,

$$w = \frac{.57 \; d^2 \, b}{N^2} \text{ in pounds} \qquad (15)$$

Weight of rim in sectional wheels with joints between arms :

$$w = \frac{.86 \; d^2 \, b}{N^2} \text{ in pounds} \qquad (16)$$

Total weight of wheel : for solid wheel,

$$W = \frac{.76 \; d^2 \, b}{N^2} \text{ to } \frac{.86 \; d^2 \, b}{N^2} \text{ in pounds} \qquad (17)$$

For segmental wheels with joint between arms :

$$W = \frac{1.05\, d^2 b}{N^2} \text{ to } \frac{1.3\, d^2 b}{N^2} \text{ in pounds} \quad . \quad . \quad (18)$$

This investigation has not the advantage of extreme accuracy, but it tends to show a relation which ought to be observed. The constants used may be modified by further experience.

The value of $F = 6,000$ lbs., is taken at one-sixth of Rankine's ultimate breaking strength of cast iron when under transverse strain. The factor of safety is, therefore, 6. It should be remembered that in wheels this factor can be rapidly diminished ; thus the strain varies nearly as the rim velocity squared, and if the velocity is doubled, the strain is quadrupled, and the factor of safety is reduced from 6 to less than 2.

[*Editor's Note:* The discussion has been omitted.]

21

Reprinted from pages iii, 1–4, 8–26, 30, and 41–42 of *Report AECL-5116*, Atomic Energy of Canada Ltd, 1975, 65pp.

KINETIC ENERGY STORAGE OF OFF-PEAK ELECTRICITY

by

L. A. Simpson

with I. E. Oldaker

and J. Stermscheg

A B S T R A C T

The concept of using large flywheels to store off-peak electricity has been considered. The development of high strength composite materials has made possible improvements in the energy storage capacity of such devices. The problems involved in designing large flywheels and their economic advantages over alternative means of energy storage are discussed. The economic arguments are based on the present or near future capabilities and costs of structural composite materials. The flywheel costs turn out to be considerably higher than for many alternative schemes including advanced batteries, gas turbine generators and pumped storage schemes.

Atomic Energy of Canada Limited
Whiteshell Nuclear Research Establishment
Pinawa, Manitoba, ROE 1LO
September, 1975

1. INTRODUCTION

Because of the high capital cost of nuclear reactors, electric utility companies prefer to operate them continuously at maximum output as part of their base load. The peak demand on a typical utility can often be more than twice the minimum demand. Hence nuclear generators, if restricted to base load operation, can never comprise more than 50 percent of a utility's generating capacity unless some use is found for the off-peak power. One such use could be the charging of an energy storage device which could later be discharged into the grid during peak load periods, thus helping to meet the peak demand. A wide variety of energy storage techniques have been proposed over the years. These include conventional methods such as pumped hydro storage, compressed gas storage, and steam accumulators, as discussed in a parallel study[1]. There are also new concepts in various stages of development such as storage batteries, hydrogen production, and kinetic energy storage. It is this latter technique which this report assesses.

The concept of kinetic energy storage in flywheels is an old one and is found in a wide variety of devices, but recent developments in materials technology have led to suggestions of new applications. Possibly under most serious consideration is the application for mass transit vehicles. These include flywheel powered trolley buses[2] which are capable of extending their range up to six miles beyond the trolley lines, and hybrid motor-flywheel systems in automobiles [3,4] and trains. In the hybrid system, the flywheel is charged up by the motor during normal operation and absorbs the vehicle's kinetic energy during braking. This stored kinetic energy is then used to provide extra power for acceleration. This system allows a reduction in engine size since acceleration power comes from storage.

A recent article by Post and Post[5] forecasts a future flywheel technology involving automobiles driven only by flywheels with a range of 200 miles between charging stops and off-peak storage units for utilities costing only \$110/kW. While the assumptions leading to these predictions were not clearly stated, it is obvious from both this article and a patent by the same authors[6] that the material strengths assumed in their study

were several times higher than are currently available for structural purposes (e.g. flywheel tensile strengths of 7 - 10 GPa). Nevertheless, recent
progress in materials development does justify an examination of the potential
for large flywheel load-levelling devices. While several authors have suggested such an application and funds are being solicited in the United States
by at least two companies to undertake such a study (John Hopkins University
Applied Physics Laboratory, and Garrett Corporation), no detailed study has
yet been completed.

This report evaluates the potential for flywheel load-levelling
devices and identifies areas where development effort is required.

2.   SUPERFLYWHEELS

Superflywheel is a term used to describe flywheels capable of
storing very large amounts of energy per unit of mass. The elementary physics
of rotating bodies is reviewed in Appendix A. It is shown that the maximum
kinetic energy density in a rotating body is proportional to the materials
strength divided by its density. Kinetic energy density is expressed as
energy per unit mass, KE/m, hence

$$KE/m = K \frac{\sigma_{max}}{\rho} \tag{1}$$

where σ_{max} = material strength

ρ = material density

K = constant depending on geometry, $0 < K < 1$

For the particular case of the disc (see Appendix A)

$$KE/m = .606 \frac{\sigma_{max}}{\rho} \tag{2}$$

Hence, for high energy density flywheels, it is necessary to
select materials with high strengths and low densities. Many fiber-reinforced
composite materials have densities several times lower than steel, while in
the uniaxially reinforced form they exhibit strengths approaching those of

the strongest metals. This feature has been responsible for the recent interest in composites for flywheel materials. Unfortunately, the strength properties quoted by Post[5,6] were based on laboratory tests on individual fibres and do not reflect realistic strengths found in even the best composites. However, even using realistic mechanical properties, composites do appear to have certain advantages. An assessment of their potential for load-levelling flywheels is justified.

There is one important difference between a composite and a metal in this application. High strengths are achieved only in composites with uniaxial reinforcement and then only in the direction of reinforcement. Compressive and shear strengths in any direction are much lower and this type of loading should be avoided. Thus, composite flywheels must be designed so that material loading is in uniaxial tension. The disc, which is a conventional flywheel configuration, is stressed in a biaxial mode with high tensile stresses in both the radial and circumferential directions. However, there are other flywheel configurations, some of which are appropriate to composites. The energy densities for all configurations obey expressions similar to equation (1).

Some flywheel configurations are shown in Figure 1. Of these, only the last three exhibit primarily uniaxial stress distributions, i.e. the bars are stressed in tension parallel to their long dimension and the rim is stressed circumferentially. None of these configurations have K>0.5 whereas, with isotropic materials, configurations with K>0.9 are possible. This reduces to some extent the composite's advantage of a high strength to density ratio but only partially as shown in Table 1. In this comparison, a K of 0.9 is used for isotropic materials and 0.5 for composites. These represent optimum values and could be lower for practical configurations.

The rod and rim configurations are not volume efficient, that is, the energy is stored in a very small fraction of the swept volume. Rabenhorst[8] proposed a configuration where a large number of rods are built into a brush configuration and has published rough details for such an assembly including

the hub (Figure 2). This device is much smaller than that which would be
required for load levelling but is capable of being scaled up in size.

Rabenhorst[9] and Post[5,6] also suggested combining a number of
rims concentrically as in Figure 3. The as-yet-unsolved problem is connecting
the rims together and to a hub without inducing radial stresses in the rims.
Post[6] suggested that a rubber-like material with a low modulus be used, but
this idea is not proven either analytically or experimentally. Rabenhorst[9]
has recently proposed an external spoke arrangement made of composite to
connect the rims together, but the feasibility has yet to be demonstrated
and he doubts that it could be easily employed in large structures. Also,
since all the rims must be constrained to the maximum allowable angular
velocity of the outer one, the inner rims would not be fully loaded and K
would be reduced to approximately that for a pierced disc[3] (Figure 1).
Rabenhorst[9] has also suggested using ballast such as lead powder to increase
the loading (and stored energy) in the inner rings. This would present
additional construction difficulties but, if feasible, would increase K to
a maximum of 0.5.

The brush and concentric rim designs represent the only concepts
for composite flywheels which have been published to date. The design details
are very sketchy and no flywheels of these types have yet been successfully
tested. Dugger et al[3] have tested individual rods of various composites,
however, and achieved energy densities close to those in Table 1 for some
composite materials. Most published flywheel concepts have been related to
small-scale applications such as in transportation, and reference to large
flywheels for off-peak storage has been mainly in lists of additional
potential applications for superflywheels.

[*Editor's Note*: Material has been omitted at this point.]

4.   ASSESSMENTS

A flywheel system for storing electrical energy consists of a
number of basic units:

1. A flywheel, including hub assembly and shaft.
2. A vacuum chamber with pumping system to house the flywheel.

---

* Canada Deuterium Uranium – Pressurized Heavy Water

3. A suspension system with bearing and seals.
4. A motor-generator device with the necessary inverters and converters
    to transform electrical energy to kinetic energy and vice versa.

The electrical equipment, which requires some development to
suit it to this particular role, is relatively straightforward. A reference
design is given in Appendix B, with some cost estimates. However, the other
components require extensive engineering and innovation and will be discussed
in some detail.

The first step in this assessment was to choose a reference size
for the system. Because of the high engineering and manufacturing costs of
components such as bearings, vacuum systems and motor generators, it is unlikely that small unit sizes, i.e. less than 1 MWh capacity, will be economic.
On the other hand, transmission costs will be reduced if the flywheel units
can be dispersed throughout the grid to be close to the electrical consumers.
25 MWh units should be an optimum size, giving flexibility in placement while
retaining the cost advantage of large sizes. The required storage capacity
per 100 MW of peak demand for the systems in Table 2 varies from 125 to 241
MWh (neglecting Ontario #1). Based on the figures in Table 2, a power
rating of 5 MW will be adequate to allow a 25 MWh flywheel to follow the
typical cycles.

4.1 THE BRUSH CONCEPT

Based on Table 1, the most economic flywheel material is E-fiberglass.
It has therefore been selected as the reference material for composite flywheels.

The brush configuration is the only one for which even crude design
details are available. This design, due to Rabenhorst and Taylor[8] (R and T),
was intended for discharging at a very high power rating (500 MW) into a plasma
physics device. The actual capacity is relatively small (36 kWh in the rods),
and thus it is fully discharged in a fraction of a second. Torques due to
deceleration were an important consideration but were found to be significant
only in the shaft. The stresses in the rods and hub were due primarily to the

centrifugal loading and these parts can be scaled up uniformly to a 25 MWh configuration.

In this design it has been assumed that all the usable energy is stored in the rods. In fact a small fraction of the energy will be stored in the hub but it has been assumed to be only 10 percent of the total. Thus, by operating the flywheel at a 3:1 speed ratio, this is just balanced by the 10 percent reserve capacity in the flywheel*.

4.1.1   <u>THE RODS</u>

The flywheel consists of 40 layers of fanned rods, each layer consisting of 20 rods and a hub as illustrated in Figure 5. R and T's[8] reference material was Kevlar-epoxy but since their working strength was close to that assigned to fiberglass in this study, the dimensions are not affected. (For any fixed configuration, flywheel volume depends on working strength only). However, the fiberglass brush will be considerably heavier due to its higher density. The scaled-up dimensions are given in Table 3. The figures for both Kevlar-epoxy and fiberglass-epoxy are included. The maximum angular velocity for the 25 MWh flywheel can be calculated from the equation for the energy density of a rod derived in Appendix A

$$KE/m = 1/3 \; \frac{\sigma max}{\rho} = \frac{\omega^2 R^2}{6} \tag{3}$$

where ω = maximum angular velocity

   R = flywheel radius (rod 1/2 length)

Since R for the 25 MWh flywheel is 19.6/2 = 9.8 m (Table 3), the maximum angular velocity is

$$\omega = \frac{1}{R} \; (6KE/m)^{\frac{1}{2}} \tag{4}$$

   = 123 rad/s or 1172 rpm

This is very close to 1200 rpm which is compatible with a six pole motor for charging the flywheel as described in Appendix B.

---

* Recall that the energy density is proportional to the velocity squared, hence 90 percent of the capacity is usable with a 3:1 speed ratio (Appendix A).

A 3:1 speed range allows a 90 percent utilization of the capacity of the flywheel as described earlier. Hence, the minimum velocity is 400 rpm.

The cost of material and fabrication of fiberglass depends on the methods of manufacture and the tooling required[18]. However, an estimate based on information from Fiberglas Canada Ltd., and taking into consideration the very large valume requirements, would be about \$2.20/kg in 1970 dollars. Using a modest average annual increase of 5%, the effective 1980 cost is about \$3.60/kg. At this price the 374 Mg of fiberglass for the rods would cost 1.35×10^6 or about \$54/kWh. This is higher than the figure in Table 1 (\$22/kWh) because of the escalation and also because the geometrical constant, K, for the brush is 0.33 rather than the 0.5 which was used to calculate the values shown in Table 1.

4.1.2   THE HUB

The design of the hub is crucial because it must sustain not only the bearing loads of the rods but also its own body stresses. While Rabenhorst and Taylor[8] give some design details, they are not optimized and, as yet, no hub of this type has ever been built. Hence the first assumption to be made regarding the hub is that the construction of a workable design is possible. R and T[8] chose AISI 4340 steel for their hub material based on the need to withstand stresses of 1.1 GPa. An isotropic material is required since the loading is biaxial tension plus very high shear stresses in the design considered here. Using steel, about one-half of the hub load comes from the rods and about one-half from the body stresses due to the hub rotation.

The machining cost of this intricate part is difficult to estimate. R and T[8] estimated about \$40/kg for the 36 kWh flywheel (1973 dollars). However, this is based on a single unit and large-scale production might result in lower costs. James[18] quotes costs for machined metal parts to be in the range \$20 - \$100/kg. These figures pertain to smaller structures and some economy might be expected in going

to the larger sizes involved with the hub. A rough estimate by the AECL
workshops suggests that the machining costs alone would be at least $1/kg
of finished product. However, they have no experience with structures of
this size. With the stringent quality control required (for fracture
prevention) a minimum fabrication cost equal to the material cost ($1.32/
kg, 1974) was selected which would give a hub cost (Table 3) of $1.37 x
10^6 in 1980.

Obviously the cost of the hub must be kept low if the brush
flywheel is to be economic. If the fabrication costs using metal prove
to be much higher than the minimum estimate above it might be necessary
to consider a composite material. R and T[8] mention the possibility of
making a composite hub. While the body stresses would be lower because
of the lower density, the bearing loads would still be high. Dugger et
al[3] demonstrated theoretically that certain laminated components could
yield a biaxial stress capability with about 40 - 50% of the uniaxial
composite strength. However, such a strength reduction may be too much
for the hub requirements and it could be difficult to design out regions
of high shear stresses. This is a major area for further study. For
the purpose of this assessment it has been assumed that either a composite
or a metal hub can be built for a minimum cost equal to the cost of the
rods, i.e. $1.35 x 10^6.

4.1.3   THE SHAFT

The final component of the flywheel itself is the shaft. In
the R and T[8] design the shaft flares out into a flange which is bolted
through the hub to a similar structure at the opposite end (Figure 6).
The bolt holes in the hub could raise problems of stress concentration
which might be particularly serious in a laminated composite. The most
serious problem will be that of alignment of the shaft. The tooling and
machining costs will probably be the significant factor for this part.

Roughly scaling up the R and T[8] shaft assembly (except for
the shaft diameter itself which, because of lower torque requirements, is

raised only to 0.3 m), reveals a need for an additional 50 - 100 Mg of steel. Even at \$5/kg, fabricated, this would add \$25 to \$50 x 10^4 to the cost. Using the smaller figure this raises the cost of the flywheel assembly to nearly \$3,000,000, or \$120/kWh.

4.2   CONCENTRIC RIM CONCEPT

While design details for the brush flywheel are sketchy, they are non-existent for the concentric rim. Thus an accurate cost assessment is not possible. The aim of the work on this concept was, therefore, to estimate the lowest possible cost assuming all problems can be overcome.

The overriding difficulties with this concept are joining the rims together without inducing radial stresses and designing a suitable hub. However, if this configuration can be built, it has the advantage of volume economy, i.e. nearly all of the swept volume stores energy. This will reduce the physical dimensions of the containment as well as the free volume requiring evacuation.

To understand the problem of rim connection it is instructive to look at the stresses in a rotating disk of inner radius a and outer radius b

$$\sigma_r = \frac{3 + \nu}{8} \rho \omega^2 \left(a^2 + b^2 - \frac{a^2 b^2}{r^2} - r^2 \right) \tag{5}$$

$$\sigma_\theta = \frac{3 + \nu}{8} \rho \omega^2 \left(a^2 + b^2 + \frac{a^2 b^2}{r^2} - \frac{1 + 3\nu}{3 + \nu} r^2 \right) \tag{6}$$

where σ_r = radial stress

σ_θ = circumferential stress

r = radial coordinate

ν = Poisson's ratio

The stress distributions are plotted in Figure 7 for various values of a. They are normalized by dividing by $\rho\omega^2 R^2$, the hoop stress in a thin rim ($a \simeq b$). Note that as soon as a hole is put in a disc, the stress at the center is doubled. This is due to the stress concentrating properties of a hole in a biaxially stressed medium. In the unpierced disc ($a = o$) the radial and circumferential stresses are equal and a maximum at the center. As the configuration is changed to that of a rim, i.e. $a \to b$, the radial stresses become very small, i.e. the stress distribution becomes uniaxial.

The only known results of tests on a fiberglass disc were published in 1967 by Morganthaler and Bonk[19]. In this test, on a wound fiberglass disc with outer and inner radii of 10 and 3 inches respectively, failure occurred by delamination at 12 Wh/kg. One interesting point in this work was the analytical result that by tailoring the elastic moduli in orthotropic composites, the radial stresses could be reduced somewhat from those to be expected in an isotropic material. Apparently this work has not been continued.

If a number of concentric rims are rigidly connected together, they will induce radial stresses in one another, i.e. they will behave like a disc. This is because the stresses are proportional to r^2 and hence the fractional increase in radius of the outer rims will be greater than that for the inner ones. While some have suggested accommodating the changes in rim spacing by connecting them with a low modulus material, e.g. rubber, this has never been proven experimentally and raises a number of doubts:

1. Is there a material which has sufficient strength to hold a large flywheel in a rigid configuration?

2. Is there a low modulus binder which would not increase the susceptibility to unbalancing by distortion?

4.2.1   THE RIMS

A cost for the rim flywheel can be arrived at after making two assumptions:

1. The inner rims can be cheaply ballasted without affecting their mechanical strength, to raise K to 0.5 for the whole flywheel.

2. The rim connection problem can be solved with no loss of performance.

Then, based on Table 1, 250 Mg of fiberglass would be required for a 25 MWh flywheel. Assuming the same material and fabrication costs as for the brush ($3.60/kg), the rims would cost $900,000. The introduction of ballast and the connection of the rims, being very critical with regard to balancing, would at least double this cost. Hence a minimum cost for a 25 MWh concentric rim flywheel would be $1.8 x 10^6.

4.2.2   HUB FOR CONCENTRIC RIMS

A wound fiberglass rim, by virtue of the construction, will always have a hole at its center. Therefore, a hub of some kind is required to support the flywheel. Certainly first generation hubs would have to be made of metal. The first problem is matching the high radial expansion of the low modulus fiberglass to the lower expansion of the hub. Morganthaler and Bonk[19] used a spring-loaded steel hub which expanded with the inner diameter of the flywheel. While no cost information is available, the construction of a hub capable of supporting 250 Mg plus ballast would be an intricate job, and 10^6 would probably be a minimum estimate*. Thus the total estimated cost for a flywheel and hub is $2.8 x 10^6 or $112/kWh. This is very close to the minimum estimate for a brush flywheel.

4.3   ENERGY LOSSES

Energy losses will accrue from several sources. First of all, the electrical-mechanical transmission (motor-generator) will be a source

---

*   Although using 50 - 100% of the flywheel cost for the cost of this hub may seem high for a component which is somewhat smaller than the flywheel it is important to keep in mind that the flywheel costs are based on a simple, easy-to-fabricate geometry, whereas, the hub designs for both the rim and brush flywheels are considerably more detailed and will require more labour to produce.

of loss. A number of schemes based on fully developed technology have
been proposed in the literature. For the system described in Appendix B,
the estimated losses during charging and discharging are 10%. Generally,
the rule will apply that the more expensive systems have the lower losses.

Estimation of the frictional losses is more difficult since
the design of the flywheel is not sufficiently detailed. These will come
under three categories – windage, bearing friction, and seal friction.

4.3.1   WINDAGE

Numerous relationships appear in the literature governing
windage friction on rotating bodies[3,4,20]. The correct relationships
depend on the detailed configuration, such considerations as whether
flow is laminar or turbulent, and interaction of the boundary layers of
the wheel and housing.

Dugger et al[3] used the following relationship to describe
a single rotating rod of square cross section

$$P_\omega = \rho_g \, Z \, R^4 \, \omega^3 \, / \, 2.16 \tag{7}$$

where ρ_g = gas density in kg/m^3

Z = rod thickness (width) in m

R = rod ½ length (wheel radius) in m

P_ω = power loss in J/s (Watts)

For air, ρ_g = 1.22 kg/m^3 at one atmosphere at 15°C. Multiplying this equation by 800 for the 800 rods and substituting the dimensions
from Table 4 leads to the upper curve in Figure 8. This serves as an upper
bound for the brush flywheel since the resistance on a .series of rods will
be less than the sum of the windage losses for each individual rod. A
lower bound is given by the windage curve for a disc using the relationship
used by Gilbert et al[4]. The disc dimensions are taken to be those for an
envelope enclosing the brush.

$$P = .043 \left(1 + 2.3 \frac{t}{R}\right) \left(\frac{P}{T}\right)^{0.8} (R)^{4.6} \omega^{2.8} \mu^{0.2} \tag{8}$$

where t = brush height = 7m

                P = air pressure (torr)

                T = temperature = 288 K

                μ = air viscosity = 1.78×10^{-2} poise

and R, P_ω, are as in equation (7).

This curve is also plotted in Figure 8. The curves are calculated for the maximum angular velocity ω and thus represent the energy loss rate of a fully charged flywheel. The dependence on ω is very strong so actual losses will be lower and depend on the daily duty cycle of the flywheel.

Assuming a simple linear ω – time relationship over a 24 hour period, the energy loss is estimated in Appendix C to be 0.89 MW or 3.6 percent of the storage capacity when the power loss at maximum speed is 0.1 MW. Because of other sources of energy loss, this represents about the highest tolerable loss. From Figure 8, the required vacuum to keep losses below this level would be about 10^{-3} torr (1.3×10^{-1} Pa).

It is difficult to estimate the type of evacuation equipment which would be necessary to maintain this vacuum. The requirements for the concentric rim would be least since its windage behaviour would be closer to that for a disc than for the brush. Also, the large free space between the brush spokes would increase the required pumping capacity over that for the concentric rim flywheel. The approximate chamber volumes are 2000 m^3 for the brush, and 150 m^3 for the rim flywheel. The type of equipment required depends on leak rates at the seals and on the ultimate vacuum required. If 10^{-3} - 10^{-4} torr (1.3×10^{-1} - 1.3×10^{-2} Pa) is acceptable, then several Roots Blowers backed by oil-sealed mechanical pumps will be sufficient[21]. The cost of such a system might be $50,000 - $100,000. If diffusion pumps are required, the costs will escalate rapidly.

The vacuum housing is also difficult to cost without a detailed design. $175/$m^3$ represents a typical construction cost for a high quality building[22]. This gives a housing cost of $250,000 for the brush and $25,000 for the rim flywheel. These figures are probably too low because of the high quality construction required to ensure vacuum tightness and also the strength requirements to contain a flywheel failure.

4.3.2   BEARINGS AND SEALS

The flywheel mass is enormous and must be supported by a low friction bearing if losses are to be kept low. The results of a study of conventional bearing systems are shown in Appendix D. With conventional design, frictional (including pumps) losses are high (up to 15 MWh per day), which represents a daily energy loss of 60 percent of capacity. The cost for lubrication pumps alone is $150,000. The total conventional suspension costs will be substantially more when the cost of bearing parts, which must satisfy extremely severe tolerances, is included. Hence, there is some incentive here to look at magnetic suspensions. On this scale, extensive development would be required to prove feasibility and costs would probably be higher than for conventional bearings.

At least one rotating seal would be required between the flywheel and motor generators and with conventional high pressure bearings (see Appendix D) several would be required. These again are a source of friction although probably not as serious as the bearings. Possibly the best candidate would be a magnetic fluid seal.

Thus, the energy losses in the suspension will be most serious unless an inexpensive magnetic suspension can be built. Even if these losses could be reduced to 10 percent, the total efficiency of the system, allowing 10 percent losses in the electrical portion and 10 percent due to windage, seals and power for the vacuum system, would be near 70 percent. Since this figure is also the reference value for the Li-S battery program[10], it has been chosen for the reference calculation.

4.4   <u>COST OF KINETIC STORAGE</u>

It is possible to calculate the cost of flywheel energy storage several ways and obtain differing results. Much depends on the particular situation of a given utility. One approach is to assume that a certain amount of reserve capacity is available during off-peak periods and the cost of charging a storage device is simply the cost of extra fuel. This is the approach taken by Unsworth in his assessment of conventional energy storage methods[1]. One might argue, however, that this method is subsidized since the availability of spare capacity means that the equipment has been costed at a capacity factor less than the maximum and consumers are being charged at a higher rate than for a fully loaded plant.

In this report the full cost of the charging device (a nuclear plant) is included in the cost of peak power. Thus the alternatives for generating the power above P_B in Figure 4 are a storage device plus increased nuclear baseload $P_N - P_B$, a gas turbine or a load following nuclear plant, both rated at $P_{max} - P_B$ and operated at a load factor $(P_N - P_B)/(P_{max} - P_B)$.

The detailed cost equations are derived in Appendix E. The basic assumptions are 70 percent in-out efficiency for the storage device, 30 year station life, 10 percent fixed charge rate, and 5 percent per annum escalation of costs to 1980 dollars for all items.

It is also assumed that the storage systems will be utilized to their full capacity on a daily basis. In practice this will not be the case since cycle demands vary from day to day. Thus as the number of hours of utilization per year decreases, the unit energy costs increase. Note (Figure 4) that because the costs of storage devices depend on storage capacity rather than on power capacity, the ground rules for determining capacity factors differ from those used for generating devices. Thus while the cycle in Figure 4, if reproduced daily, could be followed by a flywheel-nuclear system operating at 100 percent load factor, the load factor for a gas turbine or load following nuclear plant will be $(P_N - P_B)/$ $(P_{max} - P_B)$.

Using the costs and equations of Appendix E, the unit energy cost for a flywheel-nuclear storage system is plotted against flywheel capital cost in Figure 9. The costs of energy from gas turbine plants, nuclear load following stations and battery systems are also plotted.

At $100/kWh, the flywheel is just competitive with the battery and gas turbine but is more expensive than nuclear load following. While no extra costs have been included for the nuclear load following capability, both direct and indirect costs have been included in this example. This was not the case for the battery system because of the difficulty of estimation. The indirect nuclear costs are also included in the cost of the charging facility for the storage device.

It seems clear that a cost of $100/kWh is not credible at present for a flywheel system. We have shown that the minimum cost for just a flywheel and hub assembly would be $110 - $120/kWh. The costs of housing, suspension, bearings, shaft, etc. have only been crudely estimated here but would raise this figure by at least $20/kWh. These costs were arrived at by assuming no great technical difficulties in design and construction and utilization of composites to a working strength of 1.4 GPa. Indirect costs such as engineering, contingencies, spares and construction indirects, have been neglected, as have operation and maintenance costs. The large flywheel is a high technology concept and, therefore, indirect costs could significantly increase its cost.

The choice of working strength may well have been optimistic. Ultimate strengths for uniaxially reinforced fiberglass are seldom higher that about 1.8 GPa. Assuming some improvement in properties by 1980, this might rise to 2.1 GPa although obtaining this in large structures is not an easy task. The chosen working stress of 1.4 GPa is 2/3 of this value. Fatigue data on composites are limited but some authors feel that a suitable fatigue limit for fiberglass is 1/3 UTS*[4,24]. If this limit were applied to flywheels, it would halve the energy density and hence double

---

* Ultimate Tensile Strength

the flywheel capital cost. Graphite epoxy composites have superior fatigue
resistance but cost five times as much as fiberglass (Table 1) so nothing
is gained here. The fatigue properties of Kevlar are not well known but
should be somewhere between those of glass and graphite composites (fatigue
resistance appears to increase with elastic modulus).

Thus it appears that the capital cost of flywheels will certainly be in excess of $200/kWh using near term materials. However, any
discussion of costs may well be academic since it remains to be proven
that a large, high energy density flywheel can actually be constructed.
To develop large flywheels for off-peak energy storage by the early 1980's
would require a massive development effort at least equal to that in progress
for storage batteries. Composite flywheels would first have to be proven
for small scale applications such as transportation. The companies with
the greatest expertise in this area, e.g. Garrett, Lockheed, still prefer
metal flywheels and such an effort is not foreseeable at this time. Once
a composite flywheel is built and successfully operated on a small scale,
one will be in a better position to assess their application in electrical
grids.

Increases in interest rates will improve the economics for gas
turbines relative to the other three load following systems in Figure 9.
The latter three systems are capital intensive and their relative economic
position will not change significantly with a change in interest rates.

At this juncture, the question might arise as to the potential
for metal flywheels, since the materials properties are well understood and
the high efficiency (K value, equation 6) does compensate to some extent
for their high density.

The best candidate (Table 2), 4340 steel, is twice as expensive
as fiberglass, ignoring fabrication costs which will be at least 1 to 3
times the material cost. Also the maximum thickness requirement for adequate
heat treatment will restrict the size such that approximately 200 flywheels
would be required to store 25 MWh. Going to a vastly smaller unit size

would greatly increase the cost of containment, evacuation, mechanical-electrical transmission and so forth. Thus it does not appear that any flywheel system will be viable for large-scale load following.

While nuclear load following looks relatively inexpensive in this study, it remains to be proven that CANDU reactors can perform this role without increasing their cost significantly.

4.5   COMPARISON WITH CONVENTIONAL ENERGY STORAGE DEVICES

Unsworth[1], in a parallel study, has assessed several conventional energy storage techniques including pumped hydro storage, steam storage, storage of boiler feedwater, and compressed air storage in caverns. It is therefore appropriate to compare the results of his studies with flywheel energy storage. To do this, some adjustment of the costing method is required. Unsworth assumed that the cost of off-peak power for charging a storage device was simply the cost of fuel for a nuclear reactor with some reserve capacity. Appendix F shows the calculations used to cost flywheel energy storage using Unsworth's method. The result is compared with his data in Figure 10. The only change made to Unsworth's data is to escalate the costs at an annual rate of 5 percent to 1980 dollars. The results are plotted against the load factor of the generating device (motor generator in the case of the flywheel) for 300 days per year operation. While this is the conventional way to compare generators, storage systems are measured in terms of energy capacity rather than power capacity and hence are not very sensitive to load factors except at low values.

Using these costing methods, the flywheel system (at $200/kWh) is several times more expensive than any of the conventional methods. It is twice as costly as the conventional peaking gas turbine when costed this way.

This very high cost is initially surprising considering that only the fuel cost is assessed for charging the flywheel. However, the

unit cost of the flywheel is only part of a total system in the costing
method used in this report. The total energy production of the system is
$L = 24(P_N - P_B)$ (see Figure 4) of which less than 30 percent (EPOS/L) is
stored in the flywheel. Hence the cost of the flywheel is averaged out over
the system which is generating about three times as much energy as the flywheel actually stores. Since the unit energy cost for the flywheel alone is
much higher than for the base-loaded nuclear-charging system, the total
system approach yields a lower cost.

It is not the purpose of this report to decide which costing
approach is more realistic. This choice will be dictated by an individual
utility situation.

5. <u>CONCLUSIONS</u>

1. No proven concept for a composite superflywheel exists at present.
2. Optimistic predictions regarding the cost of storing off-peak
   electricity in flywheels have been based on exaggerated assumptions
   of material strengths.
3. The engineering of uniaxial composites into large superflywheel
   structures is a complex problem and may well be impossible without
   some sacrifice in energy density.
4. Assuming a superflywheel could be constructed, the energy density
   achievable with present day materials is not sufficient to make a
   utility storage system economically attractive.
5. The likelihood of materials becoming available with the mechanical
   properties and low cost required to make large scale kinetic energy
   storage viable is remote.

6.   AREAS FOR FUTURE STUDY

As mentioned previously, initial development efforts should be
directed towards use of composites in small scale applications. This
would entail the development of a detailed design capable of being scaled
up two or three orders of magnitude in size. Because of the difficulties
in using uniaxially reinforced materials, this is no easy task and success
would constitute a great breakthrough. However, even with success, if the
costs of construction are not minimal, it will be difficult to ensure an
economic device.

A doubling of the working stress would improve the prospects
for success. However, this is not likely to occur in the near term at low
cost. High quality control would be required to ensure flaw-free
construction. The suggestions of utilizing the laboratory strength of
flame-polished quartz and other glasses in large structures must await a
technical breakthrough of unprecedented magnitude. The two most important
problems are those of handling and fabricating large structures (a scale
up of 10,000 times) without losing structural perfection (handling can lower
the strength by an order of magnitude). The production of reliable 7 GPa
strength material will have benefits far beyond the flywheel industry but
must be classified as a breakthrough which is not likely to be realized in
this century.

Besides these two areas for development, much work is required
to minimize energy losses by breakthroughs in suspension and bearing
design.

<u>REFERENCES</u>

1.   G. N. Unsworth, *A Review of Pumped Energy Storage Schemes*, Atomic
     Energy of Canada Limited Report, AECL-4926, 1975.

2.   L. J. Lawson, *Kinetic Energy Wheel (KEW) Propulsion Systems*, Presented
     American Transit Assoc. Mid Year Meeting, Washington, D.C. May 20-23,
     1974.

3.   G. L. Dugger, A. Brandt, J. F. George, L. L. Perrini, D. W. Rabenhorst,
     T. R. Small and R. O. Weiss, *Heat Engine/Mechanical Energy Storage
     Hybrid Propulsion Systems for Vehicles*. John Hopkins University
     Applied Physics Laboratory Rept. CP011, March, 1972.

4.   R. R. Gilbert, J. R. Harvey, G. E. Heuer and L. J. Lawson,
     Flywheel Feasibility Study and Demonstration, Lockheed Missiles
     and Space Co. Rept. LMSC D007915, April 30, 1971.

5.   R. F. Post and S. F. Post, *Flywheels*, Scientific American, <u>229</u>
     #6, (1973) pp. 17 - 23.

6.   R. F. Post, *Inertial Energy Storage Apparatus and System for Utilizing
     the Same*, U.S. Patent No. 3,683,216, August 8, 1972.

7.   E. E. Sechler, *Elasticity in Engineering*, Dover, New York, 1968, p. 162.

8.   D. W. Rabenhorst and R. J. Taylor, *Design Considerations for a 100 -
     MJ./500 MW Superflywheel*, Applied Physics Laboratory, John Hopkins
     University Rept. No. T G 1129, December, 1973.

9.   D. W. Rabenhorst - *The Multirim Superflywheel*, Rept. No AD/A-001-081,
     August, 1974. John Hopkins University Applied Physics Lab.

10. M.J. Kyle, E.J. Cairnes and D.S. Webster, *Lithium Sulfur Batteries for
     Off-Peak Energy Storage*, Argonne National Laboratory Rept. ANL - 7958,
     March, 1973.

11. J. T. Brown and J. H. Cronin, *Battery Systems for Peaking Power Generation*,
     San Francisco, August, 1974, IEEE

12. J. M. Kay and A. A. Fulton, *Combined Use of Nuclear Power and Pumped
     Storage Hydro Stations*.

13. H. Y. Watt, Private communication to G. N. Unsworth.

14. J. H. Shortt, *Power Generation Economics*, Gas Turbine International,
     <u>15</u>, #1, (1974) pp. 32-36.

15. W. J. Walsh, J. W. Allen, J. D. Arntzen, L. G. Bartholme, H. Shimotake,
     H. C. Tsai and N. P. Yao, *Development of Prototype Li/S Cells for
     Application to Laod Levelling Devices in Electrical Utilities*, IEEE,
     9th Intersociety Energy Conversion Engineering Conference, San Francisco,
     California, August, 1974.

16.   S. P. Mitoff and J. B. Bush, Jr., *Characteristics of a Sodium-Sulfur Cell for Bulk Energy Storage*, General Electric Company, Corporate Research and Development Rept. No. 74 CRD 145, August, 1974.

17.   Fiberglass Reinforced Plastics, Book 2, a Manual published by Fiberglas Canada Ltd. 1970.

18.   R. C. James, *The Engineering-Marketing Interface in Developing Applications for Advanced Composites, Composite Materials in Engineering Design*, Proc. 6th St. Louis Symposium, 11-12 May, 1972, American Society for Metals, Netals Park, Ohio.

19.   G. F. Morganthaler and S. P. Bonk, *Composite Flywheel Stress Analysis and Materials Study, Advances in Structural Composites*, 12th Annual Symposium, Society of Aerospace Material and Process Engineers, 1967.

20.   J. W. Daily and R. E. Nece, *Chamber Dimension Effects on Induced Flow and Frictional Resistance of Enclosed Rotating Disks*, J. Basic, Eng. <u>82</u> (1960) pp. 217-232.

21.   H. A. Steinbeiz, *Handbook of High Vacuum Engineering*, Reinhold, New York, 1963.

22.   R. S. Godfrey, Building Construction Cost Data 1973, R. S. Means, Tuxbury, Mass. 1973.

23.   S. Shimamura, H. Furve and M. Nuka, *Flexual Properties of Hybrid Composites*, Proceedings 1974 Symposium on Mechanical Behaviour of Materials, Kyoto, Japan, August, 1974.

[*Editor's Note*: Tables 1, 2, and 3 and figures 2 through 10 have been omitted.]

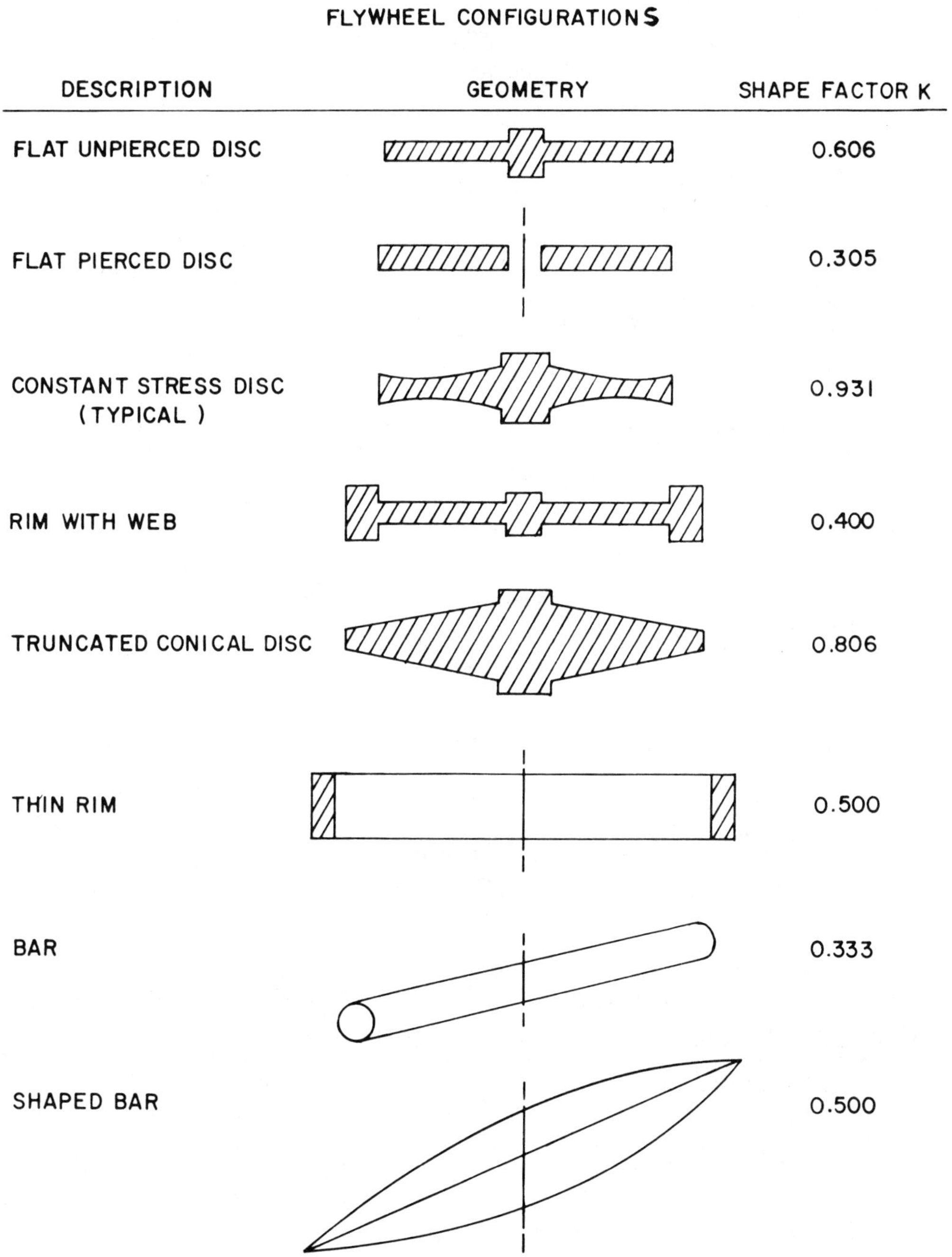

FLYWHEEL CONFIGURATIONS

DESCRIPTION	GEOMETRY	SHAPE FACTOR K
FLAT UNPIERCED DISC		0.606
FLAT PIERCED DISC		0.305
CONSTANT STRESS DISC (TYPICAL)		0.931
RIM WITH WEB		0.400
TRUNCATED CONICAL DISC		0.806
THIN RIM		0.500
BAR		0.333
SHAPED BAR		0.500

FIGURE 1 - TYPICAL FLYWHEEL CONFIGURATIONS SHOWING SHAPE EFFICIENCY FACTOR, K

APPENDIX A

PHYSICS OF ROTATING BODIES

The kinetic energy of any body rotating about a fixed axis is given by

$$KE = \tfrac{1}{2}I\omega^2 \tag{A1}$$

where

I = moment of inertia about axis of rotation

ω = angular velocity about axis of rotation.

For a flat disc of radius R, thickness t, and density ρ,

$$I = t\frac{\pi}{2}\rho R^4 = \tfrac{1}{2}mR^2 \tag{A2}$$

where

m = mass of disc.

The kinetic energy density (per unit mass) is

$$KE/m = \frac{\omega^2 R^2}{4} \tag{A3}$$

The stress analysis for the rotating disc (see equations 5 and 6) indicates that the maximum stress occurs at the center and is given by[7]

$$\sigma_{max} = \frac{3 + \nu}{8}\rho\omega^2 R^2 \tag{A4}$$

where

ν = Poisson's ratio

By equating σ_{max} to the maximum allowable stress in the disc and combining equations (A3) and (A4), an expression for the maximum energy density is obtained.

$$KE/m = \frac{2}{3 + \nu}\frac{\sigma_{max}}{\rho} = 0.606\frac{\sigma_{max}}{\rho} \tag{A5}$$

for $\nu = 0.3$.

For the rod or brush flywheel, neglecting the small effect of curvature of the rods in the brush, the maximum stress which occurs at the center of a rod is

$$\sigma_{max} = \frac{\rho\omega^2 R^2}{2} \tag{A6}$$

where $R = \frac{1}{2}$ x rod length (i.e. wheel radius).

The kinetic energy in a rotating rod is

$$KE = \frac{1}{2}I\omega^2 = \frac{m\omega^2 R^2}{6} \tag{A7}$$

where I = moment of inertia about axis of rotation = $\frac{mR^2}{3}$.

Hence the energy density is

$$KE/m = \frac{\omega^2 R^2}{6} = \frac{1}{3}\frac{\sigma_{max}}{\rho} \tag{A8}$$

which demonstrates the origin of the shape factor $K = 1/3$ for a rod.

The energy density for a thin rim may be obtained from equation (6) where $a = b = r$.

$$\sigma_{max} = \rho\omega^2 r^2 \tag{A9}$$

$$KE = \frac{1}{2}I\omega^2 = \frac{1}{2}mr^2\omega^2 \tag{A10}$$

From (A9) and (A10)

$$KE/m = \frac{1}{2}\frac{\sigma_{max}}{\rho} \quad . \tag{A11}$$

Thus, $K = 0.5$ for the thin rim.

[*Editor's Note:* Appendixes B through F have been omitted.]

22

Reprinted from pages 76–88 of *1975 Flywheel Tech. Symp. Proc.*, U.S. Energy Research and Development Administration and the Lawrence Livermore Laboratory of the University of California, 1975, 294pp.

IMPROVED FUEL ECONOMY IN AUTOMOBILES BY USE OF A FLYWHEEL ENERGY MANAGEMENT SYSTEM

A. A. Frank and N. Beachley
University of Wisconsin

ABSTRACT

A high-performance, low fuel consumption automobile is being built that will contain a flywheel package powered intermittently by a standard internal combustion engine and connected to the road through two transmissions arranged in series. The design of the 23-inch diameter, steel alloy flywheel is based on current technology. The flywheel will run in a vacuum and will have useable energy storage of 2/3 hp-hr, maximum windage of 1 hp at maximum speed, a maximum torque transmission capability of 250 foot pounds, overspeed protection and protection against locked bearings. Special controls and instrumentation peculiar to this design will be necessary, since the torque-converter transmission will be used for both propulsion and regenerative braking. Computer simulation of operating aspects shows that a standard 1976 car modified to the new design should have 58% better gasoline mileage. If certain other design improvements can be realized, ultimate gasoline mileage could be 117% better.

INTRODUCTION

A standard automobile engine driving a modern, contemporary automobile cannot operate at its best thermal efficiency at all times. The engine must perform over a wide torque-speed range, yet in standard city driving it must operate in a so-called idle-mode much of the time. In today's society fuel economy is becoming increasingly important, and non-productive modes, such as idle, and inefficient modes such as low-torque, high-speed operation, should be eliminated.

Also, safe acceptable performance in city driving requires that the brakes of an automobile be used to slow it down. This kind of braking is inefficient because it turns kinetic energy into heat. If one were not interested in performance, it would be possible to drive a vehicle without brakes, using only aerodynamic and rolling friction to slow down the car. But traffic as we know it today will not tolerate such a pattern of driving. Regenerative braking that re-uses this energy will be required if maximum fuel efficiency is to be achieved.

A flywheel power plant system is being built at the University of Wisconsin that we believe will retain the desirable features of present-day automobiles, but will improve fuel economy considerably and will minimize pollutant emission. The system will be installed in a 3,000-pound vehicle to be completed in the summer of 1976.

SYSTEM DESCRIPTION

The configuration of the proposed flywheel engine vehicular system is shown in Figure 1. This particular configuration has been chosen because of practical considerations and efficiency.

The design features of this automobile are that it must have equal or better performance than the standard unmodified vehicle in all respects. The performance criteria are: acceleration from 0 to 60 mph, top speed, gradability and braking performance. These constraints dictate that the engine in this vehicle be approximately the same size as the engine in the standard vehicle.

The system configuration consists of two transmissions in series: a standard four-speed manual shift transmission and a continuously-variable transmission unit. The continuously-variable transmission is designed to be integral with the standard rear axle assembly, necessitating four-wheel independent suspension. The reason for using two transmissions is to provide a system which can operate at its best efficiency with off-the-shelf components.

The flywheel package is based on current, proved technology. An installation sketch of the flywheel system is shown in Figure 2 and Figure 3 shows the flywheel configuration and the vacuum housing plus the steel protective system. The flywheel runs in a vacuum at 11,000 rpm. Wheel disc diameter is approximately 23 inches. Full safety protection is provided.

The basic flywheel system specifications are: useable energy storage: 2/3 hp-hour, maximum windage: 1 hp at maximum speed, over speed protection, and locked bearing protection.

The flywheel design is based on currently sponsored government research. Material is steel alloy, and the maximum torque transmission capability is 250 ft-lbs. The basic continuously-variable transmission (CVT) features are: ratio range - 3.5:1, power split principle for good efficiency, and torque control (that is, the driver's foot commands torque and the ratio is adjusted to provide the demanded torque). An internal spur gear differential is used for power splitting. Torque capability of the CVT is ± 400 ft-lbs. Top speed is designed for 80 mph vehicle speed. The CVT configuration is shown in Figure 4.

Figure 5 shows the vehicular controls, that is, the acceleration and braking controls of the CVT. The control system uses pressure feedback torque control for both acceleration and braking. This is accomplished by feedback of hydrostatic system pressures to a control valve which controls the continuously-variable ratio pistons. The particular design uses a single valve to accomplish both acceleration and braking. Maximum control torque is plus or minus 400 ft-lbs. This kind of control is necessary to minimize problems with stability.

In addition to the CVT controls there are other special controls and instrumentation required for driving this particular vehicle. These are shown in Figure 6. In addition to the standard operating equipment in a vehicle (i.e., the clutch, brake, shift lever, accelerator pedal, steering wheel and speedometer) there is an engine start clutch lever, a gear shift meter, gear shift indicators, a gear shift warning light, a flywheel speed meter,

a vehicle velocity function meter, a minimum charge warning light, a maximum charge speed warning light, and an engine start light. It should be pointed out that all these controls are easily automated, thus a production vehicle would only have accelerator and brake pedals and perhaps a flywheel charge meter.

The system has been designed using both an all-digital program and a real-time hybrid computer version of the program to test driver reaction to all the additional instrumentation and controls. All component losses have been calculated on a continuous basis in the simulation program. They have been modeled from test data obtained from component manufacturers.

The results of the simulations show that the standard 1976 vehicle can be modified to provide a 58% improvement in gas mileage with an ultimate developed potential of 117% improvement. An energy comparison table is shown in Table 1. Here we see three columns: the first column indicating the standard 1976 2.3 liter vehicle and its various energy consumptions and mileage on the bottom line. The second column (middle column) is the proposed vehicle discussed here and its various energy losses and utilization.

It is important to note that the flywheel and continuously-variable transmission use up about as much energy as was used up in the standard vehicle for deceleration and braking over the EPA-CVS cycle. It is further important to note that the idle and coast conditions use up as much fuel as is required to supply energy for the flywheel gears, the charge pump, excess brakes, engine clutch, engine inertia, and engine start. Thus, there is very little difference between the total amount of work done by the vehicles. The gain, however, is in the fact that the brake specific fuel consumption can be reduced from an average of .808 lbs. of fuel per hp-hour to .50 lbs. of fuel per hp-hour.

It should be pointed out that both vehicles are assumed to be of the same inertial weight. This is possible in a production vehicle because of system integration and component development.

The third column is important to concentrate on because it shows what the potential of the concept is. Note:

1. The four-speed transmission can be eliminated and its function included in a specially-designed continuously-variable transmission.

2. The flywheel system parasitic losses can be reduced by half by advanced design bearings and use of lighter weight wheels.

3. The losses in the continuously-variable transmission can possibly be reduced by half by application of new hydrostatic design concepts.

4. The charge pump can be designed, by use of a variable delivery system, to have approximately 1/6 of the loss now presently experienced.

These four improvements would allow one to obtain approximately 52 mpg for a 3,000-lb. car over the EPA-CVS city cycle as compared to the standard vehicle's 24 mpg for a 117% improvement.

We must look at the possible expected loss of fuel economy as a result of future new emissions standards. As the standards become more restrictive, the estimates for a standard vehicle and the hybrid flywheel vehicle are as shown in Table 2. It should be noted that a degradation in mileage occurs in both vehicles. These estimates are made by the automobile manufacturer involved in this particular project.

A continuous time simulation of the flywheel vehicle over the driving cycle is shown in Figure 7. The top trace shows the driveshaft torque. The 2nd trace shows the flywheel speed and the 3rd trace shows the engine speed. It should be noted that the engine is on at the beginning to give the flywheel its initial charge, and then comes on four more times during the cycle. The 4th trace is the vehicle speed over the EPA-CVS city driving cycle. The last trace shows the engine torque.

Note that the engine torque is not constant, but rather follows the wide-open throttle engine torque-speed characteristics. Note also that there is considerable energy left in the flywheel at the end of the cycle. A number of possible alternatives have been suggested to use this energy. One way may be to have a homing mode in which the last flywheel charge from the engine brings the flywheel to a speed corresponding to the distance required to travel which will be much less than the maximum. Another possible alternative is to use the flywheel energy at the end of the cycle to charge the battery of the vehicle. If we keep the battery only partially charged while driving, this technique seems entirely practical.

The second trace shows the effect of regenerative braking, in which the flywheel speed increases as the vehicle is braked. Regenerative braking continues until the transmission is no longer able to provide the proper ratio, which occurs at approximately 5 mph. Below this speed the regular vehicle brakes must be used.

There are four current program objectives:

1. To perform component tests to obtain verification of component data and to obtain new data where data are not currently available.

2. To investigate control techniques, stability of operation, ease of operation and computer control aspects of the vehicle control system.

3. To evaluate engine operation impacts on emissions, on temperature and heat control and on insulation requirements.

4. To compare theoretical and actual performance with respect to mileage and emissions.

After initial assessment, transmission and engine modifications may be required. The computer program will be used to assess the necessary modifications.

After the demonstration vehicle is constructed and tested, future programs will include continued investigation of advanced hydrostatic transmission concepts such as the use of balanced force systems to minimize bearing load and reduce parasitic losses, and of techniques to reduce pumping and flow losses. The objective will be to achieve maximum efficiency with a wide ratio range. The flywheel vehicle requires a ratio range determined by the maximum vehicle speed and minimum flywheel speed. Typically a ratio range of 20:1 is required. By use of some special techniques this can be reduced to about 10:1 as in the vehicle being constructed.

Other transmissions will also be investigated, such as an electric transmission, a multi-gear torque converter transmission, traction drive transmissions, and possibly other drive systems.

All of these continuously-variable transmission systems will be studied by simulation and designed for maximum fuel economy, mechanical simplicity and reliability. The objective will be a fully automatic transmission system. Vehicle size effects will be investigated with respect to 1500 lb., 3000 lb., and 10,000 lb. vehicles. System weight, of course, will be a primary parameter and must be considered in any design evaluation. Reliability must also be given a high priority.

The design and construction of an advanced continuously-variable transmission will then be undertaken. This will be based on a computer-optimized design using realistically measured dynamometer data for all components. The purpose of the construction exercise is to assess the control aspects, shifting criteria, clutch losses, ease of control, reliability and the feasibility of a fully automatic system.

Flywheel subsystems will continue to be investigated with respect to such vehicles. Of course advanced flywheels, that is composites, will be evaluated with respect to overall system impact. Other flywheel configurations will be considered. For example, a variable inertia flywheel whose use would eliminate the need for a continuously-variable transmission appears to be an interesting idea. However, the complexity of such a wheel and the weight impact may offset entirely the elimination of the CVT, thus the overall system must be studied for feasibility. It is anticipated that a flywheel subsystem will be incorporated into the continuously-variable transmission to give a single mechanical package. The flywheel subsystem, however, would be subcontracted.

Finally, the system will be installed in a vehicle and evaluated. Driveability of this automobile system will be assessed, measured fuel consumption and emissions data taken, and vehicular weight and cost impact will be evaluated.

CONCLUSIONS

It has been demonstrated by computer techniques that a vehicle containing an integral high-speed flywheel can have greatly enhanced fuel economy within emission constraints. Further, current vehicle performance levels can be completely retained.

The program described will evaluate practical consideration of flywheel systems in automobiles. It will concentrate on the most important aspect of the flywheel system, which is the transmission that connects the flywheel to the road. The main objective of the program is system overall efficiency. Thus, parasitic losses in all components are being minimized by careful system design. It should be noted that if the system is put together without extreme care to control the parasitic losses, it is very easy to obtain a vehicle with worse mileage performance than the standard unmodified vehicle. Finally, a carefully designed flywheel-engine system provides a real alternative for future transportation.

<u>Acknowledgements</u>

The following members of the University of Wisconsin research team have contributed to the results described in this paper.

Richard R. Radtke

Peter Ting

Tom Hausenbauer

Douglas Brooks

This project is sponsored by the United States Department of Transportation, Office of University Research, Contract No. DOT-OS-30112.

ENERGY COMPARISON TABLE

Table 1.

EPA-CVS CYCLE ENERGY COMPARISON

ITEMS	STANDARD 1976 2.3 LITER VEHICLE (HP-SEC)		1976 FLYWHEEL 2.3 LITER VEHICLE (HP-SEC)	POTENTIAL FROM CONTINUED R & D (HP-SEC)
Road Load	3700		3702	3702
Rear Axle	470		536	536
Transmission	648		648	200**
Deceleration and Brakes	2555	Flywheel	855	400
		CVT	1809	900**
Total (+) Work	7373	FW Gears	172	172
		Charge Pump	634	100
Idle & Coast Fuel 0.25 lb	1111*	Excess Brakes	50	50
		Engine Clutch	99	99
		Engine Inertia	93	93
		Engine Start	66	66
Total Work	8484		8664	6318
Fuel for (+) Work	1.655 lb		1.202 lb	0.876 lb
Fuel Total	1.905 lb		1.202 lb	0.876 lb
(+) BSFC	0.808 lb/HP-HR		0.50 lb/HP-HR	0.50 lb/HP-HR
Mileage	24.0 MPG		38.0 MPG	52.0 MPG
Improvement			58%	117%

*Equivalent Work Computed at 0.808 lb/HP-HR

**A Single CVT Package Will Replace Both Units

Table 2.

EXPECTED LOSS IN FUEL ECONOMY AS A FUNCTION OF EMISSION STANDARDS

	Emission Standards, gms/mi	2.3 Liter Standard Manual 24.0 mpg, percent loss	2.3 Liter Hybrid Vehicle 38.0 mpg, percent loss
1976 49 states	HC 1.5 CO 1.5 NO 3.1	---	---
1977 49 states	HC 1.5 CO 15 NO 2.0	4 - 5%	5 - 10%
1976 California	HC .9 CO 9.0 NO 2.0	11 - 12%	5 - 10%
1977 California	HC .4 CO 9.0 NO 1.5	24 - 33%	10 - 15%
1978	HC .41 CO 3.4 NO 0.4	Unknown	Unknown

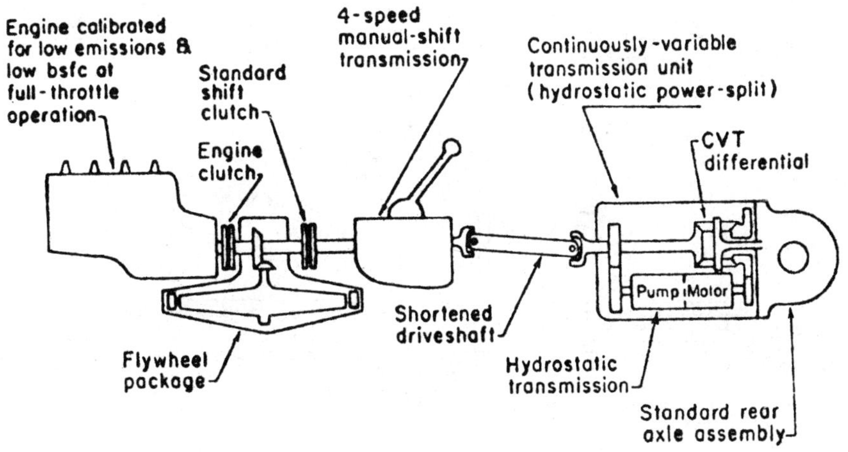

SYSTEM CONFIGURATION

Fig. 1.

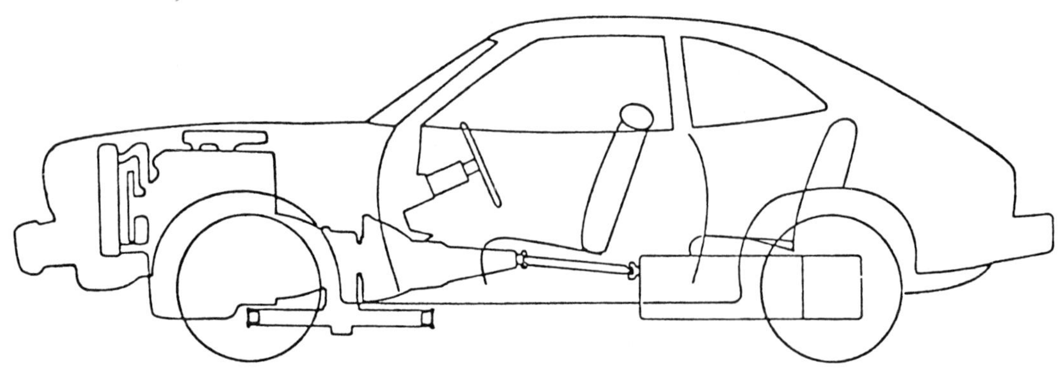

INSTALLATION SKETCH

Fig. 2.

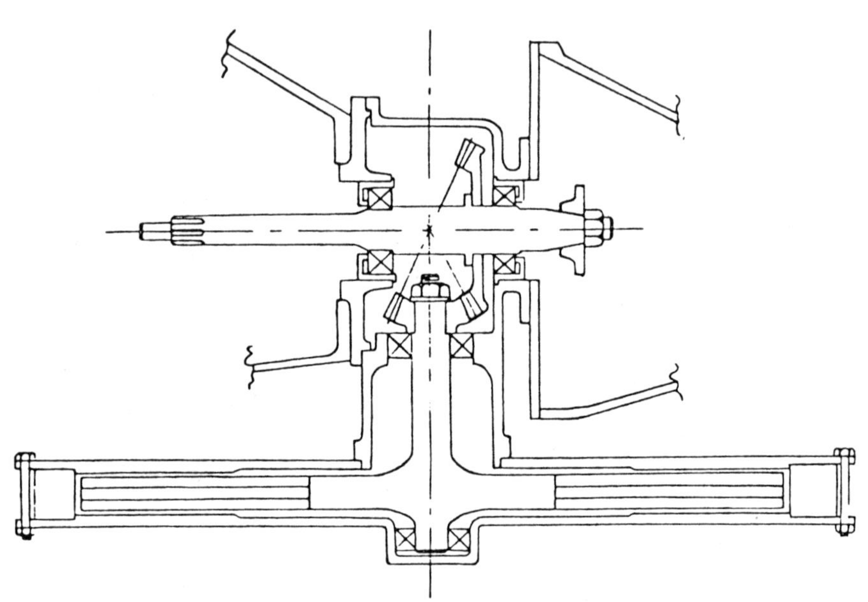

FLYWHEEL CONFIGURATION

Fig. 3.

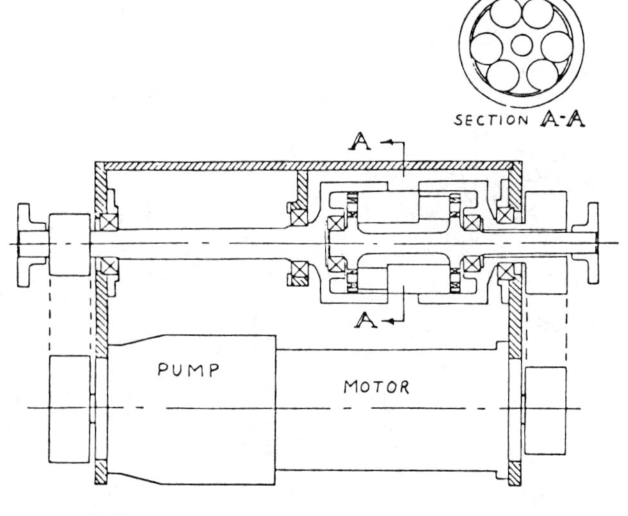

CONTINUOUSLY VARIABLE TRANSMISSION CONFIGURATION (CVT)

Fig. 4

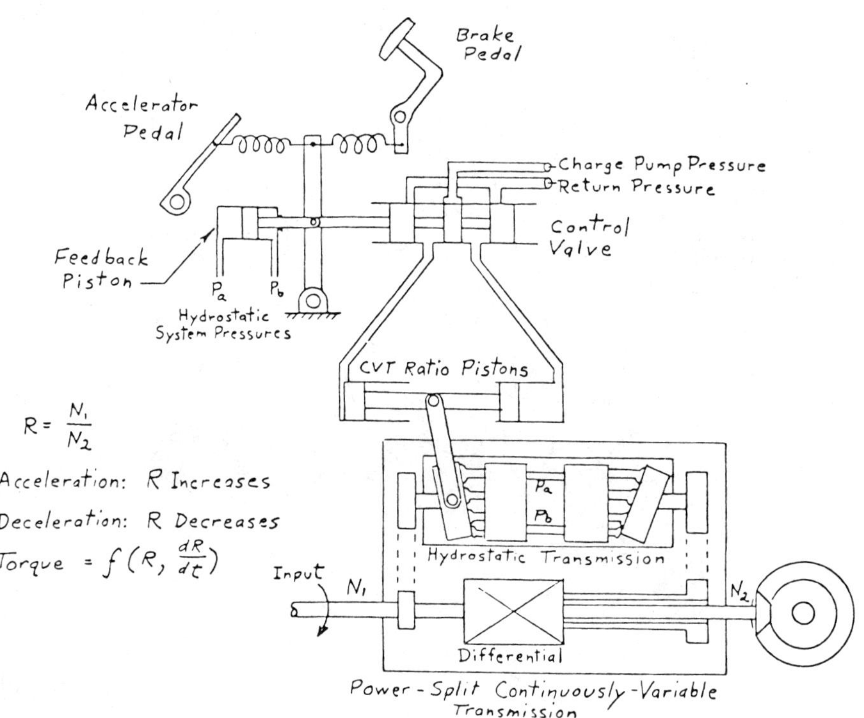

VEHICLE ACCELERATION AND BRAKING CONTROLS

Fig. 5.

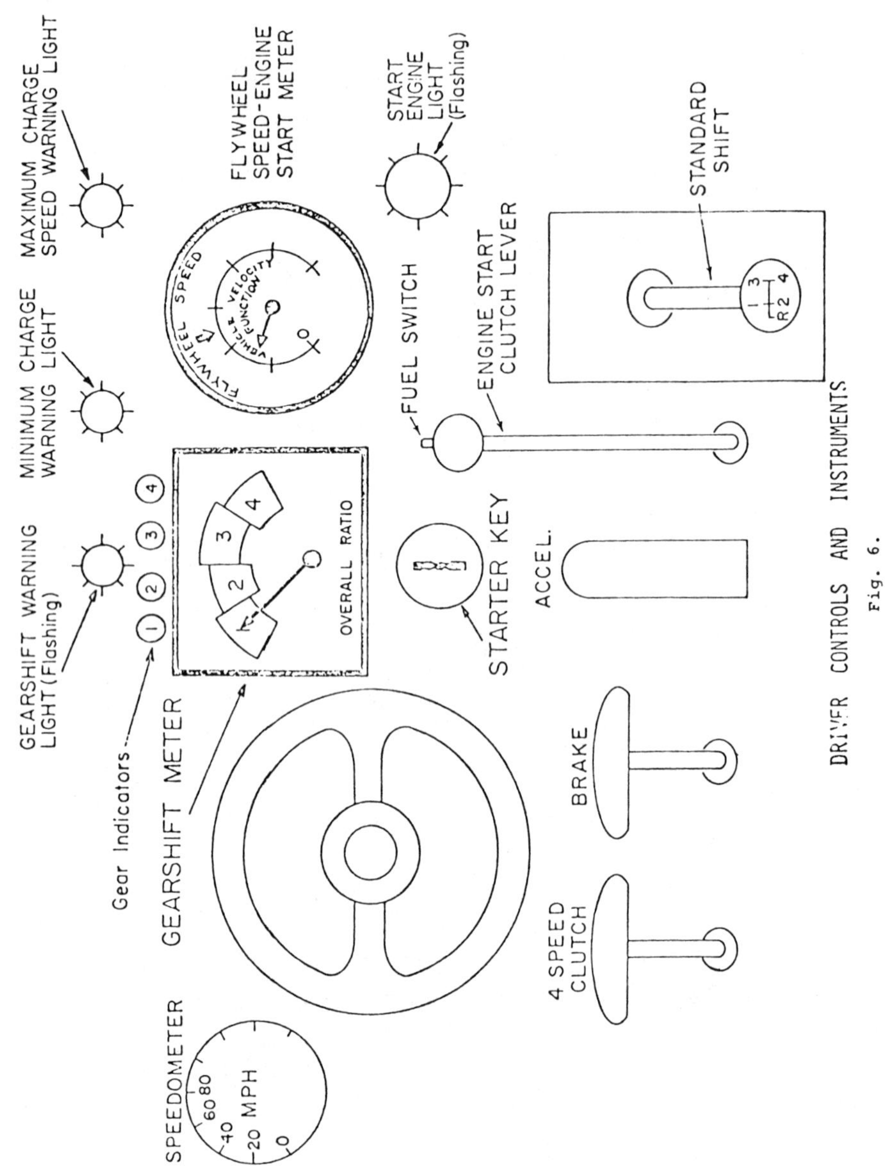

DRIVER CONTROLS AND INSTRUMENTS. Fig. 6.

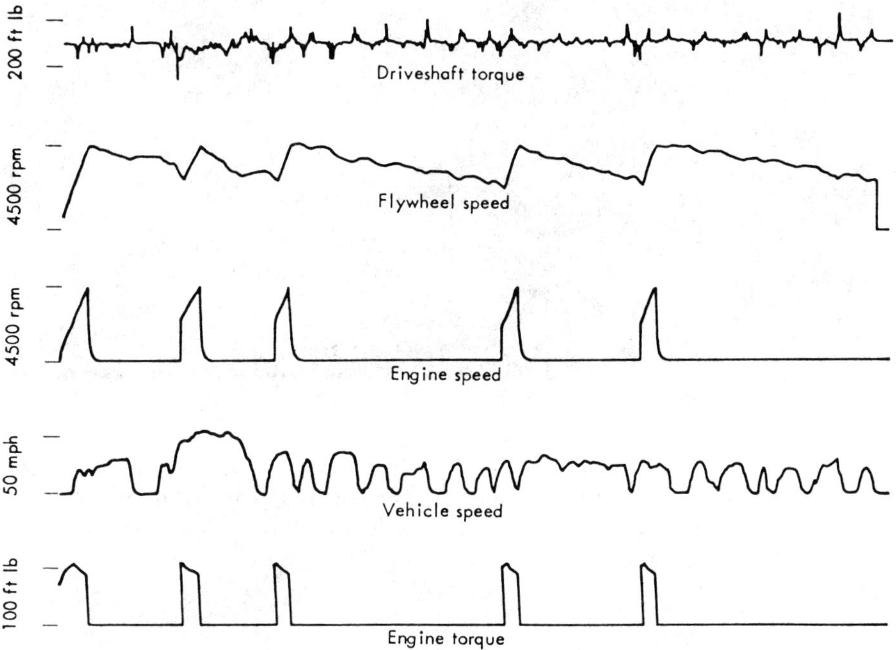

Fig. 7. Recorder traces of a flywheel
vehicle simulation over the
EPA-CVS city driving cycle

[*Editor's Note*: Photograph of the interior of the University of Wisconsin flywheel-assisted vehicle supplied by the authors. The flywheel housing is to the left and the special controls are in the center of the vehicle and on the control panel.]

23

Reprinted from pages 229–233 of *1975 Flywheel Tech. Symp. Proc.*, U.S. Energy Research and Development Administration and the Lawrence Livermore Laboratory of the University of California, 1975, 294pp.

ENGINEERING DESIGN DATA FOR COMPOSITE MATERIALS[*]

Linda L. Clements

Lawrence Livermore Laboratory, University of California

Livermore, California 94550

ABSTRACT

LLL has been studying some basic mechanical properties of filament-wound composites. These include elastic and ultimate properties in tension and compression for both longitudinal and transverse modes as well as ±45°-tensile shear properties.

Measurements of the properties of composites of Kevlar 49 in an epoxy matrix are complete and are reported here, along with limited comparative data from Kevlar 49 in a second epoxy matrix. Some preliminary data from measurements of E-glass/epoxy specimens are also included.

INTRODUCTION

Designing a flywheel requires a certain, important minimum of realistic engineering data. Chiao and Hamstad[1] concluded that for filament-wound composites the minimum should include both elastic and ultimate properties in tension and compression for both longitudinal and transverse modes, as well as in-plane shear data.

Recent studies at LLL have concentrated on measuring these properties for composites of Kevlar 49[**] and E-glass fibers in an epoxy matrix. Table 1 lists the three composite systems studied.

TEST DATA FROM THE KEVLAR 49/EPOXY A COMPOSITE

For Kevlar 49 in epoxy A, all of the above mechanical characterization tests have been completed. Figure 1 compares the resulting stress-strain curves schematically. As can be seen in Figure 2, in which the data are normalized to 65 vol% fiber, properties in tension and compression differ considerably. The ultimate stress and strain values obtained from elongated-ring longitudinal-tensile tests (1500 ± 230 MPa, 1.71% ± 0.25%)[†] are several times greater than those from flat-plate "Celanese" longitudinal compression tests (254.6 ± 3.5 MPa, 0.478% ± 0.026%). Furthermore, the elastic modulus in compression (73 ± 8 GPa) is significantly less than that in tension (87.6 ± 3.2 GPa), although the Poisson's ratio for compression (0.44 ± 0.09) is comparable to the tension result (0.39 ± 0.12). We feel that refinement of the compression test is necessary in order to verify these results.

As shown in Fig. 3, transverse properties of flat-plate specimens also differ, depending on whether they are measured in tension or compression. Here, however, the Celanese compression ultimates (53.0 ± 3.4 MPa, 1.41% ± 0.12%) are several times greater than the tension results (12.35 ± 0.38 MPa, 0.283% ± 0.011%). However, the secant modulii at 0.1% strain for tension (4.65 ± 0.17 GPa) and compression (4.5 ± 0.6 GPa) are comparable. Again, compression-test refinement is necessary.

---

[*]This work was performed under the auspices of the U. S. Energy Research & Development Administration under contract No. W-7405-Eng-48.

[**]Reference to a company or product name does not imply approval or recommendation of the product by the University of California or the U. S. Energy Research and Development Administration to the exclusion of others that may be suitable.

[†]All limits are 95% confidence limits.

Table 1. Composite Systems Studied.

Fiber	Nominal vol% fiber	Resin	Cure
Kevlar 49 – 1420 denier	65%	Epoxy A: Dow XD7818/ Jeffamine T403 (100/49)	16 h at 60°C 3 h at 90°C
Kevlar 49 – 1420 denier	65%	Epoxy B: Dow DER332/ Jeffamine T403 (100/39)	24 h at 60°C
E-glass – Type 30	70%	Epoxy B: Dow DER332/ Jeffamine T403 (100/45)	24 h at 60°C

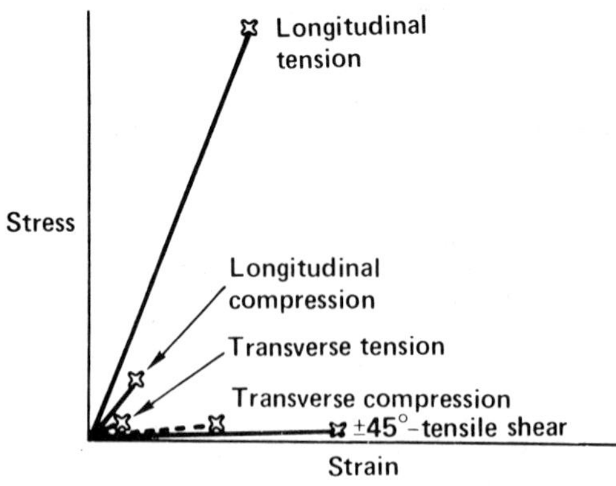

Fig. 1. Schematic comparison of various stress-strain curves obtained for Kevlar 49 in epoxy A.

Shear properties were obtained from the ±45°-laminate tensile test. While the exact significance of the ultimate properties measured in this type of test has not been established, the test does give good comparative values for different composite systems.[2] As shown in Fig. 4, the shear modulus obtained for a 58.3 vol%-fiber composite is 1.877 ± 0.028 GPa, and the ±45°-shear ultimates (taken at the inflection of the stress-strain curve) are 37.96 ± 0.36 MPa and 2.56% ± 0.06%.

KEVLAR 49/EPOXY B COMPOSITE

Limited data for comparison only was obtained for Kevlar 49 in epoxy B. Table 2 shows the ultimate strengths of elongated rings of Kevlar 49 in epoxy A and epoxy B. The epoxy B matrix seems to produce a composite with higher ultimate strength.

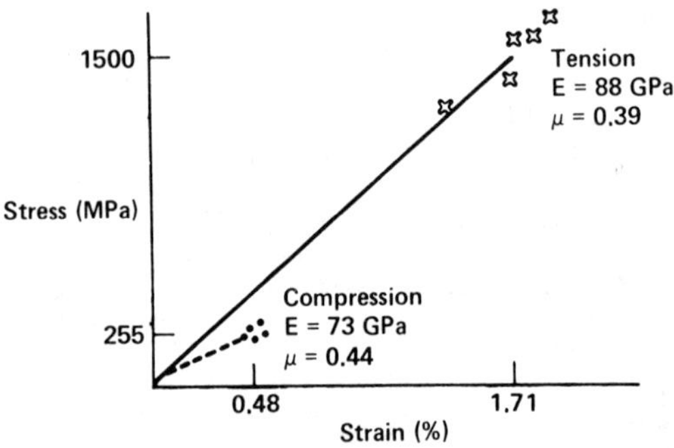

Fig. 2. Longitudinal properties in tension and compression for 65 vol% Kevlar 49 in epoxy A.

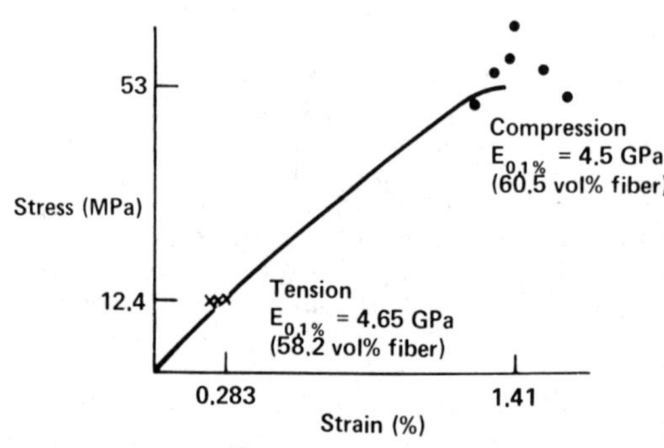

Fig. 3. Transverse properties in tension and compression for Kevlar 49 in epoxy A.

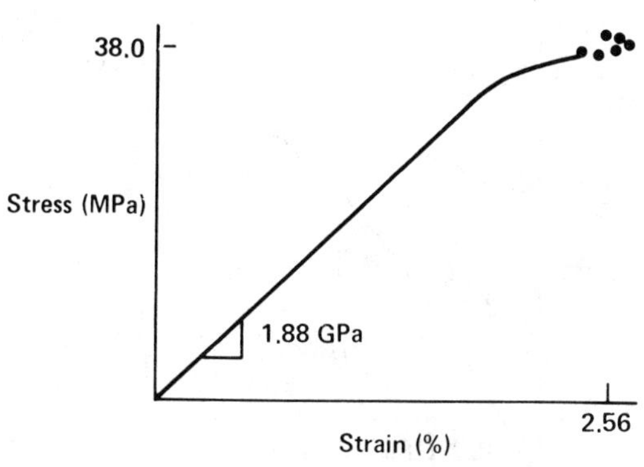

Fig. 4. Shear properties from ±45°-
tensile specimen for 58.3 vol%
Kevlar 49 in epoxy A.

The ±45°-tensile shear properties of
the two composites are compared in Fig. 5.
Although the shear modulus for the epoxy B
composite (1.85 ± 0.12 GPa) is comparable
to that for the epoxy A composite, the
ultimate stress and strain for the epoxy B
matrix (31.5 ± 1.1 MPa, 1.94% ± 0.09%)
are significantly less.

E GLASS/EPOXY B COMPOSITE DATA

Only preliminary results are available for E-glass in epoxy B. However, as
shown in Fig. 6, these include almost the
entire spectrum of test modes. Figure 7
presents the longitudinal properties,
normalized to a 70-vol%-fiber composite.
We see that the ultimate strengths in
elongated-ring tensile tests (1240
± 80 MPa, 2.30% ± 0.16%) are about twice
those measured in compression (570
± 120 MPa, 1.11% ± 0.27%). The tensile
modulus (69 ± 11 GPa) is greater than that
in compression (53.5 ± 4.3 GPa).

Figure 8 gives the transverse tensile
properties. The secant modulus values are
high (16.5 ± 4.9 GPa), but the ultimate
stress and strain are extremely low (5.9
± 1.0 MPa, 0.040% ± 0.018%). Since the

Table 2. Longitudinal tensile strengths
of 65 vol% Kevlar 49 in two
different matrices.

Matrix	Ultimate stress (MPa)
Epoxy A: XD7818/T403	1500 ± 230
Epoxy B: DER332/T403	1760 ± 110

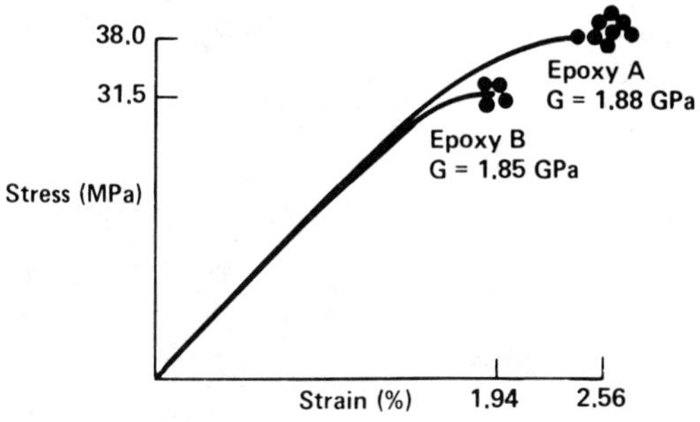

Fig. 5. Shear properties from ±45°-
tensile specimen for Kevlar 49
in epoxy A and epoxy B.

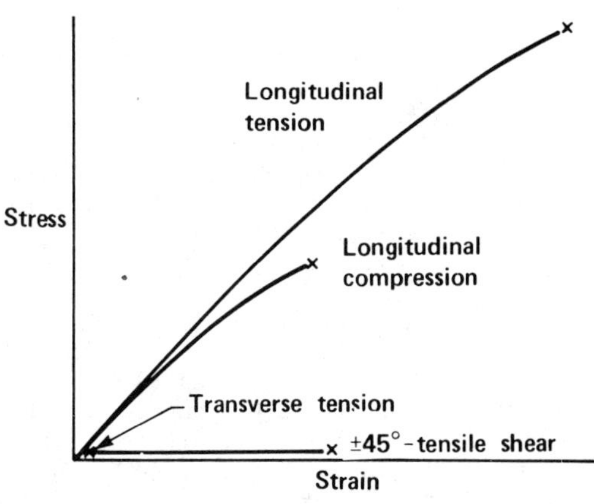

Fig. 6. Schematic comparison of various
stress-strain curves obtained
for E-glass in epoxy B.

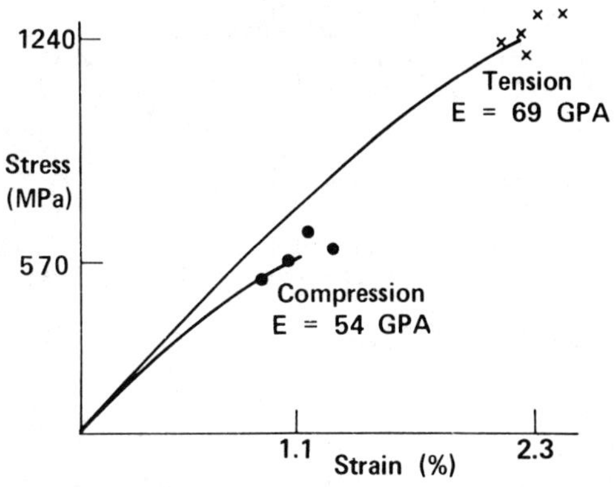

Fig. 7. Longitudinal properties in tension and compression for 70 vol%
E-glass in epoxy B.

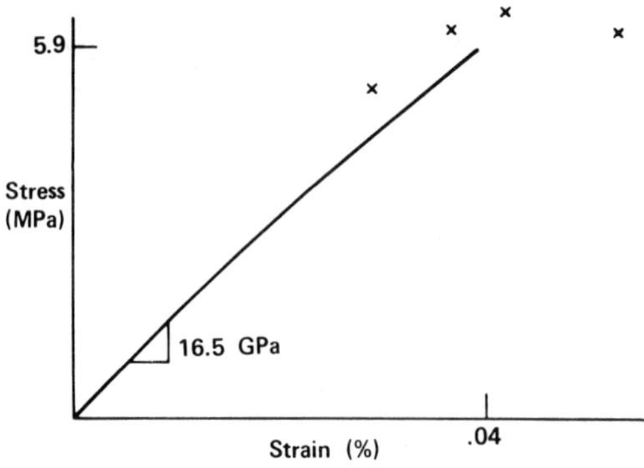

Fig. 8.   Transverse tensile properties for 67.6 vol% E-glass in epoxy B.

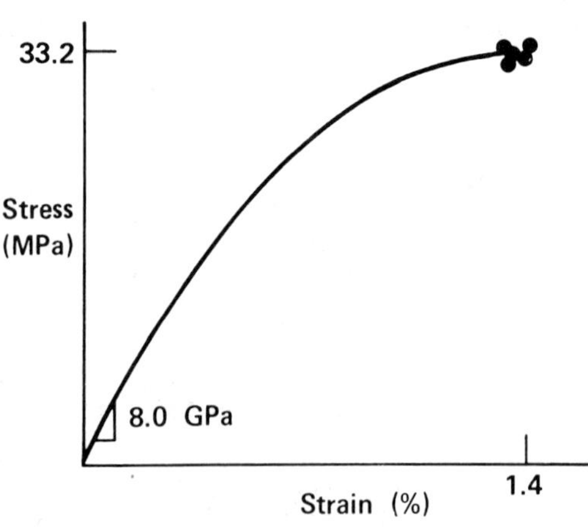

Fig. 9.   Shear properties from ±45°-tensile specimen for 69 vol% E-glass in epoxy B.

transverse tensile test is extremely sensitive to edge effects, the low values could be caused by an edge problem in the flat-plate specimen configuration. Additional transverse tensile measurements of filament-wound tubes will be made to verify these results.

Shear properties from ±45°-tensile specimens of E-glass in epoxy B are shown in Fig. 9. The shear modulus (8.0 ± 0.9 GPa) is quite high, compared to the Kevlar 49 composites. The ultimate stress and strain values in Fig. 9 require further refinement.

Kevlar 49 in epoxy B, but only prelimization for flywheel will include:

- Continuation of E-glass work, including more study of the 70-vol% fiber composite, study of 65- and 75-vol%-fiber composites, and determination of physical properties.

- Both mechanical and physical characterization of composites of other fibers (i.e., S2-glass, Kevlar 29) and other resins (i.e., more flexible resin).

- Study and refinement of test techniques, particularly those used for compression testing.

SUMMARY AND FUTURE PLANS

In summary, mechanical characterization data for engineering design were obtained in detail for Kevlar 49 in epoxy A. Limited work was done on

REFERENCES

1. T. T. Chiao and M. A. Hamstad, "Testing of Fiber Composite Materials," Lawrence Livermore Laboratory, UCRL-76365, Livermore, CA, Jan. 9, 1975.

2. C. C. Chiao, R. L. Moore, and T. T. Chiao; to be published.

24

Reprinted from pages 361–374 of *Symp. Fusion Tech., 8th, Proc.*, Commission of the European Communities, 1974, 1016pp.

LARGE FLYWHEEL POWER SUPPLY FOR FUSION EXPERIMENTS IN THE

MAX-PLANCK-INSTITUT FÜR PLASMAPHYSIK, GARCHING, GERMANY

by

A. Knobloch Max-Planck-Institut für Plasmaphysik, Garching
M. Kottmair Max-Planck-Institut für Plasmaphysik, Garching
W. Schlüter Siemens A.G., Erlangen
G. Vau   Siemens A.G., Erlangen

[*Editor's Note:* The photographs that comprise figures 4, 6, and 7 have been omitted.]

Introduction

The second range of pulsed plasma machines based on
the Stellarator and Tokamak concept represents a major
field in preliminary experimental steps in the development
of fusion reactors. The experimental set-ups under construction and in the planning stage at the Max-Planck-

305

Institut für Plasmaphysik in Garching necessitate electric
power of 100 to 200 MW and useful energy of up to 1500 MJ.
The characteristic pattern of the major electrical variables
during a load cycle can be seen from Fig. 1, which ex-
emplifies the power supply of the main field windings and
the helical windings for the Wendelstein VII Stellarator
experimental set-up in Garching.

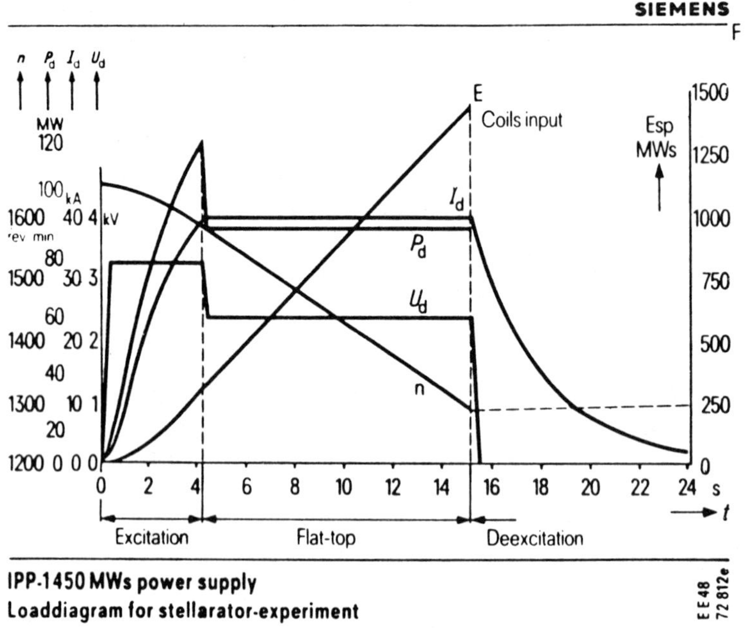

Under normal circumstances it is not possible to obtain
such power direct from the public supply system without
adopting special measures.

The magnitude of the voltage drops, changes in frequency and system harmonics caused by the pulsed power in
the system on the one hand and the resulting retroactive
effect on parallel loads and power station generators on
the other necessitate highly effective decoupling of the

power supply from the system and from parallel loads as
well. The two basic solutions for designing pulsed power
supplies are shown in Fig. 2. An assessment of the advantages and disadvantages of the two solutions, i.e. one
with an interposed energy store and the other with power
drawn direct from the system, depends very largely on local
factors such as the available system and the terrain.

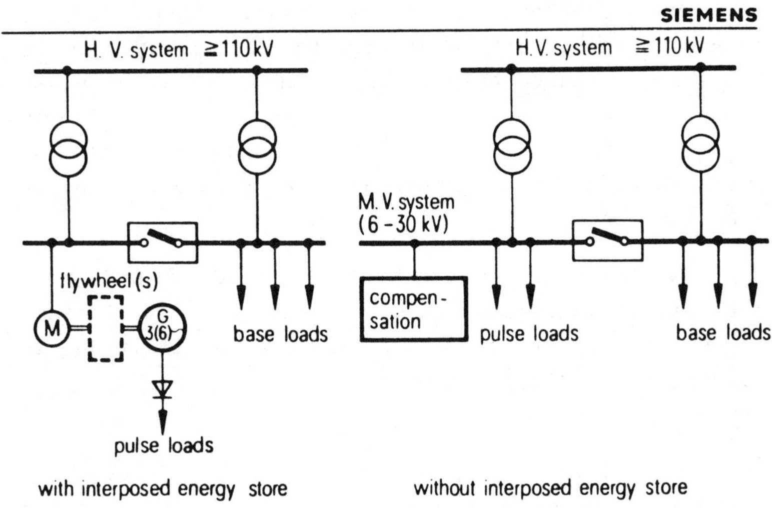

Possible power supply arrangement for pulse loads

The operation of the experimental set-ups which are
now being planned and rigged up at the Institut für Plasmaphysik necessitate power pulses of approximately 150 MW
with rise times within a matter of seconds and a flat-top
duration of 10 to 15 seconds. It is not possible at the
present time to obtain such pulse power direct from the
110 kV system available in the Garching area, which has a
short-circuit rating of approx. 2 GVA. An analysis of
various power supply equipment revealed that the employ-

ment of a motor-generator set represents the most economic
and reliable solution. A major advantage of this solution
lies in the relatively low drive power of 5.7 MW which can
be supplied from the existing 10 kV system without additional outlay, the maximum power fluctuations being
approximately 3 MW.

Further advantages are:

1. The main generator, which is of salient-pole design,
    permits the generator voltage to be largely matched to
    the load data because lower voltages can be attained
    without any intermediate transformer; this is due to
    the fact that the radial dimensions of the generator
    permit parallel-connected external part windings to
    be employed.

2. The selected generator frequency of 100 Hz and the cur-
    rent/voltage regulation of the load circuit via the
    generator excitation ensure favourable conditions for
    smoothing the d.c. current.

3. The use of a static cyclo converter in the rotor
    circuit of the slipring drive motor to control the
    speed and power enables a speed range to be attained
    which is ·higher than the synchronous speed and it also
    permits operation of the complete power supply at unity
    p.f. across the input without any additional power-
    factor correction equipment.

The experimental set-up is described in line with the
basic circuit diagram (Fig. 3).

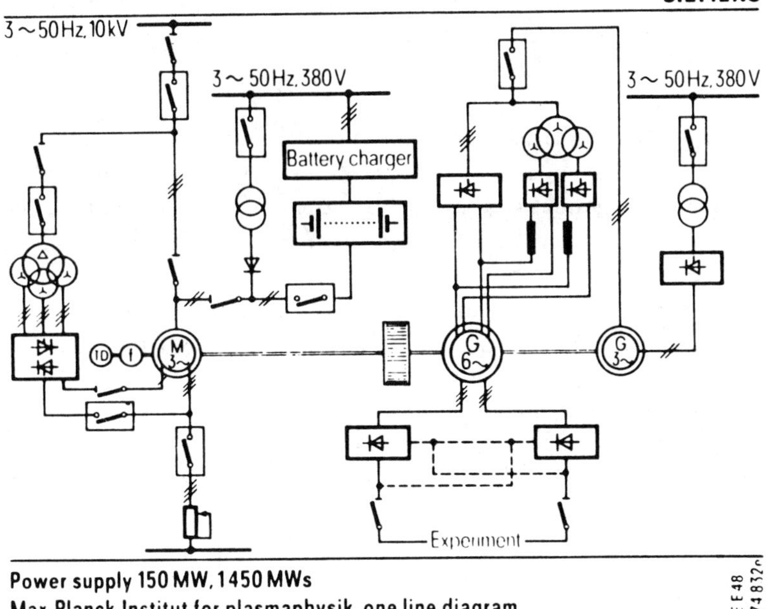

Power supply 150 MW, 1450 MWs
Max-Planck-Institut for plasmaphysik, one line diagram

Technical description

MG set

The MG set consists of a drive motor, flywheel, main generator and - for exciting the latter - a shaft-mounted
generator as well. The total length of the MG set is
approximately 22 m. The foundation frame is supported
resiliently on springs and vibration dampers.

The 4-pole 5.7 MW drive motor is of the induction type
with slipring rotor. Variable-speed or power control
of the drive is employed to decouple the pulse load from
the system. A static converter in rotor circuit enables
energy to be either tapped off or fed to the rotor, i.e.
subsynchronous and supersynchronous speeds can be set.
The required energy of 1450 MJ is drawn from the kinetic
energy of the flywheel by reducing the speed from 1650
to 1275 rev/min. The pulse load thrown on the system is
thus only that part of the power of the drive motor which
is proportional to the slip.

The main generator has two in-phase windings which each
produce a rated voltage of 2.75 kV. The peak power is
167 MVA and the respective current 2 x 17.5 kA.

All the winding ends are brought out. The generator is
excited by a converter which is fed from a special shaft-
mounted generator in order to relieve the system from the
surge-like excitation power of the order of 1.4 MVA.

The 8-pole generator is driven at twice the 50 Hz speed,
so that a frequency of 100 Hz + 10% and -15% is attained.

The shaft-mounted generator supplies the excitation power
for the experimental set-up. A partial current flows
through the auxiliary brushes and sliprings of the main
generator during the interval in each cycle in order to
ensure that the contact layer is maintained between the
brushes and the sliprings.

The remanence voltage is simultaneously reduced to zero
by reversing the voltage across the excitation winding in
order to make the experimental set-up off circuit during
the intervals. The shaft-mounted generator is also excited via a converter.

Rectifier

A diode rectifier is connected pro system on the output
side of the main generator. Each of these rectifier
groups, which are in three-phase bridge connection, consists of six cubicles. Fig 4 (below) shows a partial view.

The supply can be matched to changing load data by connecting the two groups either in series or parallel.
Each rectifier cubicle contains 5 air-cooled stacks
fitted with 3 diodes each of type SSi PO 1 100. Each
leg of a bridge thus contains four diodes connected in
series and ten connected in parallel.

The magnitude of the d.c. voltage per group is 3.3 kV
with a peak current of 22.5 kA for a flat-top duration of
10 seconds. Each leg is protected by fuses for selective
shutdown on an internal short circuit. I_S limiters are
used to protect the rectifier against a short circuit on
the bus or experimental set-up. The mechanical stress
placed on the bus in the event of such a short circuit is
thus reduced simultaneously.

Speed and Power Control

The frequency, amplitude and phase angle of the three-
phase supply fed to the rotor of the drive motor via the
cyclo converter can be varied. The rotor frequency
emitted by the converter represents the difference between the system frequency and the particular rotating-
field frequency, so that the motor behaves like a synchronous machine, but without the disadvantages of the
latter, such as a tendency "hunt" and possibly to fall
out of phase. The amplitude and phase angle of the rotor
three-phase supply can be adjusted by two independent
control circuits for active and reactive power. The output variables represent two partial setpoints of the
supply system which are phase-displaced by 90°. For
continuous pulse operation, the current setpoint in the
active axis corresponds to the constant active power
consumed while that in the reactive axis corresponds to
the magnetizing reactive power of the induction motor.
The motor is thus magnetized from the rotor, the reactive
power from the system is zero and the power factor at
unity. A variable-speed control circuit is superimposed
on the reactive power control circuit; this speed-control
circuit not only reduces the speed to the desired working one but also the active power as well if the motor
is operated in the "waiting" position at the lower speed
limit.

Voltage regulation

It is only possible to adjust the load current by altering
the generator voltage through the excitater. Voltage regulation with excitation current cut-off is therefore employed to attain a certain current pattern, particularly
during a sudden transition from initial excitation to
flat-top. Appropriate over-excitation ensures favourable
transient behaviour, i.e. rapid adjustment of the generator
voltage to the desired setpoint.

The most important technical data are tabulated in Fig. 5 below.

SIEMENS

DC - VALUES		MOTOR - GENERATOR - SET	
Peak current	45 kA	Rated - speed	1500 rpm
Peak voltage	3,7 kV (at no load)	Speed range	1650 - 1275 rpm
Peak power	150 MW	Moment of inertia	1000 tm^2
Number of groups	2	Total weight	4,0 MN
Repetition time	≥ 6 min	Useful energie	1450 MWs
LOAD		**GENERATOR**	
Inductance	135 mH	Voltage	2 x 2,75 kV
Resistance	30 m Ohm	Current	2 x 17,5 kA
Time constant	4,5 s	Frequency	100 Hz + 10% / - 15%
Rise time	2,4 s	Peak power	167 MVA
Duration of flat top	10 s	Subtransient reactance	0,15 p.u.
		MOTOR	
		Voltage	10 kV
		Power factor	1
		Rated output	5,7 MW

IPP Garching principal technical data and parameters of the power supply

E E 48 74849e

Distribution of the d.c. power to several experimental set-ups

The power supply system dealt with here is fed to several experimental set-ups in the Institut für Plasmaphysik (e.g. Wendelstein VII and the Tokamak "Asdex" set-up) which - for various reasons - are not located immediately next to the power supply. The problem of transmitting the energy (max. 45 kA and 6.6 kV) over distances of 200 to 300 metres from the power supply plant to the set-ups thus arises. Special measures also had to be adopted to contend with the peak short-circuit current of the generator, which approaches a value of 1000 kA. Various forms of d.c. lines were analysed. The most favourable

solution proved to be an arrangement with parallel aluminium
buses with a rectangular cross section of 250 x 80 mm which
are laid in reinforced concrete ducts. To reduce the mechanical forces developed on a short-circuit, the I_S
limiters already mentioned - with parallel-connected extra
fast response fuses - were fitted between the generator
and rectifier. The short-circuit current is thus limited
to about 300 kA and elaborate bracing of the d.c. buses
avoided because the mechanical forces developing on a
short-circuit are absorbed by the duct walls. The losses
during a pulse at 45 kA are 7.2 kW per metre of double-
laid bus (on an average 288 W/m), so that natural cooling
suffices. With a maximum bus length of 350 m the line
losses are approximately 1.6% of the peak power and for a
200 m length 0.93%.

Flywheel

Since the flywheel is a major component it is dealt with
more closely. The flywheel has an overall weight of 223
tons, a total length of 5.7 m and a diameter of 2.9 m; it
can either be made of one piece or of several. Analyses
have shown that for a rated speed of 1500 rev/min a flywheel made of several forged discs is a better solution
both with respect to reliability as well as to the possibility of rejection because of defects. The multi-piece
flywheel consists of 4 forged discs stacked side by side
with an internal borehole. A centre bolt, which is
threaded at each end, passes through this borehole. The
two shaft ends are designed in the form of nuts which -
via the centre bolt - press the 4 discs together with
appropriate pretension. Apart from the volume of the
flywheel the energy which can be derived from the latter
depends solely on the speed range and the strength of the
material. The speed determines the magnitude of the
alternating stresses to which the material of the flywheel is subjected. It therefore determines the life of
the flywheel. The dimensioning in turn depends on the

endurance limits of the material selected. Owing to the
type of operation, the flywheel material is more susceptible to faults in the area closer to the axis than similar
forgings which are subjected to constant stressing. The
change in speed which occurs with every stress pulse leads
to tensile stress which changes as the square of the speed,
which is the result of the natual centrifugal force of the
flywheel. In constrast to workpieces subjected to constant
stressing, the flywheel must therefore be dimensioned in
line with the fatigue strength, allowing for ultimate
stress, and not exclusively with regard to the yield point

The concept of material fatigue strength also offers the
possibility of performing ultrasonic quality tests during
commissioning and subsequent operation.

Transporting such a flywheel naturally presents problems.
Fig. 6 shows the flywheel on the way to Garching.

It was moved on a supporting frame by rail and road transport. The next picture - Fig. 7 - shows the flywheel being mounted in the bearings at Garching by two 120 ton
cranes.

Fig. 8 shows the complete MG set during erection.

Commissioning

Erection and commissioning of the entire power supply were
finished on schedule. In particular, the ultrasonic measurements taken on the flywheel do not vary from the calibrated
measurements; the tests were conducted after commissioning
at different speeds, including maximum speed. The scheduled
load tests will be conducted when the rated load values or
available.

<u>Facilities for extending the plant</u>

The entire plant has been designed so that the power supply
section can be enlarged at any time without difficulty.

Such an enlargement would involve extending the plant in
order to increase the power rating and also uprating of
the useful energy available by connec-ting one or several MG
sets in parallel.

Fig. 9 shows some of the possible ways in which the MG
sets could be combined, the amount of energy which can be
drawn off at present being governed by the flywheel weight.
The maximum weight of a steel flywheel which it is now
possible to make is approximately 300 tons. Air-cooled
generators for supplying fusion experimental set-ups can
be made with individual ratings of more than 400 MVA.

SIEMENS

- Variants	Number of MG-sets	Number of flywheels	principal arrangement	stored/ useful energy MWs	remarks
1	2	2 x 2	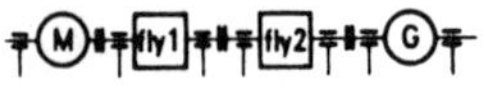 M—fly1—fly2—G	12220/ 4900	only toroidal field
2	2	2 x 2	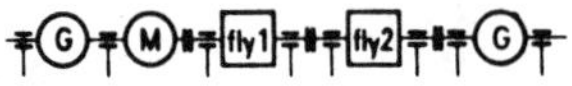 G—M—fly1—fly2—G	13750/ 5500	toroidal +poloidal field with separate generator
3	1	3	 M—fly1—fly2—fly3—G	12220/ 4900	only toroidal field
4	3	3 x 1	 M—fly—G	13750/ 5500	only toroidal field

Variants for power supplies with MG-sets with flywheels

EE 48 74847e

Part V

THERMAL ENERGY STORAGE

Editor's Comments
on Papers 25 Through 29

The technology of thermal energy storage has only recently been developed to a point where it can have a significant effect on modern technology. Many historical instances of the use of thermal energy storage come to mind, however. The major non-technical use of thermal storage was to maintain a constant temperature in dwellings, usually to keep them warm during cold winter nights. Large stones, blocks of cast iron, and ceramics were used to store heat from an evening fire for the entire night. The materials and techniques for this type of energy storage have remained essentially unchanged for several thousand years.

Dwellings need not be heated by fires to make effective use of thermal storage. In southern temperate zones, where the sky is clear in winter, the houses are designed so that the thick walls, frequently made of mud or adobe blocks, can retain the heat from the sun throughout the night.

With the advent of the industrial revolution, thermal energy storage existed as a by-product of the energy production and conversion process even though some thermal energy storage existed in the boilers that produced steam for steam engines.

A variety of new techniques of thermal energy storage has become possible in the past four or five decades as the industrial countries have become highly electrified. The two major applications for thermal storage today are at the site of electric power generation and in single- or multiple-family dwellings. Heat storage at power plants typically is in the form of steam or hot water and is usually for a short time (1 hr). Very recently other materials such, as oils having very high boiling points, have been suggested as heat storage substances for the electric utilities. Other materials that have a high heat of fusion at high temperatures, mainly special salts, have also been suggested for this application.

Another application of thermal energy storage on the electric utilities that has received widespread attention in Europe is to provide hot water throughout the day by heating rocks or ceramics at night and then heating water by passing it through tubing in contact with the rocks.

Perhaps the most promising application of thermal energy storage is for solar-heated residential and commercial structures. Occasionally these buildings are also cooled via solar energy. Almost any materials can be used for thermal energy storage. Solar-heated houses, for example, have used rocks, water and other liquids, hydrides, oil, and heat-of-fusion salts for heat storage.

Two particular problem areas for thermal energy storage systems are the heat exchanger and, in the case of phase-change materials, the method of encapsulation. The heat exchanger must be designed to operate with as low a temperature difference as possible to avoid inefficiencies. In residential heating systems, however, the heat exchanger is not so critical as it is in high-temperature applications such as electric power plants.

Proposed applications for thermal energy storage include on-board storage for vehicle propulsion via a Sterling or Brayton cycle engine, diurnal storage for solar electric power plants and for commercial process heat, the accumulation of ice in winter to be used to cool houses in summer, and recuperation of waste heat from industrial processes for use in residential heating. Another application of thermal storage may extend this technology to the energy-conversion and energy-transmission sector. In this technique heat is used to drive a reaction with two or more end products. These products may or may not need to be separated, but they are stored and transmitted to a site where the inverse reaction is allowed to proceed and thereby produce heat again. Thus, one has in effect ambient temperature storage of heat. Reactions based on methane, sulfur dioxide, and hydrides have been proposed for this type of storage.

The potential use of oil as a diurnal energy storage medium for large nuclear and fossil-fired power plants is described in Paper 25. The advantage of this application of energy storage is that it separates the thermal power system, which would be fatigued by frequent thermal cycles, from the electrical generator, which is not susceptible to damage from fluctuating power demands. The thermal power system is permitted to run at nearly constant power, and any unneeded energy is fed into the thermal storage. The electrical generators draw energy from the thermal storage as needed to meet the fluctuating demands for electrical power. Paper 26 examines the underground storage of pressurized hot water in association with a thermal power plant. The water must be stored deep underground so that the hydrostatic pressure will keep the hot water from boiling. Paper 27 contains a great deal of information on the possibilities of heat storage. The excerpt included here gives some details of the characteristics of heat of fusion materials and includes a rather complete bibliography on thermal energy storage. Paper 28 looks at a very interesting energy-storage concept in which heat pumps are used in winter to heat houses, and the resulting low temperature discharge, which is actually colder than ambient, is used to freeze water that is stored until summer, when it is used for air-conditioning. The many unique features and the application to a complete annual system make this technique quite interesting and instructive to study even though at present it does not appear to be economical. Paper 29 describes the use of metal hydrides (which were discussed in Part III where the storage of hydrogen was the goal) as a thermal storage medium where the heat of formation of the hydride is used thereby avoiding large temperature differences and the subsequent losses.

BIBLIOGRAPHY

Anon. 1974. *Phase-Change Materials Handbook.* NASA Tech. Brief B72-10464.

Balcomb, D., J. C. Hedstrom, S. W. Moore, and B. T. Rogers. 1975. *Solar Heating Handbook for Los Alamos.* Los Alamos Sci. Lab. LA-5967.

Borucka, A. 1975. Survey and Selection of Inorganic Salts for Application to Thermal Energy Storage. *ERDA Rept. ERDA-59.*

Bramlette, T. T., R. M. Green, J. J. Bartel, D. K. Ottesen, C. T. Schafer, and T. D. Brumleve. 1975. Survey of High Temperature Thermal Energy Storage. *Sandia Lab. Rept. SAND-75-8063.* (Contains a compete set of references.)

Droege, J. W., D. W. Locklin, E. D. Grasselli, and J. A. Eibling. 1958. *A State-of-the-Art Survey of Methods of Heat Storage for the Heat Pump.* Report to the Joint AEIC-EEI Heat Pump Committee.

Gilli, D. V., and K. Fritz. 1971. Nuclear Power Plants with Integrated Steam Accumulators for Load Peaking. *IAEA Symp. Nucl. Power, Oct. 1970, Proc.,* pp. 79–93.

Glenn, D. R. 1976. Technical and Economic Feasibility of Thermal Energy Storage. *General Electric Corp. Annual Rept. C00-2558-1.* Energy Res. Dev. Admn.

Hale, D. V. 1971. *Phase Change Materials Handbook.* NASA Rept. NASA-CR-61363.

Janz, G. L. 1967. *Molten Salts Handbook.* Academic Press, New York.

Kovack, E. G., ed. 1976. *Thermal Energy Storage.* Rept. NATO Sci. Committee Conf. Science Affairs Division, NATO, Brussels, Belgium.

Lane, G. A., J. S. Best, and E. C. Clarke. 1975. *Solar Energy Subsystems Employing Isothermal Heat Sink Materials.* Project Repts. Dow Chemical Corp. Supported by NSF-RANN and ERDA.

Libowitz, G. C. 1974. Metal Hydrides for Thermal Energy Storage. *Intersoc. Energy Convers. Eng. Conf., 9th, Proc.,* Am. Soc. Mech. Eng., pp. 322–325.

Maru, H. C., J. F. Dulles, and V. S. Huang. 1976. Molten Salt Thermal Energy Storage Systems. *Salt Selection Inst. Gas Tech. Rept. C00-2888-1.*

Naumes, P. A., and J. L. McCabnia. 1966. Solar Thermoelectric Power Conversion Coupled with Thermal Storage for Orbital Space Applications. In *Space Power Systems Engineering,* ed. G. C. Szego and J. E. Taylor. Academic Press, New York, pp. 851–868.

Phillips, W. F., and R. A. Pate. 1974. A Hot Liquid Energy Storage System Utilizing Natural Circulation. *1974 ASME Winter Meet. Paper 74-WA/HT-16.*

Schroeder, J. 1974. Energy Storage in the Form of Heat. *Umschau* 5:155–156.

Southeastern Conf. Appl. Solar Energy, 2d, Proc. 1976. Baton Rouge, La.

Sundheim, B. R., ed. 1964. *Fused Salts.* McGraw-Hill, New York.

Telkes, M. 1955. Solar Heat Storage. In *Solar Research,* ed. F. Daniels and J. F. Duffie. University of Wisconsin Press, Madison, Wisc.

Wessling, F. C., Jr. 1974. Thermal Energy Storage in Adobe and in Stone Structures. *1974 ASME Winter Meet. Paper 74-WA/HT-15.*

Wilden, M. W., S. F. Gilman, E. R. McLaughlin, and F. H. Bridgers. 1976. Experimental Results for a Heat Pump System with Thermal Storage. *Penna. State Univ. Rept. C00-2704-3.*

25

Reprinted from pages 598–605 of *Intersoc. Energy Convers. Eng. Conf., 11th, Proc.,*
American Institute of Chemical Engineers, 1976, 1802pp.

STORAGE IN OIL OF OFF-PEAK THERMAL ENERGY FROM LARGE POWER STATIONS

Edward W. Nicholson
Exxon Enterprises Inc.
New York, NY

Robert P. Cahn
Exxon Research and Engineering Co.
Linden, NJ

Abstract

A novel method of storing and recalling off-peak thermal energy from large power stations using a high-boiling refined oil as storage medium is described, and the economics of the system are developed and compared with competitive energy storage techniques. The effect of storage medium cost is analyzed, and potential methods of improving the economics of the overall system are outlined. Specific applicability to various nuclear reactor types is discussed, and other potential uses of the method are presented.

INTRODUCTION

A method has recently been disclosed[1] for storing heat derived from off-peak operations of nuclear or fossil-fuel fired electricity generating stations in oil prior to its conversion into electricity. The thermal energy thus stored can be recalled with high efficiency during peak demand periods. Atmospheric pressure tankage can be used to store the hot and the cold oil. This system was developed in the course of extensive studies by Exxon Enterprises Inc. on alternative techniques for energy storage in electric utility operations.

Storage is achieved by increasing the steam extraction from the turbine during off-peak periods resulting in a drop in power output. The steam heats oil through heat exchangers as the oil passes from cold oil tanks to hot oil tanks. During peak demand periods, the oil flow is reversed and the stored heat is used to preheat boiler feed water, releasing extraction steam to do useful expansion work in the turbine, raising power production to the required level. While these electrical output changes are taking place, the power plant's boiler runs under constant conditions.

A somewhat similar scheme is being studied in France[2] by Electricite de France in conjunction with several companies, but they are considering a more expensive material for thermal storage capable of withstanding higher temperature conditions than the medium being studied for the present applications. The cost and availability of the thermal energy storage material has a major effect on the economics and range of application of such a scheme.

This will be brought out in the present paper, which will cover a number of possible schemes for optimizing this storage concept. In addition, various hardware considerations will be discussed, indicating how the system is particularly well adapted to the present generation of nuclear reactors with their relatively moderate temperature levels. Use of the system in connection with supercritical fossil-fuel fired units, however, could also be of interest.

THE EXXON HEAT STORAGE SYSTEM

1. Flow Plan and Method of Operation – In the Exxon Heat Storage (EHS) system, a highly refined, high-boiling hydrocarbon oil is used to store excess thermal energy produced during low-load periods in a power plant which is running at constant thermal output. Water is used as the transfer medium to transport the heat between the powerhouse and the oil storage.

While many possible transfer configurations are possible, a typical scheme is shown in Figure 1.

EXXON HEAT STORAGE SYSTEM

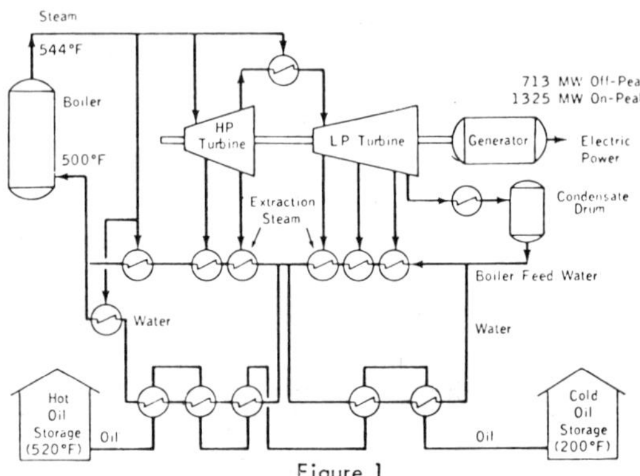

Figure 1

During periods of low load, a higher than normal water stream is sent through the boiler feed water (BFW) exchangers. This, in turn, leads to an increase in the steam extracted from the turbine at all levels, resulting in a drop in turbine-generator output. The preheated water in excess of the constant BFW requirements, after some additional heating with high pressure steam, is used in countercurrent heat exchangers to reheat oil which is being pumped from a cold oil tank to a hot oil tank. It should be noted that this can be run as a completely steady state operation, depending on the oil flow rate, as long as cold oil is available for this purpose. All conditions of operation can be held at constant, predetermined levels, i.e., hot oil temperature, BFW temperature, power output, etc.

When the oil flow is stopped, water flow through the BFW preheaters reverts back to BFW demand level, and the power plant output rises to its normal level for as long as desired.

As power demand increases, BFW flow can be diverted from the normal extraction-steam fed BFW preheaters into the oil-water exchangers. Then, by pumping oil through the exchangers from the hot oil storage tank to the cold oil tank, an increasing fraction of the BFW preheat duty can be taken over by the hot oil. As a result, steam extraction rates from the turbine decrease, with a consequent increase in turbine-generator output. Again, output can be held at a constant level, or varied as desired, as governed by the fraction of the BFW diverted to the hot oil/water exchangers.

This system is entirely flexible, and electrical output can be varied from maximum (by heating boiler feed water with hot oil) to minimum (by reheating oil with hot boiler feed water) while the boiler is operating under completely steady conditions. The rate and temperature of the BFW to the boiler are held constant, and high pressure steam production will not vary.

2. Major Alternates - There are many alternate ways of interconnecting the powerhouse and the oil storage facilities. The most direct is to heat the oil during the off-peak periods with extraction steam in countercurrent heat exchangers adjacent to the turbines, and then to use these same exchangers to heat the boiler feed water with the hot oil passing in the reverse direction. This scheme is probably lowest in cost, but it carries with it potential problems of contamination and other hazards which make it a less desirable engineering choice.

At the other end of the spectrum, the powerhouse and the oil storage facilities can be interconnected by closed loops of circulating heat transfer fluids, preferably water. This, of course, essentially eliminates all possible contamination, but introduces additional heat exchangers, temperature losses and consequent inefficiencies.

The specific scheme selected for the present study and illustrated in Figure 1 represents a reasonable compromise which allows for physical separation of powerhouse and oil facilities without incurring excessive costs and efficiency losses.

3. Capacity and Efficiency of System - In its most direct application, the heat stored in the oil is used to heat the boiler feed water to the desired boiler feed temperature. Depending on this temperature, up to about 25% of the output of a large nuclear station can be stored and recalled by this method. Of course, it is really not the output, but the input to the turbine-generator which is being stored.

In the previous paper[1], the relationship between boiler feed water temperature and fraction of storable power was covered in some detail. As will be seen from Figure 2, this fraction is very sensitive to BFW final temperature, rising from below 15% at 425°F to 25% at 500°F. In addition to BFW heating, however, excess off-peak heat can be stored for other purposes. One of these is interstage steam reheat between the high and low pressure

turbine stages. Another purpose could be to use the stored heat to generate intermediate pressure steam for additional power generation during high-load periods, or to drive power plant auxiliaries during that time. Both of these are ways to increase the fraction of storable energy.

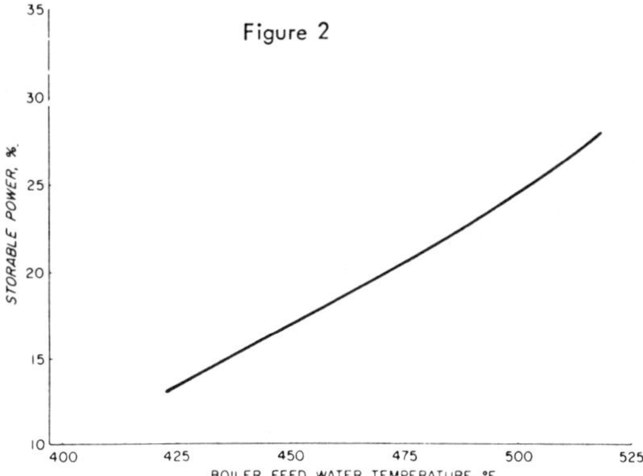

The fraction f of storable energy, together with the efficiency e, determine the peak-to-valley ratio which can be satisfied with this scheme when coupled with a constant thermal output nuclear or fossil-fuel station. If fraction f of a powerhouse's output is stored, only 1-f will show up during the off-peak period. When demand is high, the stored energy fraction f can be recalled with efficiency e, raising the plant's output to 1 + fe, so that the peak/valley ratio becomes (1 + fe)/(1 - f). For the EHS system, an overall efficiency of 73.4% has been calculated, giving a peak/valley output ratio of over 1.8/1 at a 500°F BFW temperature. Adding in storage for interstage steam reheat raises this ratio to above 2/1. Generation of intermediate pressure steam can raise this ratio to any value desired because, if desired, all the heat generated in the boiler can be stored in the oil.

ECONOMIC AND DESIGN CONSIDERATIONS
1. Results of the Exxon/General Electric Study - A joint study was arranged by Exxon with the Steam Turbine Generator Products Division of the General Electric Company to evaluate the feasibility and economics of the concept of storing excess thermal energy from a large power station in a well-refined petroleum fraction. Exxon handled the design and cost estimating of the oil and oil-handling facilities, including the large heat exchanger train required. GE studied the feasibility, efficiency and cost of the large turbines required and the other powerhouse and concomitant electrical facilities. The results of this joint study, as applied to a 1043MW light water reactor (LWR) nuclear power plant of the pressurized water type (PWR), are summarized in Tables I, II, and III.

Table I

Exxon Heat Storage System -- Study Basis

Type of power plant	Nuclear--LWR-PWR type
Base case capacity, MWe	1043
Live steam, P psia/T°F	1000/545
BFW temperature, °F	500
Oil temp., hot/cold, °F	520/200
Thermal energy stored	BFW preheat

Table II

EHS -- Performance

Capacity of nuclear station without added storage facilities	1043MW
Base capacity of system with storage facilities	1066MW
Minimum output during charging of storage	713MW
Maximum output during discharge of storage	1325MW
Energy charged into storage	3533MWh
Energy delivered during discharge over 10 hours	2594MWh
Thermal efficiency for storage cycle	73.4%

Table III

Total EHS System Investment

Incremental Costs Above 1040 Megawatt Nuclear Power Station
(1975 Dollars)

	$M
Power-Dependent Facilities (260 MW)	
Turbine and other powerhouse equipment oil/H_2O heat exchangers, piping, pumps, safety equipment	61.2
Storage-Dependent facilities (2600MWH)	
Tankage, Blanketing	12.1
Oil (667,000 bbl. @ 69¢/gal.) (non-depreciable asset)	19.3
	31.4
Total Depreciable Capital	73.3
Total Capital	92.6

Table I lists the major design basis items selected for this study. The calculated performance is summarized in Table II, showing that when the storage is being "charged", power output would drop by about 1/3 and, when the storage is being discharged, power output would rise by about 1/4 over normal, at an overall efficiency of 73.4%. A surprising finding was that the nuclear station became more efficient when the thermal energy storage was incorporated into its design, because the larger turbine and condenser required for the peak operation allowed a lower exhaust pressure and consequently more efficient operation when the station ran without storage. This 2.3% improvement in net output is allowed for in the overall storage cost computation.

The cost estimate for the plant has been divided into two parts: power-dependent facilities which are a function of the output (and input) rate, and energy-dependent items which depend on the quantity of energy stored and recalled. The latter will depend on the maximum number of hours/cycle at which the unit must be capable of operating at its rated capacity. All costs, including that of the oil, are in 1975 dollars assuming 1975 plant completion, and include a 25% contingency to allow for uncertainties in estimating the costs of a new process.

In order to permit comparison of the oil-based scheme with other storage techniques, the above costs in Table III have been converted into specific storage system investments ($Cp/kW, $Cs/kWh) in Table IV. In computing the power-dependent cost of $182.8/kW, a $600/kW credit has been taken for the 2.3% increased base power output provided by the larger turbine and condenser required in the storage system design. These added costs, of course, are included in the totals.

Table IV

Specific Storage System Investment

$$C_t = C_p + C_s T$$

Power-dependent costs (C_p)
$$C_p = \frac{\$61.2M - (1066 - 1043MW)(\$600/kW)}{(1325.4 - 1066MW)} = \$182.8/kW$$

Storage-dependent costs (C_s) for 10-hr cycle

Depreciable capital $= \dfrac{\$12.1M}{(1325.4 - 1066MW) \times (10\ hours)} = \$4.66/kWh$

Oil (nondepreciable asset) $= \dfrac{\$19.3M\ (0.78)}{(1325.4 - 1066MW) \times (10\ hours)} = \$5.80/kWh$

$$\overline{\$10.46/kWh}$$

Total cost for 10-hour system (C_t) $= \$287.4/kW$

Two points must be emphasized before these numbers are discussed further. The investments do not include interest during construction, usually a sizeable item in connection with nuclear power plants. Petroleum refinery construction experience, however, indicates that oil storage and handling facilities can be erected in two years or less.

Inasmuch as these facilities are operationally separate and physically removed from the nuclear power plant, their construction can be scheduled for the last part of the total construction period. Hence, it is believed that the interest during construction contributed by the thermal energy storage facilities should not exceed 10% of the total investment shown. The other point is that the design evaluated for the oil thermal storage scheme was selected on the basis of judgement and experience and has not been optimized yet. There are many possibilities for decreasing the cost of this system, and some of these suggestions will be dealt with further below.

The comparative cost of the EHS scheme vs. above-ground and underground pumped hydro is shown in Table V. Thermal storage in hot water is not shown because reliable cost figures for a large installation of this magnitude are not available. Based on a recently completed study by Public Service Electric and Gas Company (New Jersey) for EPRI/ERDA[3], however, it is doubtful that hot water storage will become competitive until low-cost pressure vessels can be developed and demonstrated.

Table V

Exxon Heat Storage Versus Pumped Hydro

Capital Costs - 1975 Dollars

	Pumped Hydro		Exxon
	Above-ground	Under-ground	Heat Storage
Power related, $/kW	120-180	140-190	183
Storage related, $/kWh	2.5-12	4-14	10.5
Total (10-hr system), $/kW	145-300	180-330	287

The values in Table V show that thermal storage in oil has a good chance of being competitive with pumped hydro in the long range as suitable storage sites for hydro storage systems become less available and more expensive to develop. A significant advantage of the EHS system is that it is not site dependent, whereas the only other energy storage systems which can be installed by utilities in the near future -- pumped hydro and compressed air -- cannot be considered unless appropriate topographical and underground geological structures are available. The only restriction on the location of the oil system is the limitation of any thermal storage unit -- it must be situated reasonably near the power house.

Regarding feasibility of the EHS system, the joint Exxon/GE study, while preliminary in nature, utilized the long experience of Exxon Engineering in designing large and safe oil handling facilities, and drew on GE's extensive experience in the design and application of large steam turbines and their auxiliaries. All operations and equipment are based on already existing technology, or only slight extensions of present experience. The authors concluded that no insurmountable obstacles nor lengthy research and development programs stand in the way of reducing this concept to commercial reality.

2. Effect of Oil Cost on Economics and Overall Application - The cost used in this study for the oil, for a 10-hour storage cycle, amounts to approximately 20% of the total new investment (Table III). When allowance is made for the investment credit for the additional power produced due to the improved efficiency, the oil cost actually represents 25% of the total. If longer storage cycles are considered, this percentage would, of course, be larger.

The oil cost that has been taken for this study represents a highly-refined, low-vapor-pressure petroleum fraction (plus a 25% contingency allowance). This oil is known to have the requisite thermal stability and other desirable properties. Research is under way at Exxon, however, to find other, more cost-effective oils which have the necessary qualities but would be available at a lower cost.

It is entirely possible that a less expensive oil could be added continuously to the storage facility at some fixed rate to ensure that the quality of the oil inventory remains at the required level.

More expensive oils and synthetic organic liquids commonly used for heat transfer purposes would extend the temperature range over which the fluid could be used, but would severely adversely affect the overall economics.

One of the important advantages of the EHS system as designed for LWR applications is that it is inherently limited to relatively mild thermal conditions for the oil, and a thermal stability much greater than that obtainable with the type of oil specified herein would not be useful or justified. Of course, there may be applications in the future which will require more stable oils as higher temperature nuclear cycles begin to be applied in the utility industry.

Table VI summarizes the effect of oil cost on the overall project cost. A quick comparison with the competitive technologies in Table V shows the importance of keeping the oil cost down or getting it even lower. With a thermal storage fluid costing twice the 69¢/gal. used in the current estimate, this system would not be competitive with pumped hydro or compressed air, and might be edged out by hot water storage. On the other hand, if oil costs could be lowered to one-half, which would bring them into the range of a low-sulfur fuel oil, then the system would be in the cost range of current pumped hydro installations.

Table VI

Effect of Oil on EHS System

Basis: 2,594 MWh/250 MW storage unit 1975 dollars
Oil considered non-depreciable asset in Cases 1 and 2

Case No	Oil Cost ¢/gal.	Base for Selection of Oil Price	Likeli-hood	Total Specific Cost $/KW
1	69	Base Case (incl. 25% contingency)	Good	287
2	35	(a) Low cost oil developed or, (b) National Stockpile pays half of oil cost	Fair	258
3	-	National Stockpile pays total oil inventory cost	Low	229
4	140	Low cost synthetic heat storage fluid is used	Very low	380
5	250	Medium cost synthetic heat storage fluid is used	Unlikely	499

Table VI also shows the effect of zero oil cost, which is not entirely unrealistic if this material could be considered a part of the National Strategic Petroleum Stockpile. This is especially appropriate if the oil cost is in the range of a low-sulfur utility fuel. While it is not implied that the oil would be free to the utility and fully charged against the National Stockpile, it basically does not matter how the charges are divided: If the oil can be "used" for both purposes, the net credit can be applied to the energy storage system.

If oil in the unit must be withdrawn for fuel purposes in case of a national emergency, the effect will only be to shorten proportionally the available storage time per cycle until an inoperable condition is reached. Inasmuch as it is expected that this will happen only in times of national emergency, the use of electricity by consumers can be regulated then much more effectively than during normal times in order to achieve the necessary load leveling.

The amount of oil required for thermal storage is about 11 gal./kWh. When this is burned in a modern fossil fuel station, 140-150 kWh of electricity can be produced. In a combustion turbine, about 100 kWh of peak power can be generated with the oil required to store 1 kWh. This gives an idea, not only of how often the oil must cycle before its thermal storage capacity is equal to its energy content as a straight fuel, but also how effectively this material can be used as a utility powerhouse fuel in case of emergency.

One last item in connection with the National Stockpile: Assuming that it can be arranged for the oil to be considered part of this strategic reserve, and this is still very problematical, then some of the storage tank costs might be allocated as well.

327

3. Potential Improvements and Optimizations - As previously mentioned, the arbitrarily selected basis for the thermal storage design case summarized in Tables I-IV can be optimized with a reasonably good chance of improvement. A number of possible approaches to this optimization are discussed below, together with rough estimates of the potential effects of such changes on the specific storage costs Cp and Cs. Only approximate changes are shown since more accurate numbers can only be determined through a definitive engineering design, preferably applied to a specific project and location. The estimates are shown in Table VII, and they are over and above any changes in the oil cost discussed in the previous section.

Table VII

Potential EHS Improvements

Item	Possible Effect on Storage Cost*		
	ΔC_p, $/kW	ΔC_s x 10h, $/kW	ΔC_t, $/kW
(a) Optimize Turbine Extraction Pressures for Storage Case	(13-26)	-	(13-26)
(b) Addition of Motor Drives to Powerplant Auxiliaries	(5-10)	-	(5-10)
(c) Optimize Heat Transfer Surface vs. Exchanger ΔT	(10-15)	5-10	(5-10)
(d) Storage of Hot Water Coupled with Oil Storage	(5-10)	-	(5-10)
Total possible credit	(30-60)	5-10	(25-50)

Each $/kW saved in nuclear portion of powerhouse saves $4/kW in storage costs

* Numbers in parentheses represent reductions in cost

(a) Improved Selection of Turbine Extraction Points - In the base case, the design of a more or less conventional LWR nuclear plant turbine was modified to suit the extraction rates required for the oil storage design. However, the BFW going to the oil/water exchangers must be heated to a much higher temperature (even using throttle steam) than is the normal practice in a nuclear power plant. These new requirements impose a different boundary condition on the number and levels of extraction steam streams from the turbines so that a very worthwhile optimization can be accomplished here. The net effect of such an optimization will probably not only be an improved efficiency for the thermal storage scheme, but also an increase in the base efficiency of the overall thermal cycle, whether storage is being used or not. It is estimated that the improvement would amount to 1/2 - 1% of the total powerhouse output, which, when credited at $600/kW nuclear capacity, could amount to a credit of $13 - 26/kW in the Cp term.

(b) Addition of Motor Drives to Powerplant Auxiliaries - Normally, powerplant auxiliaries are mainly steam-turbine driven. They include the boiler feed water pumps, the condenser water pumps and any fans in forced draft cooling towers, as well as the primary coolant circulating pumps in pressurized water reactor units. Now, if these auxiliaries (including the oil circulating pumps in the EHS case) are provided with electric motor driven spares, then the steam turbine drives can be used during the peak operation, while the electric motor drives would be running during off-peak periods.

The principal effect would be that the main turbine and generator would be designed for less of a swing between peak and off-peak operation than indicated by the net plant output. If this results in some improvement in turbine efficiency, a savings of $5-$10/kW can be anticipated in Cp.

(c) Optimization of Heat Transfer Surface - Over one-half of the total Cp cost term is for heat exchange equipment between extra extraction steam and water, and between water and oil. This has not yet been optimized, particularly the effect of ΔT across the exchangers on equipment cost, oil inventory cost, and overall system efficiency.

While an accurate assessment of this has not yet been made, a rough estimate indicates a potential reduction in Cp of $10-15/kW. There would, of course, be some loss of thermal efficiency.

Oil cost is a major variable in this optimization. As the ΔT across exchangers increases, the surface requirements go down, but system efficiency deteriorates. As a result, more oil has to be used per kWh stored, and this increases costs by $5-10/kW for a 10-hour storage capacity. If the oil is cheap or even "free", much lower efficiencies may be justifiable than indicated by the present study.

(d) Storage of Hot Water Coupled with Oil Storage - The storage of water can probably be advantageously combined with oil in a thermal energy storage scheme. Up to its boiling point, water can be stored in atmospheric pressure tankage, so that the temperature range from condenser conditions up to about 210°F can be covered with water cycling between these two limits. While this, of course, adds hot and cold water tankage, the net effect is a reduction in overall heat transfer surface requirements and a small, yet not insignificant increase in storage capacity per unit volume of oil and a similar increase in the fraction of storable power. Contrasted to the oil-alone base case, some of the lower portion of the oil temperature range can be taken over (and extended downward) by water, and this will result in exchanger surface savings, because it is precisely in this range that the oil is most viscous and has the lowest specific heat.

It is estimated that the storage of BFW preheated to 210°F or so may save in the order of $5-10/kW in Cp.

Another modification worth considering is the possibility of storing a nominal amount of BFW in a high pressure drum at BFW final temperature. The advantage of this scheme is that it could allow rapid load-following and simplified control in the nuclear unit, regardless of the thermal inertia of the oil/water exchanger system. The amount of BFW which would have to be stored need only be a small fraction of the total storage cycle requirements, and therefore the high cost of hot water pressure storage would not be a significant factor. On the other hand, the easy control and rapid response of the hot water system may result in appreciable savings in the nuclear portion of the plant. As has been pointed out in a previous publication[1], each $1/kW saved in the powerhouse (on 1066 MW) translates to $4/kW saved when applied to storage (259kW storage rate). No estimate of possible savings has as yet been made for this modification.

(e) <u>Summary of Possible Savings</u> – In summary, the possible net savings through optimization studies could be in the range of $25-50/kW. In addition to this, the previously discussed effects of oil cost could have a major influence. Furthermore, the inclusion of the EHS system, ensuring operation of the nuclear reactor at constant conditions throughout the year despite wide variations in electrical output, may result in significant savings in the nuclear portion of the plant. These could include reduction in nuclear reactor control facilities normally required for load following, and improved fuel utilization. These possibilities need to be addressed in any definitive engineering evaluation.

4. <u>Hardware and Siting</u> – A variety of considerations can affect the selection of equipment and process schemes for any given application. Without going into detail regarding specific component performance, an attempt will be made here to discuss a number of important criteria which influence the design.

(a) <u>One vs. Two Turbines</u> – The difference between maximum and minimum extraction rates imposes a significant design problem on a single turbine and would require compromises and possible performance debits. Therefore, it may be more economical, in the long run, to consider two turbines, with one more or less designed to operate as a peaking turbine when more motive steam becomes available for power generation. Of course, this is the most likely way in which this scheme could be retrofitted into an existing power plant; in this case, the second turbine with its own generator would pick up the incremental steam capacity made available to the powerhouse as a result of the installed thermal storage.

Another consideration in favor of a second turbine is the possibility that the stored thermal energy in the oil is used for the actual generation of additional intermediate pressure steam rather than just for BFW preheat. In that case, this additional steam can be used to drive the second turbine, or at least be one of several motive steam feeds powering this turbine.

As discussed previously in Section 3, the second turbine need not be an electric power generator; it can actually drive powerplant auxiliaries and thereby indirectly increase the available power output of the installation.

On the other hand, the use of one large, flexible turbine has the advantages of process simplicity, less space requirement and possibly lower total installed cost when the relative costs of all piping and controls are included. Specific design studies of the alternatives probably will be necessary to reach a final decision.

(b) <u>Type of Nuclear Plant</u> – BFW temperature and BFW chemistry are two considerations which indicate that the oil storage scheme may be more directly applicable to pressurized water reactors (PWR) than to boiling water reactors (BWR). However, it is only after a detailed review of all the design features and through a future definitive design that the differential cost, if any, of one system or another can be ascertained.

The BFW temperature in BWR nuclear plant designs is lower than desirable for maximum thermal energy storage capacity and efficiency. This apparent handicap, however, can be used to advantage by utilizing some of the high level heat available from storage for the purpose of interstage steam reheat and intermediate pressure steam production in the case of the BWR. This could be particularly interesting in the two-turbine application.

Water chemistry considerations enter in because it is vital that carbon steel rather than stainless be used as the material of construction for the extensive oil/water heat exchanger train. Otherwise, the investment for the exchanger surface would be excessive. While this may favor some PWR designs over the BWR, there is nevertheless a good possibility that powerhouse and oil storage can be interconnected by entirely closed circulating loops of heat transfer fluid, such as water, for both PWR and BWR. In that case, the oil-water exchangers can be made of carbon steel because the material depends only on the water chemistry of the circulating loop fluid, not that for the BFW.

Applicability of the concept to the CANDU-type of heavy water reactor needs to be checked. Inasmuch as BFW temperatures in this system are somewhat lower than in the LWR's, the fraction of the plant's output which can be stored when using BFW preheat only is reduced. This, however, can probably be made up by intermediate pressure steam production with the heat stored in the oil. The lower temperature, of course, means even less of an oil stability problem, and the possibility that a lower cost oil will suffice.

(c) <u>Nuclear vs. Supercritical Fossil-Fuel Plant</u> – The major justification for energy storage in utility systems resides in substituting low-cost, off-peak nuclear energy for expensive combustion turbine fuels otherwise needed during peak demand periods. Storage of off-peak energy produced from coal firing to displace distillate fuels in combustion turbines has somewhat lower economic benefit. Nevertheless, this may be justified in some cases for supercritical coal-fired stations. The high pressure, supercritical plants are seriously affected by large changes in load on a cyclic basis, resulting in reduced efficiencies and significantly increased maintenance costs. These factors may provide sufficient additional credits to justify energy storage despite the reduced fuel credit.

Because supercritical coal-fired stations operate at much higher temperature than nuclear units, the amount of BFW circulated per kWH generated is less, of the order of about one half. Therefore, the storable energy is also cut in half when its use is limited to BFW preheat, and maximum oil temperatures are restricted to those considered safe and operable for the oil.

It may again be possible, however, to supplement storage of BFW preheat with thermal energy stored for the purpose of generating intermediate pressure steam during peak demand periods.

A further consideration is that the first thermal energy storage system using oil should be installed in conjuction with a fossil-fuel fired plant. This will permit testing all aspects of the installation, including its response to various upsets, in a non-nuclear installation so that the appropriate safeguards can be specified for a later unit coupled to a nuclear plant.

(d) Retrofit Installation - The EHS system can be considered for retrofit installations, particularly if the boiler portion of the plant has been derated relative to the turbine-generator portion. The capacity of the turbine would have to be checked, especially in the low-pressure stages, for the much higher throughput during peak periods where extraction steam is minimized. Retrofitting can be considered for any installation, however, space permitting, if a peaking turbine is provided to handle the additional steam quantities which will be made available due to the thermal storage.

(e) Siting at Nuclear Parks - When a thermal storage plant is located, as it must be, next to a power station, it necessarily acquires the service factor of that unit. The service factor of an oil storage system could be improved by tying it into a nuclear park which contains more than one nuclear unit, usually of the same kind and size. In that case, it may be possible to tie the thermal storage into one of the two turbine halls, and then cross-connect the two nuclear boilers to the turbine hall with the storage tie-in. In that case, the storage can be used as long as at least one of the nuclear steam systems is operative.

5. Other Applications of the Exxon Heat Storage System
There are several other potential applications of the EHS system in addition to those described above.

(a) Industrial Operations - The use of nuclear reactor systems to provide both electricity and industrial process steam by drawing intermediate pressure steam from the turbines has been proposed by Dow Chemical Company [4] and others. In the arrangements utilized, however, process steam must be drawn off continuously as needed, including hours when electrical demand is at peak levels. Through incorporation of the thermal energy storage system described earlier, excess thermal energy could be stored in oil during off-peak periods and then recalled to produce some or all of the process steam needed during the periods of peak electricity demand. This possibility has also been suggested by Tillequin. [2]

(b) Solar Thermal Systems - In high temperature solar thermal systems for electricity generation or process steam production, an EHS system can be incorporated to store the lower level heat needed for BFW heating during periods when direct solar energy is not available. The higher level heat must be stored in a molten salt of liquid metal, but these storage media are not usable at BFW temperatures, and the addition of the EHS system for the BFW substantially increases the overall system efficiency.

It is also possible to consider an EHS system to provide part or all of the BFW preheat from solar energy in a nuclear or fossil-fuel fired electricity generating station. In this case, energy received by a solar thermal collector while sunlight is available would be stored in oil for use throughout the day to heat boiler feed water.

(c) Compressed Air Storage Systems - Compressed air storage systems for electric utilities in which the adiabatic heat of compression is stored as well as the energy in the compressed air have been described [5]. Such systems eliminate the undesirable need for combustion turbine fuel.

Glendenning [5] described systems in which the adiabatic heat of compression is stored in underground pressurized containers filled with ceramic or other solid heat storage media. For systems in which two or more stages of compression are used, resulting in moderate interstage temperatures, oil could be used as the heat storage medium in an EHS-type system.

Normal hydrocarbon oils, however, could not be used in the same heat exchangers with high pressure air because of the safety problems. Either a non-flamable fluid could be used, or a separate water heat transfer loop between the hot air and the oil stream could be incorporated.

No economic assessments of these possible additional applications have yet been made.

CONCLUSIONS
The Exxon Heat Storage system for storing off-peak thermal energy in a stable, high-boiling, refined petroleum fraction and recalling this energy for use during peak demand periods appears to have potential value to electric utilities.

- The EHS system permits running nuclear or fossil-fuel fired plants at constant, rated output continuously while varying net electrical output over a range of about 2/1.

- Economic assessments made so far show EHS to be reasonably competitive with pumped hydro storage, and possibilities for improving its competitiveness through optimizations and modifications have been identified.

- EHS is not site-dependent, as pumped hydro and compressed air systems are.

- The EHS system is based on existing technology and experience. No long and expensive research and development programs are needed in order to design commercial facilities, and a large-scale demonstration could be undertaken immediately.

The EHS system may also be useful in storing thermal energy in industrial operations, solar thermal systems and in connection with compressed air storage systems.

REFERENCES

(1) R. P. Cahn and E. W. Nicholson "Storage of Off-Peak Thermal Energy in Oil", IEEE Power Engineering Society Summer Meeting, Portland, OR, July 1976.

(2) J. Tillequin "Power Modulation for Nuclear Power Stations Supplying Electric Power or Heat Distribution Networks", Transactions of the American Nuclear Society, April 21-25, 1975, The European Nuclear Conference, Paris, France, pages 736-738.

(3) "An Assessment of Energy Storage Systems Suitable for Use by Electric Utilities", Public Service Electric and Gas Company study for ERDA (Contract No. E(11-1) - 2501) and EPRI (Research Project 225), 1976, T. R. Schneider, Project Manager.

(4) G. L. Decker, Chemistry and Industry (London), No. 17, pages 722-725, September 6, 1975

(5) D. J. Littler, "UNIPEDE Report on Electrical Energy Storage" IERE Meeting, Tokyo, May 14-19, 1975. See Appendix on "Compressed Air Storage" by I. Glendenning.

26

Reprinted from pages 586–590 of *Intersoc. Energy Convers. Eng. Conf., 11th, Proc.*, American Institute of Chemical Engineers, 1976, 1802pp.

UNDERGROUND STORAGE OF OFF-PEAK POWER

S. L. Ridgway and J. L. Dooley
R & D Associates, Marina del Rey, California U.S.A.

Abstract

Underground storage of pressurized hot water in association with a steam power plant as a means of off peak power storage is shown to be an alternative to pumped hydroelectric storage. Steam is flashed from the hot water and directed to peaking turbines for handling the peak load; at low load times, surplus steam is condensed in the water to recharge the system. The saturation pressure of the hot water is borne by the overburden pressure of the formation in which the storage cavity is prepared. For cavities operating in the pressure range from 70 to 140 bars the effective energy storage is 18 to 21 kwhr per cubic meter of cavity volume, and the loss in useful energy due to storage ranges between 5 percent and 9 percent. The estimated cost of a facility to provide 1000 Mw for 10 hr is $309 million.

INTRODUCTION

Electric utilities would appreciate the opportunity to store energy generated in time of low demand for delivery at times of peak demand, for the purpose of lowering the capital cost of the plant. Planners of energy supply systems whose energy source is variable, such as sun and wind power systems must have a large, inexpensive energy storage method if those systems are to have any utility.

In this thermal energy storage scheme, the excess capacity of the steam generation system of the power plant would be used at times of slack demand to charge with steam a cavity containing pressurized hot water. At times of peak demand, steam flashed off this hot water would drive supplementary turbines and their associated electric generators. Underground storage of hot water has some important advantages as a means of achieving the benefits of off-peak energy storage. These are as follows:

1. Water has a high heat capacity, low cost, and can serve as both working fluid and storage medium without the need for heat exchangers.

2. Water is well understood and accepted both by the utility industry and by the public.

3. Recovery of the energy stored can be as high as 90 to 95 percent.

4. The system can accommodate suitably rapid load changes.

5. The geology beneath many power plants would be suitable for this type of storage which is particularly of interest where pumped hydro-electric storage, is not applicable.

Steam power plants have always used the heat stored in the boilers and settings, and have often added accumulators holding hot water for energy storage to take care of short term fluctuations in heat supply, or in power demand. However, storage in this form of the large amount of energy needed to carry a utility peak load at

5:00 p.m. from fuel burned or nuclei fissioned at 2:00 a.m. has not previously seemed attractive due to the cost of the pressure vessel needed to contain the pressurized hot water. For several years, work has been underway on the possibility of harnessing fusion energy by underground containment of thermonuclear explosions. This work for Project PACER has created a new body of information on methods of construction, cost, and safety of deep underground containers whose walls are supported against internal pressure by the overburden pressure of the earth itself. Such containers have a wider application than containing explosions, since they can be designed to hold hot pressurized water for energy storage, and it is this development that makes hot water interesting for off peak power storage.

THE HOT WATER ENERGY STORAGE SYSTEM

To a conventional thermal power plant is added a large underground cavity that contains pressurized hot water at its saturation temperature. Depending upon the maximum steam pressure of the associated thermal power plant, this cavity is constructed at a depth between 1200 ft for a 1000 psia nuclear plant to 2400 ft for a 2000 psia fossil fueled plant. The depth is chosen so that the overburden pressure is somewhat over the maximum cavity pressure to insure that the rock is always in compression, and provide a margin of safety against the consequences of accidental venting of the cavity into the geological formation. If the rock pressure is greater than the steam pressure, compressive forces between the rocks will keep the rock arrangement intact; if the steam pressure is greater than the lithostatic pressure the steam could open up a leak by displacing rocks along the leak path and lead to possible venting of the cavity. The cavity has a thin steel liner for sealing against the escape of water or steam, but the pressure load is taken by the rock out of which the cavity was excavated.

Submitted Through AIChE HT&EC Division for 1976 Intersociety Energy Conversion Conference

In Figure 1 is shown the general arrangement of the components of the base load
plant with the additions that are to provide the peak load capability.

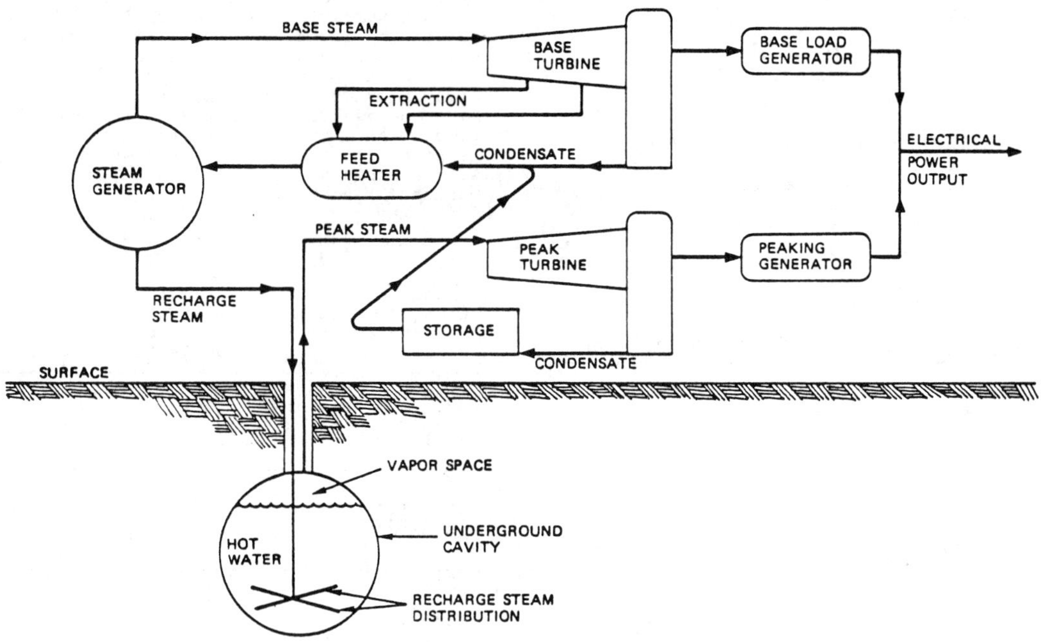

Figure 1.   Thermal Energy Storage System Schematic.

Steam turbines and electric generators are
added to give the desired increase in capability. They are connected to operate on
the steam available from the cavity, and
provision is made to store the condensate
from these turbines. During periods of
slack demand, steam is directed from the
steam generator unit to a set of pipes
below the water surface in the cavity, and
steam is bubbled into the water in the
cavity to reheat it to the charged condition. During this recharge operation, the
stored condensate is pumped to the feedwater heaters, and heated by extraction
steam from the base load turbines.

SYSTEM THERMODYNAMICS·

The general principle of operation of this
method of storage is to postpone the conversion of heat energy to electrical
energy as contrasted with pumped hydroelectric storage, where electric power is
generated at the slack period, and then
converted to a form suitable for storage
and reconversion into electric power.
The appropriate measure of the efficiency
of this thermal method of energy storage
is to compare the electrical output energy
actually realized as a result of the discharge of the thermal store in the cavity
with that electrical energy that would
have been obtained by using the recharge
steam directly in the turbines. The
energy passes only once through the turbine
and the alternator in each case, so their
efficiencies cancel out in such a comparison. The essential point in the thermodynamic analysis is to find how much the
available energy in the steam supply has
been degraded in the cavity charge-discharge
cycle by irreversible processes. The second desired result of the thermodynamic
analysis is to determine the effective
storage of the cavity per unit volume as a
function of the operating parameters of
the system.

Consider a mass of water, w, with specific
internal energy u_f which is in equilibrium
with its vapor at a pressure p. Allow an
infinitesimal amount to evaporate into
steam. Applying the law of conservation
of energy, we write

$$w\,du_f = (h_g - h_f)\,dw,$$

where h_g is the specific enthalpy of the
vapor, and h_f the specific enthalpy of the
liquid.

This states that the change in internal energy of the liquid provides the heat of evaporation needed to form the element of steam. This may be rewritten as

$$dw/w = du_f/(h_g - h_f) .$$

Integrating between two states of interest we have

$$\ln \frac{w_2}{w_1} = \int_1^2 \frac{du_f}{(h_g - h_f)} .$$

Since the process takes place along the liquid-vapor equilibrium, the integral on the right hand side can be evaluated numerically by reference to the steam tables, since u_f is a suitable determinant of the thermodynamic state and therefore of $h_g - h_f$. This integral, which we have chosen to call the flashing integral, was evaluated at 10°C intervals over the temperature range from 200° to 360°C, and is tabulated in Table I.

Table I. Values of the Flashing Integral.

Temperature	$\int_T^{360} du/(h_g - h_f)$
360	.00000
350	.103440
340	.178025
330	.237993
320	.289286
310	.334129
300	.374482
290	.411318
280	.445389
270	.477210
260	.507181
250	.535593
240	.562686
230	.588644
220	.613624
210	.637745
200	.661110

The method of analysis of the energy yield per unit cavity volume, and the intrinsic (rate independent) cycle irreversibility proceeded as follows.

The initial temperature and a cavity filling factor (97.2%) was chosen. For computing convenience, a nominal cavity volume of 1 liter was selected. From steam table values of the specific volume of the vapor and the liquid at the starting temperature, the mass of water, the mass of steam, and the total mass is determined. From the value of the flashing integral between the starting and the finishing temperatures determined by differencing the values from Table I, the amount of water remaining after the discharge is calculated. Its volume is determined, the cavity volume remaining for steam calculated, and the masses of water and steam remaining after the discharge are determined. The amount of steam that has passed through the turbine and been condensed is now known. The total energy and entropy of the system before and after the discharge are compared. Under the isentropic expansion assumption in the turbine, the entropy decrease of the system is rejected

in the condenser cooling water. The corresponding energy discarded in the cooling water is, therefore, the absolute condenser temperature times the entropy decrease. Subtracting this energy from the system energy decrease gives the shaft work delivered by the turbine. This shaft work is then multiplied by the turbine efficiency to get the actual work out, and from this value is derived the storage per unit volume. The results of these calculations for a high pressure, high temperature case as might be associated with a fossil fueled power plant, and a lower pressure, lower temperature case as might be associated with a nuclear power plant are given in Tables II and III.

Table II. Steam and Energy Yield in the Discharge of a 1 Liter Cavity from 330°C.

Dis-Charged Temperature	Steam Yield (grams)	Thermal Energy Extracted (Joules)	Ideal Work Out (Joules)
320	25.98	65460	25510
310	49.37	124980	48460
300	70.66	179540	69280
290	90.12	229700	88140

Table III. Steam and Energy Yield in the Discharge of a 1 Liter Cavity From 290°C.

Dis-Charged Temperature	Steam Yield (grams)	Thermal Energy Extracted (Joules)	Ideal Work Out (Joules)
280	22.17	57930	21500
270	42.78	111440	41230
260	62.01	162470	59340
250	80.02	209840	75980

Recharge is accomplished by returning steam and water to the cavity to restore the original cavity energy and mass. For the high temperature case, assuming that 128 bar steam at 550°C was available from the steam generator unit, and 230°C feedwater was available from the feedwater heater, recharge from the 290°C discharged state to the 330°C charged state required 62.5 grams of steam and 27.6 grams of water. The available energy in the recharge materials is 5900 joules larger than 88,140 joules obtained on discharge--a cycle loss of 6.7 percent. Desuperheating the steam with the feedwater is the origin of a loss of 4.3 percent and the remaining 2.4 percent loss is due to the cycle. The loss in the low pressure case is 3.0 percent. This is lower than in the high pressure case because of the lack of a desuperheating penalty. The steam required was 81.09 grams, and 1.07 grams of water needed to be removed to achieve simultaneous mass and energy balance. The cycle considered was from 290°C to 250°C. Results are summarized in Table IV.

Table IV. Energy Storage and Cycle Losses.

Charged Temperatures	Storage (Joules/cm³)	Cycle Loss (Percent)
330°C	75	6.7
290°C	65	3.0

Two other losses are of significance, pipe friction, and pressure loss in flashing. These combine to be no more than 2 to 3 percent with reasonable provision of pipe to carry up the steam to the surface, and underwater solid surface to nucleate bubble formation.

CAVITY MECHANICS

The cavity would be excavated by conventional mining techniques. The depths involved are not deep as mines are concerned, and good rock will allow the cavity to be excavated in it. Attention must be paid before and during excavation to the character of the original stress distribution in the rock, and to the presence of a sufficiently large homogeneous mass of competent rock to host the cavity. When the initial horizontal stress in the rock is at least as large as half the vertical stress, only compressive stresses are induced in the cavity wall by excavation. At the proposed depths these stresses are within the compressive limits of competent rock. If the horizontal initial stress is quite small, excavation of a spherical cavity would induce tensile stresses in the roof and floor, and possible failure. In this case, the cavity could have an ellipsoidal shape to assure compression in the wall.

An extensive program of rock bolting, drain provision, and grouting of weak spots with high strength concrete laid on steel mesh attached to the rock bolts should yield a smooth, strong surface for the support of the cavity liner. The best choice for a cavity liner seems to be wrought iron, or a very ductile corrosion resistant steel with similar characteristics. Its purpose is to seal the rock from the high pressure hot water and steam. After the rock cavity is prepared, rock bolted, and smoothed, the liner would be installed close to the wall. It would be welded up out of steel plates, and the welds would all be carefully inspected by x-ray and other techniques. As the fabrication progressed and the welds passed inspection, the void between the liner and the wall would be filled with grout.

When the lined cavity is finished it is at atmospheric pressure, and ambient temperature. It will need to be brought to a state of high temperature and pressure.

Application of the internal pressure relieves some of the tangential compression in the wall, and applies a radial pressure. The rock in the wall will be extremely stable under this condition.

Thermal stresses present a problem for the steel liner, and the cavity wall. Somewhat surprisingly, the thermal displacement of the host rock at the cavity wall is zero. The mild steel liner (with a linear expansion coefficient of .65 x 10^{-5}/°F) in being heated 500°F, would expand 0.00325 if not restrained. The liner is clamped against the rock wall by the internal pressure in the cavity, and compressive stresses are set up in the liner which cause a counteracting strain. Mild steel will yield under the thermal stress induced during startup, but would cycle daily afterwards in the elastic range.

For some rocks it appears that the thermal stress at the cavity wall, which is given by

$$\sigma_{th} = \frac{\alpha E\, T_i}{1-\nu},$$

with α the coefficient of thermal expansion, T_i the temperature increase at the cavity wall in starting up, and ν Poisson's ratio, exceeds the compressive strength. This inference depends upon using a Youngs modulus E for the rock wall characteristic of single compact laboratory specimens. Most host formations would be expected to contain at least a volume percent of porosity and cracks that would provide expansion room and thus appreciably ameliorate the thermal stress problem. The question requires careful study when a specific site with a specific underground formation has been selected.

ECONOMICS

Analysis of the expected heat to electrical energy conversion efficiency by layover of the heat in the storage cavity compared to direct use of the steam as it is generated indicate that there will be about a 5 percent loss in available energy for the storage facility attached to a low pressure saturated steam nuclear plant, and about a 9 percent loss for a storage system connected to a high pressure (2000 psi) high temperature (1060°F) fossil fueled plant.

Preliminary underground capital cost estimates have been made for a range of cavity sizes which determines the available storage; and for a range of power levels, which determines the size of the pipes, valves, and fittings that carry the peaking steam to the surface turbine, and return the recharged steam to the cavity. The depth dependence of the cavity cost was not large, and so average values are presented in Table V.

The specific costs are clearly lower in the larger storage capacities and the higher powers. The total underground cost at low pressure for 1000 Mw of capacity and about 10,000 Megawatt hours of storage is $120,000,000. The underground part of the capacity cost is $33 per kilowatt, and

335

the storage cost is $9 per kilowatt hour.
To the underground part of the capacity
cost must be added the cost of the conventional turboelectric capacity at the
surface, which is in the range of $150
to $250 per kilowatt.

Table V.   Cavity Costs.

| Diameter | Cost | Storage | | | |
| (ft) | $x10^6 | High Pressure | | Low Pressure | |
		Gw-hr	$kw-hr	Gw-hr	$kw-hr
125	11.32	.64	17.60	.555	20.39
200	27.23	2.64	10.33	2.27	11.97
275	57.67	6.85	8.42	5.91	9.76
325	85.25	11.31	7.54	9.76	8.75

Table VI.   Power Proportional Underground
Costs.

| Power (Mw) | Pressure | |
| | 1000 psi (low) | 2000 psi (high) |
	$ x 10^6	
200	21.82	30.48
500	25.85	38.86
1000	33.32	55.03

Mining and geological consultants were used
in preparation of the underground costs.
Supplier quotes were used for most surface
equipment costs. The direct cost was increased by 50 percent to cover engineering,
administration, legal, environmental and
other costs, which included an allowance
for underground geological uncertainties.
To these accumulated costs is then added
a 20 percent finance charge for interest
on the funds used during construction.

27

Reprinted from pages 6-14–6-26 of *Report NSF/RANN/SE/GI27976/PR73/5,* Univ.
Penna. Natl. Center of Energy Management and Power, 1973, 380pp.

CONSERVATION AND BETTER UTILIZATION
OF ELECTRIC POWER BY MEANS OF THERMAL
ENERGY STORAGE AND SOLAR HEATING

H. G. Lorsch, M. Altman, and H. Yeh

[*Editor's Note:* In the original, material precedes this excerpt.]

6.5.3   Heat of Fusion

Readily available data on melting points and heats of fusion of
representative TES materials are presented in Table 6-2. The fusion
of a salt hydrate, unlike the fusion of, say water or metallic tin, is not
a simple process of solid going into liquid. In the latter case, the
crystal structure of solid tin is broken down at the melting point, and
the long-range order of the solid is replaced by the short-range order
of the liquid (Ref. 25). The constituents however are not changed. A
more complicated fusion is that of water. In this case, beside the trans-
lational freedom acquired on melting, there is a rotational degree of
freedom which is also acquired on melting, and possibly new vibrational
degrees of freedom: this is made possible in the water molecule by the
relative movement of the oxygen and hydrogen atoms.

Table 6-2
Heats of Fusion of Representative Materials (Ref's. 23 and 24)

Compound	Mol. wt.	Melting Point		Heat of Fusion		
		^oC	^oF	kcal/mole	cal/g	Btu/lb
$Ca(NO_3)_2 \cdot 4W$	210	47	117	8.13	38.7, 33.9	65.7
$Na_2CrO_4 \cdot 10W$	342	16	61	13	38, 43.8	73.8
$Ni(NO_3)_2 \cdot 6W$	265	53	127	10.5	39.6, 36.4	68.4
$Al\,K(SO_4)_2 \cdot 12W$	474	80	176	20	42.2	76
$MgCl_2 \cdot 6W$	203	115	239	8.2	39.4	70.9
$NaOH \cdot 31/2W$	103	15	59	5.5	53.4	96.1
$CaCl_2 \cdot 6W$	219	30	86	8.8	40.2	72
$Na_2S_2O_3 \cdot 5W$	248	45	113	5.6	22.6	40.7
$Zn(NO_3)_2 \cdot 4W$	235	48	118	$\sim$9	$\sim$38	68.4
$Mg(NO_3)_2 \cdot 6W$	256	90	194	9.8	42.6, 38.]	72.7
$Fe(NO_3)_2 \cdot 6W$	288	60	140	$\sim$8.5	$\sim$30	54
$Na_2HPO_4 \cdot 12W$	358	40	104	23.9	66.8	120

In salt hydrates, the "fusion" is really the sum of two
processes: dissociation into anhydrous salt and (liquid) water,
and solution of the salt into the water. This also explains the relatively small heat of fusion of the hydrates compared to pure water:
the first process (dissociation) is endothermic, while the second
process (dissolution) is generally exothermic. The two processes
thus oppose each other, and the heat of fusion is small.

This cycle can best be represented by an enthalpy diagram

(Fig. 6-4, Ref. 26) taking $CaCl_2 \cdot 6W$ as an example. Starting with the elements in their standard states, Ca, Cl_2, H_2, O_2, the heat of formation of $CaCl_2$ (solid) is 190 kcal/mole (B); the heat of formation of $CaCl_2$ dissolved in six molecules of water is 203.5 kcal/mole (C) and the heat of formation of $CaCl_2$ in an infinitely dilute solution is 208 kcal/mole (D). Also, the heat of formation of $CaCl_2 \cdot 6W$ (solid) is 212.3 kcal/mole (E). Therefore, if the initial state is $CaCl_2 \cdot 6W$ (solid), and the final state is $CaCl_2$ (solution in 6W), the enthalpy change is (C-E), i.e., $-203.5-(-212.3) = +8.8$ kcal/mole. Thus

$$CaCl_2 \cdot 6W \text{ (solid)} \rightleftharpoons CaCl_2 \text{ (solid in 6W)}$$
$$\Delta H = 8.8 \text{ kcal/mole}$$

The diagram in Fig. 6-4 can be used to obtain the heats of fusion when calorimetric data are not available. However, most of the heats of solution are measured and tabulated at 18 or $25^{\circ}C$ and, therefore, a correction (due to the difference in the specific heats of solution and solid) must be applied if the temperature of fusion is different. Generally, the specific heat of the solution is larger than that of the solid, so the heat of fusion at the temperature of fusion will be larger than that calculated from solubility data.

In the case of $CaCl_2$, the heat of fusion is given by (C-E) in Figure 6-4, i.e. 8.5 kcal/mole, neglecting the correction due to specific heat differences. This gives 38.8 cal/gm compared to 40.2 cal/gm as determined experimentally.

6.5.4   Solidification

The major problem associated with salt hydrates as TES materials is the fact that they tend to supercool considerably. This is of major importance and has not always received sufficient attention

B – A : Heat of formation of anhydrous salt
E – A : Heat of formation of hydrate
C – A : Heat of formation in concentrated solution
D – A : Heat of formation in dilute solution
C – B : Heat of solution of anhydrous salt in concentrated solution
D – B : Heat of solution of anydrous salt in dilute solution
D – E : Heat of solution of hydrate in dilute solution
C – E : Heat of fusion of hydrate
D – C : Heat of dilution
E – B : Heat of hydration

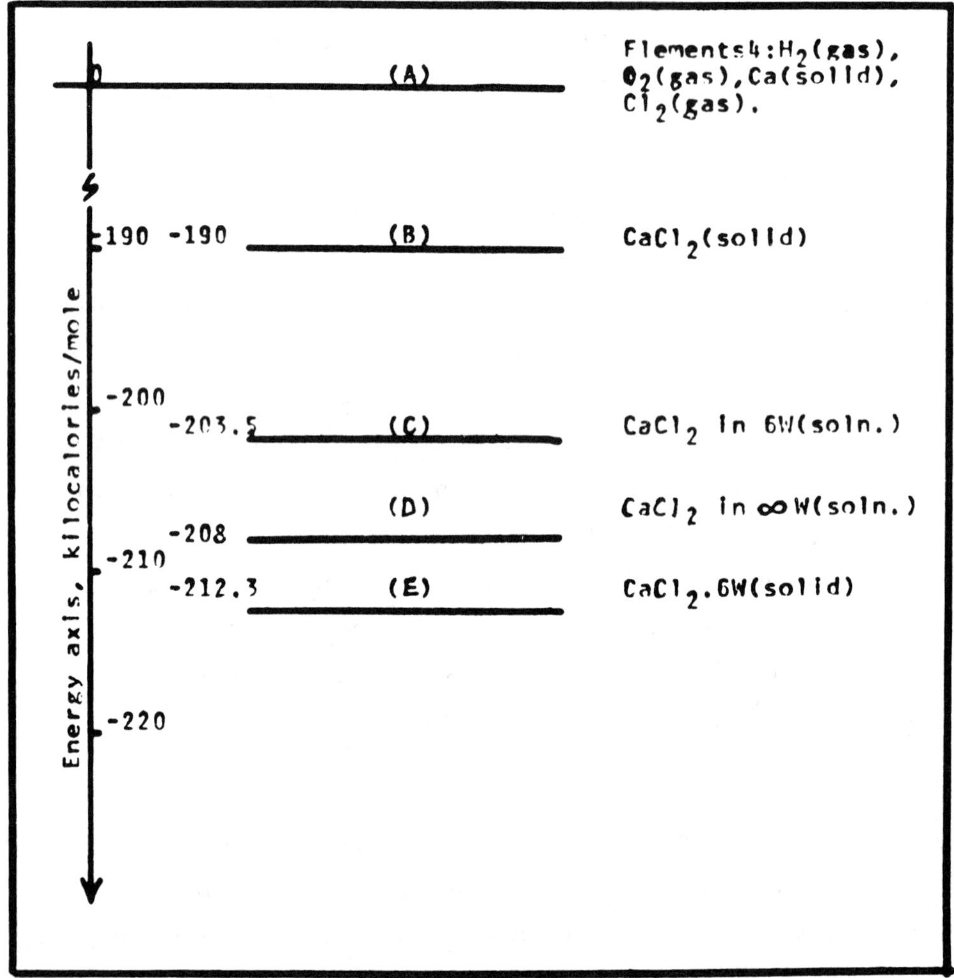

Figure 6-4
Enthalpy Diagram for the System $CaCl_2$ - H_2O

from workers in the field.

The reason for the high degree of supercooling is the fact that either the rate of nucleation (of crystals from the melt) or the rate of growth of these nuclei (or both) is very slow. This means that, as the melt is cooled, it does not solidify at the thermodynamic melting point. Thus, the value of the material for heat storage is reduced.

The reason for the strong tendency to supercool is not known exactly. It is possible, however, to show some empirical evidence connecting the tendency to supercool with the viscosity of the melt at the melting point. Materials with high viscosity in the liquid state have low diffusion coefficients for their constituent atoms (or ions) and these are unable to rearrange themselves to form a solid; instead the liquid supercools and, at sufficiently low temperatures, becomes a glass (Ref. 27). Table 6-3 shows how the viscosity affects the solidification behavior: high iscosity materials are good glass formers, and low viscosity materials solidify easily into crystals. In order to overcome these solidification problems, the nucleation rate can be increased by two methods: homogeneous nucleation and heterogeneous nucleation.

6.5.5   Nucleation

In homogeneous nucleation, the rate of nucleation of crystals from the melt is increased without adding any foreign materials. One way in which this can be done is to use ultrasonic waves. Most of the work in this field has been done by the Russian school (Ref. 31). It has been suggested that ultrasonic waves stir the liquid and thus increase the diffusion of ions in the melt; they also create cavities which act as nucleation centers and, finally, they break up the forming

Table 6-3
Viscosity and Crystallizing Behavior

Melt	Temperature*, °C	Melting Point, °C	Viscosity (poise)	Crystallization Behavior	Ref.
Iron	1595	1550	0.025	Crystallizes	28
Lead	328	328	0.028	Crystallizes	29
$Cd(NO_3)_2 \cdot 4W$	58	58	0.23	Supercools	30
$Ca(NO_3)_2 \cdot 4W$	47	47	0.75	Supercools	30
Water	0	0	0.017	Crystallizes	28
Glycerine	18	18	18	Supercools	28
$NaAlSiO_4$	1050	1050	1000	Forms stable glass	27

crystals and distribute them through the melt, creating new nucleation centers.

While no experiments have been performed on salt hydrate melts, many crystals, such as sodium thiosulfate and potassium alums, have been nucleated from more dilute solutions. Also, other classes of materials, such as metals and organic crystals, have been nucleated by ultrasonic waves.

In heterogeneous nucleation, the walls of the container, or some impurity present within the melt, act as a catalyst for nucleation by providing a substrate on which the nuclei can form. In order for an impurity to be an effective nucleating agent, it should satisfy the following

---

* Temperature at which viscosity is measured.

criteria:

(a)   Melting point higher than the highest temperature reached by
      the energy storage material in its temperature cycle.

(b)   Insoluble in water at all temperatures.

(c)   Does not form solid solutions with the salt hydrate.

(d)   Crystal structure similar to that of the salt hydrate.

(e)   Unit cell dimensions do not vary by more than 10% from those
      of the salt hydrate.

(f)   Does not chemically react with the hydrate.

These criteria show the importance of a study of the crystal structures, including atomic positions, of the salt hydrates. Some of these structures are known (Ref's. 32, 33), and these have been collected in Table 6-4.

6.6   Candidate Materials

Based on the criteria developed in Sect. 6.4, a literature survey (Ref's. 22, 47) was made to determine which of the many salt hydrates could be considered for TES. The results are given in Table 6-5. The literature survey was confined to the binary systems (salt + water). Any salts containing precious metals, such as Li, Cd, Pb, were eliminated from consideration. Salts of unstable radicals, such as chlorates and sulfites which may decompose or explode, were also eliminated.

Table 6-4
Crystal Structures of Known Hydrates

Hydrate	Structure & Lattice Constants (Angstrom)	Density (gm/cm^3)
$CaCl_2 \cdot 6W$	tet ; 7.860, 3.905	1.71
$Ca(NO_3)_2 \cdot 4W$	mon	1.89, 1.82
$Mg(NO_3)_2 \cdot 6W$	mon; 6.194, 12.71, 6.600, 93°	1.6.36
$Mg(NO_3)_2 \cdot 4W$	mon; 15.71; 9.015 12.77; 121°40'	1.52
$Na_2S \cdot 9W$	tet; 9.331, 12.85	1.427
$Na_2S_2O_3 \cdot 5W$	mon; 9.944, 21.57, 7.525; 103°55'	1.729
$Ni(NO_3)_2 \cdot 6W$	tric; 5.70; 1191. 7.65 111°, 100°30'; 78°40'	2.05
$KAl(SO_4)_2 \cdot 12W$	cub. 12.158	1.757
$(NH_4)Al(SO_4)_2 \cdot 12W$	cub. 12.240	1.64
$MgCl_2 \cdot 6W$	mon; 9.90, 7.15 6.10; 94'	1.569
$NaOH \cdot 3\,1/2W$	mon; 11.64, 11.38 6.49; 104° 7'	1.63
$Na_2S \cdot 5W$	orth; 6.475, 12.55, 9.655	1.80

Table 6-5
Possible Systems for Thermal Energy Storage

	System	Melting Point	Melting Behavior
low temperature	$K_2HPO_4 \cdot 6W$	13°C	semi-congruent
	$ZnCl_2 \cdot 3W$	10°C	congruent
	$NaOH \cdot 3\,1/2W$	15°C	congruent
	$Na_2CrO_4 \cdot 10W$	18°C	semi-congruent
intermediate temperature	$CaCl_2 \cdot 6W$	30°C	semi-congruent
	$Ca(NO_3)_2 \cdot 4W$	47°C	" "
	$K(CH_3COO) \cdot 1\,1/2W$	42°C	" "
	$K_3PO_4 \cdot 7W$	45°C	congruent
	$Mg(NO_3)_2 \cdot 6W/Mg(NO_3)_2 \cdot 2W$	55.5°C	eutectic
	$Na_2HPO_4 \cdot 12W$	35.5°C	semi-congruent
	$Na_2S_2O_3 \cdot 5W$	48°C	semi-congruent
	$Zn(NO_3)_2 \cdot 4W$	45.5°C	congruent
	$Zn(NO_3)_2 \cdot 2W$	54°C	congruent
	$Fe(NO_3)_2 \cdot 6W$	60°C	congruent
high temperature	$Al(NO_3)_3 \cdot 9W$	89°C	semi-congruent
	$Al_2(SO_4)_3 \cdot 16W$	112°C	congruent (?)
	$AlK(SO_4)_2 \cdot 12W$	80°C	"
	$(NH_4)Al(SO_4)_2 \cdot 12W$	95°C	"
	$CaBr_2 \cdot 4W$	110°C	semi-congruent
	$KOH \cdot W/KOH$	99°C	eutectic
	$MgCl_2 \cdot 6W$	115°C	semi-congruent
	$Na_2S \cdot 5\,1/2W$	97.5°C	congruent
	$Mg(NO_3)_2 \cdot 2W$	130°C	congruent
	$Mg(NO_3)_2 \cdot 6W$	90°C	congruent

Retained were systems of the following metals only: Na, K, Mg, Ca, Al, Cr, Mn, Fe, Co, Ni, Cu and Zn. Within these limitations, only those systems that exhibited a transformation within one of the three temperature ranges of Table 6-1 were further investigated. The number of systems was further reduced by retaining only those that exhibited congruent or semi-congruent behavior. In this way it was possible to satisfy the criteria of Sect. 6.4.

6.7   Sodium Sulfate Decahydrate for TES in Solar Heating

6.7.1 Basic Properties

Sodium sulfate decahydrate (Glauber salt), $Na_2SO_4 \cdot 10H_2O$, has been considered as one of the most promising low-cost materials for the storage of thermal energy, especially in solar heated houses, heat pumps, air conditioning, and similar applications. Its use in the past has been extensively described in the Ref. 1 report and in numerous articles (Ref's. 34, 35, 36). It has been applied as the heat storage material in two solar heated houses (Ref's. 37, 38). When used in closed containers, a crystal seeding or nucleating agent is required to initiate the formation of the solid phase, otherwise the material supercools.

Sodium sulfate decahydrate forms monoclinic crystals which melt incongruently (see Sect. 6.5.2) at $32.38^{\circ}C$ ($90.3^{\circ}F$) changing into a saturated solution of Na_2SO_4 in 10 mols water of crystallization. The heat of fusion is 60 cal/gram (108 Btu/lb). The specific heat and the heat of fusion have been determined repeatedly (Ref's. 39, 40, 41). In Table 6-6, the specific heat and density are compared with that of water.

Table 6-6
Specific Heat and Density of Water
and Sodium Sulfate Compounds

| Constituent | Density | | Specific heat | Relative specific |
	g/cm^3	lb/ft^3	$Btu/lb \cdot {}^{\circ}F$	heat per unit volume
Water	1	62.5	1.0	1.0
$Na_2SO_4 \cdot 10H_2O$ (solid)	1.46	91.0	0.46	0.67
Na_2SO_4 (solid)	2.66	166	0.21	0.56
Saturated solution of Na_2SO_4 at $90.3^{\circ}F$	1.33	83	0.78	1.04
Saturated solution of Na_2SO_4 at $70^{\circ}F$	1.24	77.5	0.84	1.04

The melting of sodium sulfate decahydrate is incongruent, that is, during melting some anhydrous sodium sulfate remains undissolved in its water of crystallization. Due to its higher density (2.66), Na_2SO_4 sinks in the saturated solution (density $= 1.33$ at $90^{\circ}F$). When the mixture solidifies again without mechanical mixing or stirring, dissolved Na_2SO_4 cyrstals recombine with their water of crystallization. The heavy crystals on the bottom of the container, however, recombine only with available water molecules in their immediate vicinity, forming solid $Na_2SO_4 \cdot 10H_2O$ crystals. This solid layer on the top of the anhydrous Na_2SO_4 prevents further recombination of the remaining Na_2SO_4 with the balance of the water of crystallization. Due to this effect, molten sodium sulfate decahydrate, when it solidifies – without stirring or without additives – forms three distinct layers: a bottom layer of white anhydrous Na_2SO_4 crystals embedded into crystals of $Na_2SO_4 \cdot 10H_2O$, a larger intermediate layer of translucent $Na_2SO_4 \cdot 10H_2O$ crystals, and a top layer of liquid translucent saturated solution.

During melting, the following changes occur:

Below $90.3^{\circ}F$ 100 lb $Na_2SO_4 \cdot 10H_2O$ crystals contain

            44 lb Na_2SO_4 and

            56 lb   of water.

Above $90.3^{\circ}F$ 84 lb   saturated solution is formed containing

            28 lb Na_2SO_4 dissolved in

            56 lb   water, leaving

            16 lb Na_2SO_4 sediment.

The heat of fusion required to melt this salt is 108 Btu/lb which can
be released again if the salt is homogenized during solidification by
stirring or by suitable additives. During cooling without homogenizing
or stirring, the heat released is less because part of the sediment cannot regain its water of crystallization. Some saturated solution remains
in this case upon cooling, depending upon the solubility of the salt, as
shown in Table 6-7 and graphically in Fig. 6-2.

Table 6-7

Solubility of Na_2SO_4 in Water at Various Temperatures

Temperature		Solubility (grams Na_2SO_4 per 100 grams water)	%
$(^{\circ}C)$	$(^{\circ}F)$		
0	32	4.5	4.3
10	50	9.0	8.3
20	68	19.0	14.0
30	86	41.2	29.0
32.4	90	49.7	33.2
40	104	48.1	32.4
50	122	46.4	31.6
60	140	45.2	31.1

Without an additive to induce nucleation (Sect. 6.5.5) the decahydrate phase will not form unless the mixture is cooled in a refrigerator. The agent should have the same crystal structure as the material to be nucleated and be slightly soluble or insoluble near the melting point. Borax (sodium tetraborate decahydrate) was found to be a suitable agent, when added in quantities of 3 to 5 percent to the decahydrate. This mixture shows very little super-cooling and melts at 89°F.

[*Editor's Note*: Material has been omitted at this point.]

28

Reprinted from pages 1–12 of *Report ORNL/TM-5525*, Oak Ridge National Laboratory (operated by Union Carbide Corp. for the Dept. of Energy), 1976, 59pp.

THE ANNUAL CYCLE ENERGY SYSTEM: INITIAL INVESTIGATIONS

H. C. Fischer, J. E. Christian, E. C. Hise, A. S. Holman, A. J. Miller, W. R. Mixon, J. C. Moyers, and E. A. Nephew

[*Editor's Note:* In the original, material precedes and follows this excerpt.]

1. INTRODUCTION

This report describes the findings of the initial phase (January 1, 1975 to June 30, 1975) of the program for assessing the Annual Cycle Energy System (ACES). The ACES is an integrated system which uses a heat pump and a thermal-storage bin to provide space heating and cooling and domestic hot water. The heat pump extracts the required heat from a fixed volume of stored water which is converted into ice during the heating season. During the cooling season, the stored ice provides space cooling to the building and is melted.

The primary purpose of phase 1 activities was to assemble and to verify experimentally the design data required for a real-life demonstration of the ACES concept. The intent of the planned demonstration program is to accelerate industry acceptance of the energy-saving concept by acquainting the manufacturing, utility, and building construction industries with the advantages and cost effectiveness of ACES for residential and commercial applications.

Subsequent to the initial investigations described in this report an ACES demonstration house has been constructed in Knoxville, Tennessee. The ACES development program is now continuing under joint sponsorship of the Energy Research and Development Administration and the Department of Housing and Urban Development.

Following are the activities covered in the Phase I program:

1.1 System Analysis and Performance

Analytical investigations were conducted to establish a basis for materials and components selection and to determine the requirements of controls design. Computer programs were developed to assist in the design optimization of an ACES demonstration in an actual building. Tests were conducted to determine the performance and compatibility of systems components. The experimental work required the design, construction, and operation of an ACES components test assembly. This test assembly was used to explore possible mechanical-stability problems of the ACES heat-exchanger coils, which are submerged in the ice-storage tank, and to

verify the designs of the intermediate heat exchanger and of the heat
exchanger for heating domestic hot water using refrigerant superheat.
Heat-transfer rates during ice-buildup and brine-chilling.

1.2 Program Planning

Plans were developed for the construction (in Knoxville, Tennessee)
of a single-family residential building equipped with an ACES. The
performance of the system under realistic operating conditions will be
measured to confirm the adequacy of the design procedures. Contacts
with industrial organizations interested in ACES have been maintained
and strengthened so that cooperation leading to an early commercialization
of ACES can be achieved. The coordination of research and development
efforts with industry will be maintained as an essential part of future
ACES program activities.

2. THE ANNUAL CYCLE ENERGY SYSTEM

2.1 Historical Background

The Annual Cycle Energy System (ACES) offers an attractive means of providing heat from the latent heat of fusion of water. The ice produced by a water-to-air heat pump in the winter is stored to provide "free" air conditioning in the summer. To a great extent, the ACES concept is not really new. The idea of using a heat pump to heat houses was first suggested in 1852 by Lord Kelvin (William Thompson) in a paper presented before the Royal Society. The paper, **entitled "On the Economy of Heating** and Cooling of Buildings by Means of Currents of Air," was published in the December 1852 issue of the *Glasgow Philosophical Society Proceedings*. Today, air-to-air heat pumps have become quite common in many areas of the United States.

In the United States, interest in heat pumps was revived in the late 1920s, and in 1932 a paper by F. H. Faust et al., entitled "Application of Refrigeration to Heating and Cooling of Homes," was published. The paper suggested that extracting heat from an insulated tank of water, rather than from air, would be desirable because the efficiency of the heat pump would be increased. According to the paper, ". . . It has been suggested that the latent heat of the water be extracted by freezing it. In Washington (D.C.) the amount of ice formed during the winter in heating a 14,000-ft^3, well-insulated house would be 210 tons, or 7800 ft^3, which would make a pile about half the size of the house." This idea attracted the attention of H. C. Fischer,[*] who in the mid-1950s extended the concept to include keeping the ice formed at the end of the heating season and using it later for space cooling. In this way, the heating and cooling loads of a building are balanced over a complete annual cycle, greatly reducing the total expenditure of energy.

---

[*]H. C. Fischer was a colleague of F. H. Faust at the General Electric Company in the 1950s. Mr. Fischer is presently working as a consultant to the ACES program at the Oak Ridge National Laboratory.

2.2   Concept Description

The ACES is essentially a thermal-storage system for balancing the cyclic heating and cooling loads of a building. The ACES equipment requirements consist basically of a refrigerant compressor operating as a unidirectional heat pump, an air-cooled condenser for space heating, a water-cooled condenser for domestic water heating, a chilled-water coil for air conditioning, a brine-heated evaporator, ice-freezing coils, an ice bin, and a circulation system with controls. All of this equipment either is already commercially available or can be adapted from existing units. Figure 1 shows a schematic diagram of the system.

Heat is obtained by freezing the water located in the ice-storage structure (Fig. 1) and is pumped by the unidirectional heat pump for delivery to the building. During the summertime, the melting of the ice provides air conditioning to the building; at the same time, the summer heat is stored in the water by increasing the enthalpy of the water as it changes from the solid to the liquid state. The cycle is repeated each year, and the major energy input is simply the energy required to operate the heat pump during the heating season.

Figure 2 illustrates typical operating conditions for the ACES as contrasted with the conventional air-to-air heat pump. As shown, the ACES heat pump operates between rather constant evaporating and condensing temperature limits of 20 and 105°F respectively. Thus, the system can be optimized to these temperatures to obtain a high coefficient of performance (COP). Using compressors currently available, a COP of 3.5 can be reached on the heating cycle; moreover, the use of high-efficiency compressors now coming on the market may enable a COP of 4.0 to be achieved on the heating cycle alone.

In assessing the overall performance of the ACES, however, both the heating and cooling cycles must be taken into account. Because both the heating and cooling capacity of the heat pump are utilized beneficially, the average annual COP of the system is increased. The annual COP of the ACES, neglecting system losses, is defined as

$$\text{COP (annual)} = \frac{\overset{\text{(heating)}}{\text{heat of rejection}} + \overset{\text{(cooling)}}{\text{heat of absorption}}}{\text{electrical energy input}} \ .$$

ORNL-DWG 75-3855R3

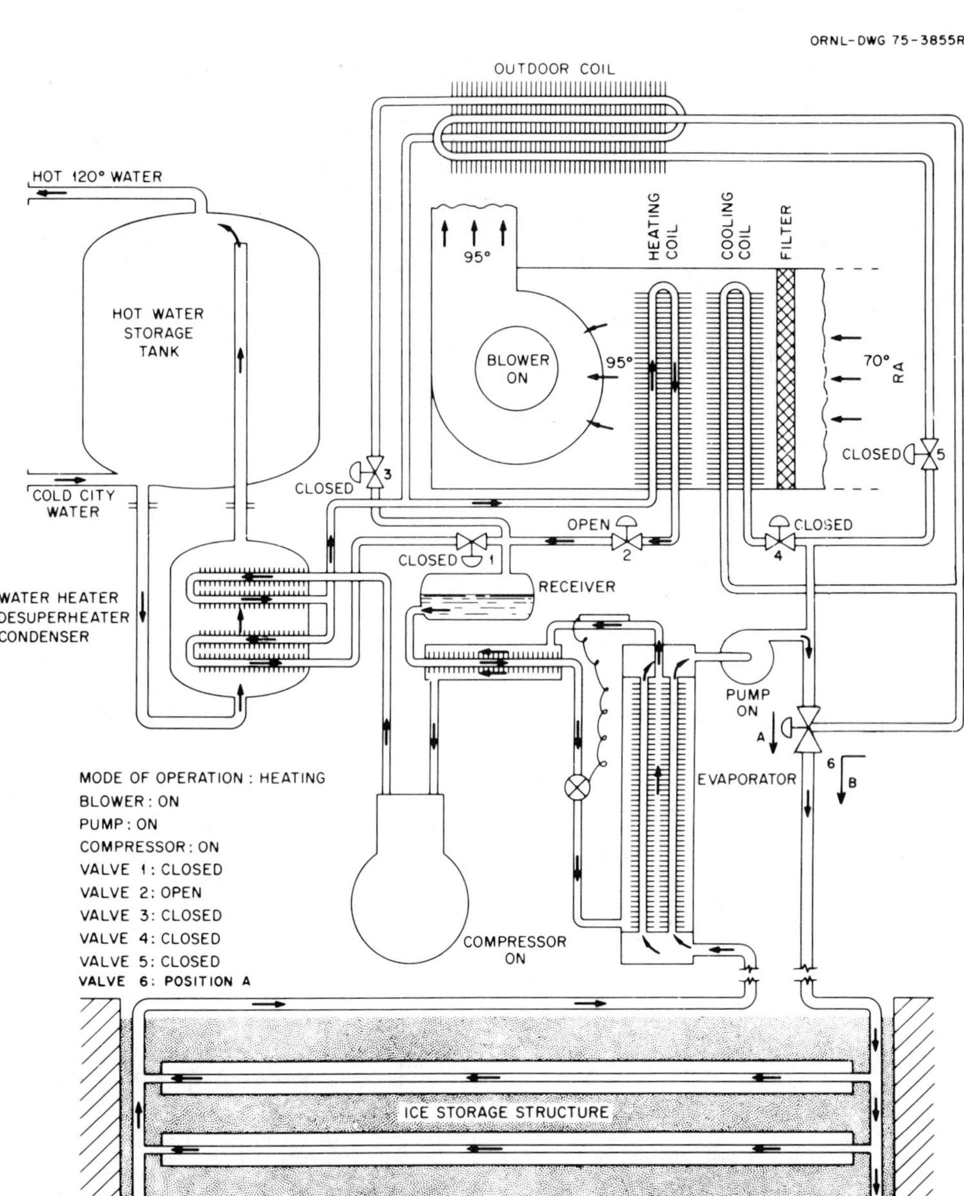

Fig. 1.   Schematic diagram of the ACES as applied to a single-family residential building.

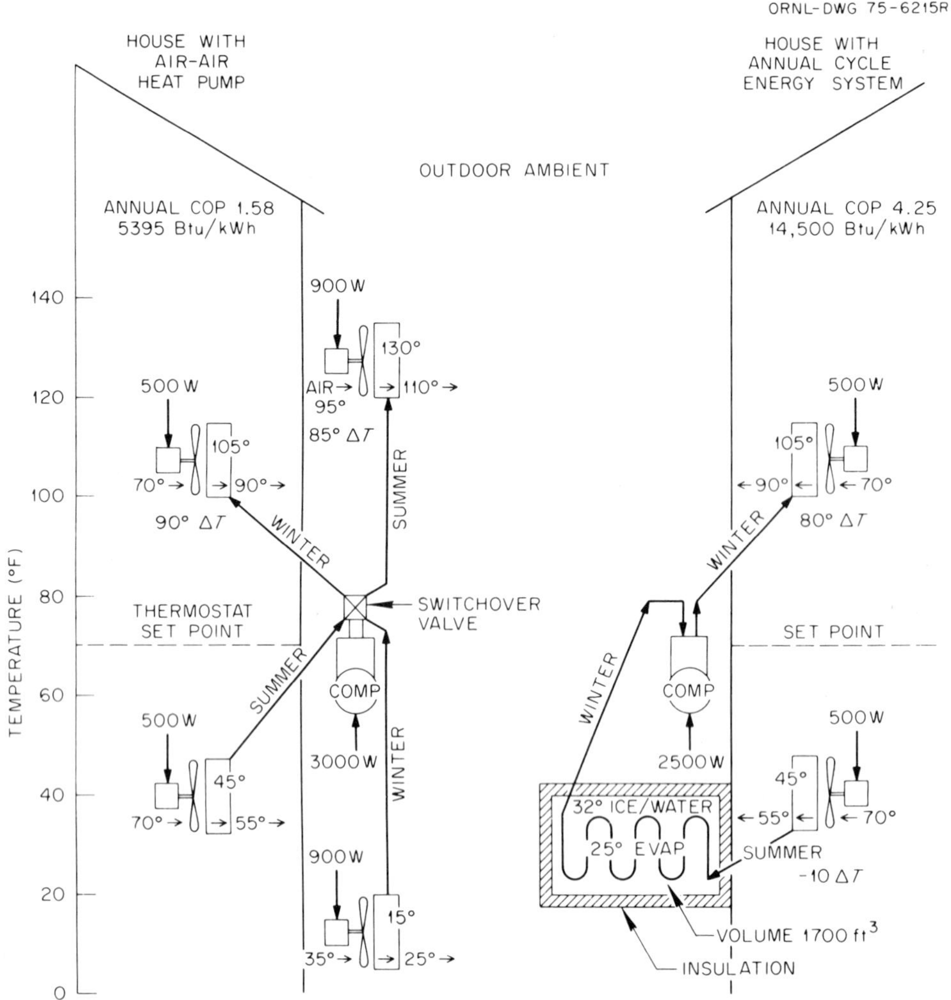

Fig. 2. Air-to-air heat pump vs ACES heat pump, showing typical operating conditions.

In estimating the actual COP of the ACES, system losses must be taken into account. These losses result primarily from pumping power requirements and from heat leakage into the ice bin. Heat leakage into the bin during the cooling season is undesirable because the amount of ice available for air conditioning is reduced. The results of preliminary investigations indicate that, depending on the size of the water-storage tank,

proper insulation of the top, sides, and bottom of the tank can reduce
the average monthly rate of heat leakage to about 3% of the tank's thermal-
storage capacity. On this basis the actual annual COP of the ACES is
estimated to be about 4.25, using present technology. With new high-
efficiency compressors, the annual COP of the ACES would be approximately 5.

2.3   Climatic Influences on ACES Design and Application

Under optimum climatic conditions, the annual COP of the ACES heat-pump system may be as high as 5 because both heating and cooling outputs of the heat pump are used. In 80% of the United States, the heating and cooling requirements are in balance or can be brought into balance for a well-insulated building. Reduction of ice-bin insulation, gain of heat from the outside air, or use of solar panels or outside air coils are possible means for compensating for heating- and cooling-load imbalances in the North. In the South, occasional supplemental compressor operation as an off-peak ice maker may be necessary in the summer. In general, the quantity of water to be frozen should be such that the amount of heat energy obtained is sufficient to supply the heating load of the building through the coldest 12 weeks of the heating season.

Obviously, in the progression toward more northerly latitudes, the winter heating loads increase and the summer cooling loads decrease. If only the heating and cooling loads of a well-insulated residential building are considered, they are found to be in balance at a latitude of about 38° N. However, if domestic hot-water heating is added to the space-heating load, the balance point shifts southward to a latitude of about 36° N (about the latitude of Nashville, Tennessee). The balance point for a poorly insulated home is farther south than for a well-insulated one. In areas north of the balance point, steps must be taken to prevent the buildup of more ice in the winter than is necessary for summer cooling. This control can be achieved by using some alternative heat source, such as solar energy, to impede the formation of ice and to bring the bin size into balance with the summer cooling requirements. Solar energy is abundantly available for this purpose during both the early (September through October) and late (March through May) portions of the heating season.

The thermal storage and solar energy collection requirements for a well-insulated 1800-ft^2 house equipped with an ACES have been estimated for different locations in the United States. Figure 3 shows the variations in required ice-storage volume, solar panel size, and supplementary

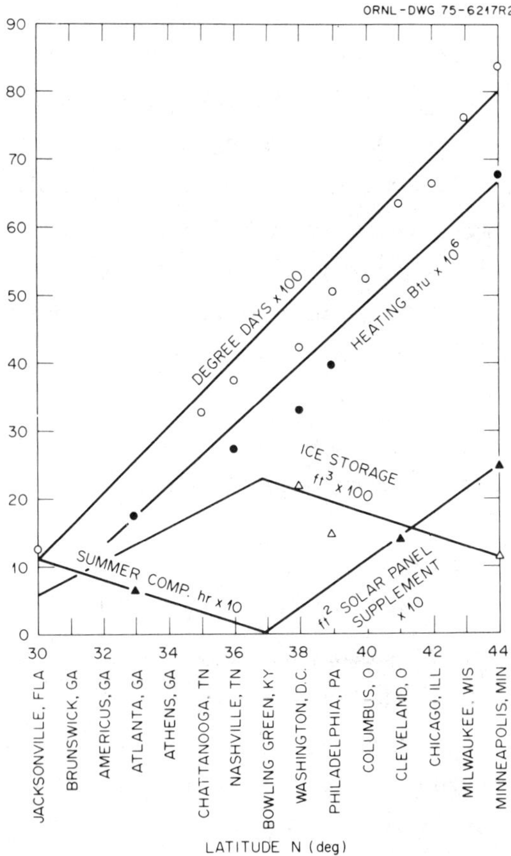

Fig. 3. Effect of latitude on heating load, ice storage, solar panel size, and supplemental compressor operation for ACES-equipped house.

summertime compressor operation for geographical locations ranging from Florida to Minnesota. In Minneapolis, at the northern end of the normal ACES territory, ACES supplies all of the air-conditioning load, all of the domestic hot-water load, and 31% of the space-heating load. The remaining 69% of the space-heating load is furnished by supplementary solar panels. In the Minneapolis region, about 250 ft^2 of solar panel in the form of a vertical solar wall or fence is required to collect the 46.6 million Btu of energy needed to balance seasonal heating and cooling

loads. The collected solar energy is stored in the ice-storage bin and
elevated to space-heating temperature by the ACES heat pump.

In Atlanta, an ACES would provide the house considered above with
all the energy required for space heating and for domestic water heating
and would meet 58% of the summer air-conditioning needs. The **remaining 42%**
of the air-conditioning load could be shifted to nighttime operation to
take advantage of better operating conditions and lower off-peak electric
rates. Whether or not ACES can be applied economically in locations farther
south than Atlanta depends on the seasonal power-rate structure of the
given locality.

2.4   Alternative Modes of ACES Operation

While the ACES is intended primarily to provide space heating and
cooling of buildings, it can also be used to produce domestic hot water.
Water heating is important because in many climatic zones the energy
required for this purpose constitutes a substantial portion of the total
energy requirement of the building. For example, calculation for a
2000-ft^2 single-family home in Knoxville, Tennessee, show that the annual
energy consumptions for space heating, for space cooling, and for water
heating are, respectively, 43.8 million, 22.7 million, and 16.2 million
Btu/year. These calculations are based on assumed heat losses through
building sections in accordance with HUD minimum property standards.[1]

In the ACES design shown in Fig. 1, high-temperature superheated
refrigerant vapor is pumped to a desuperheater condenser to produce
domestic hot water at a minimum temperature of 120°F. During mild weather
and during summer months, the compressor is operated several hours each day
to meet the domestic water-heating load. At the same time, ice is produced
and stored for use in space cooling. Either the solar panel or an air-
cooled condenser can be used to dissipate to the atmosphere the waste heat
generated by summertime compressor operation.

The primary objective in ACES design is to match the capacity of
system components with the thermal characteristics of the building. The
system must possess a high degree of operating flexibility to satisfy the

[1]U.S. Department of Housing and Urban Development, *Minimum Property Standards for One-and Two-Family Dwellings, Vol. 1,* Washington, D.C., 1973

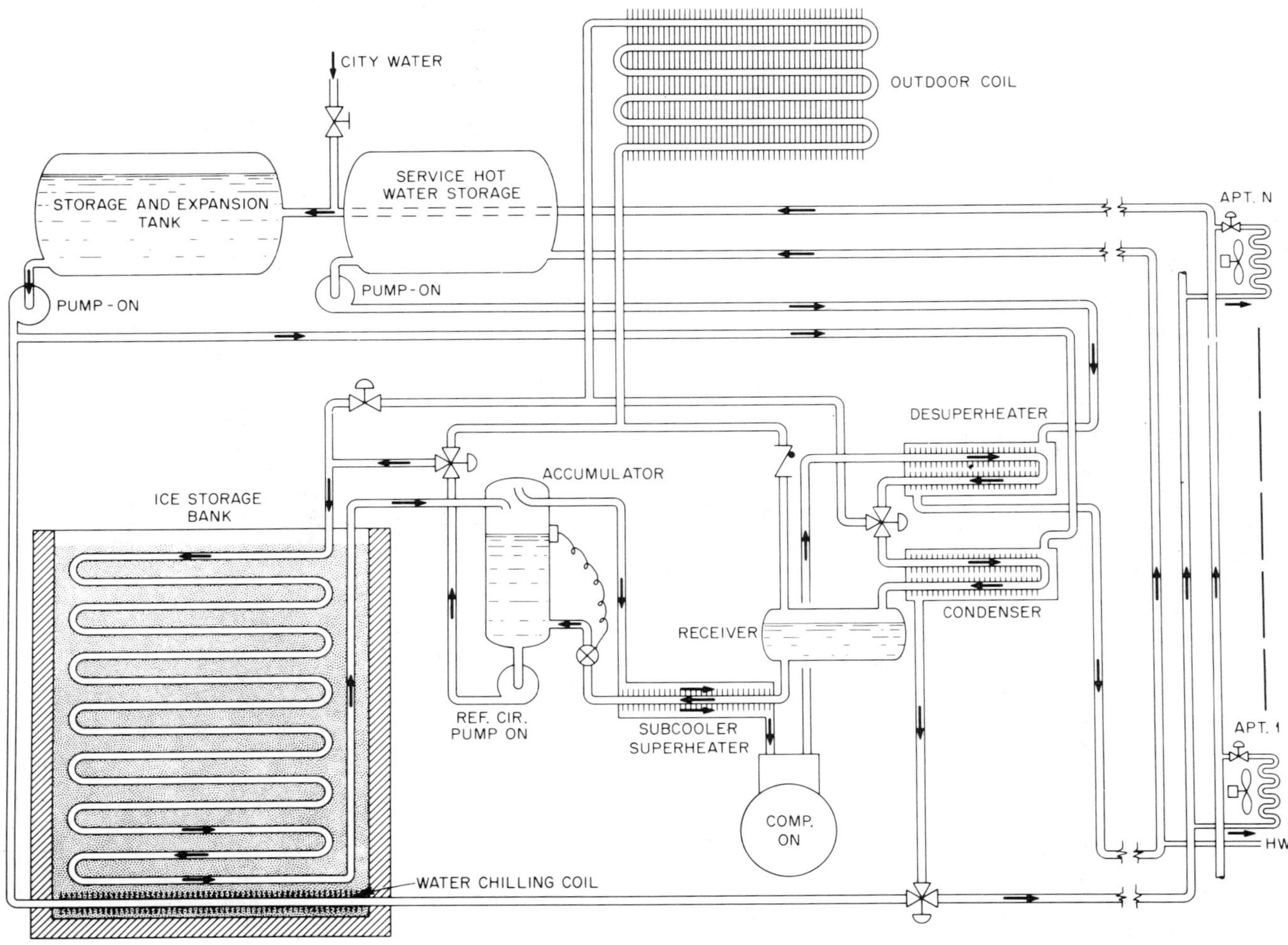

Fig. 4. Schematic diagram of the ACES central system as applied to apartment houses (heating mode).

range of heating and cooling conditions that will normally be encountered over a seasonal cycle. Figure 4 depicts an ACES for an apartment complex in which the individual apartments are served by a common storage bin located in the basement of the apartment building. In this design, liquid refrigerant from the accumulator is circulated directly through heat-exchanger coils submerged in the water-storage bin.

The various modes of operation obtainable with the ACES shown in Fig. 1 are listed in Table 1. The first operating mode listed in the table provides space heating to the building during the winter months. Compressor heat and latent heat of fusion of water are delivered to the room by a heat-exchanger coil located in the forced-air circulation duct. Ice is formed in the water bin and accumulated to provide summer air conditioning. In the primary space-cooling mode, a methanol-and-water brine is chilled by passage through the coils of the ice-bin heat exchanger and then circulated through a cooling coil located in the air-circulation duct. Forced-air circulation around the cooling coil provides air conditioning for the building. An ACES automatic control system selects the appropriate mode of operation according to need.

Table 1. Possible operating modes for the ACES

Mode of operation	Compressor	Air blower	Water pump	Valve Position[a]					
				1	2	3	4	5	6
Space heating/ice building	On	On	On	C	O	C	C	C	A
Space cooling, mode 1	Off	On	On				O	C	B
Space cooling heat rejection to outside	On	On	On	C	C	O	O	C	B
Water heating/ice building	On	Off	On	O	C	C	C	C	A
Water heating and cooling, mode 1	On	On	On	O	C	C	O	C	B
Ice melting	Off	Off	On				C	O	B
Space cooling, water heating	On	On	On	O	C	C	O	C	B
Ice buildup, heat rejection to outside	On	Off	On	C	C	O	C	C	A

[a]Valves 1, 2, and 3 control the refrigerant flow; valves 4, 5, and 6 control the brine flow. Valve position: C = closed, O = open, A and B = positions as depicted in Fig. 1.

29

Reprinted from pages 673–680 of *Intersoc. Energy Convers. Eng. Conf., 11th, Proc.*,
American Institute of Chemical Engineers, 1976, 1802pp.

AN EVALUATION OF THE USE OF METAL HYDRIDES FOR SOLAR THERMAL ENERGY STORAGE

G. G. Libowitz and Z. Blank

Allied Chemical Corporation, Materials Research Center, Morristown, N. J. 07960

Abstract

The basic properties of metal hydrides relevant to their application for storing solar thermal energy are reviewed. Several schemes are discussed and evaluated in which the enthalpy of formation of a primary hydride is used to provide heat and a secondary hydride or compressed gas is utilized for hydrogen storage. The results show that, with present technology, a metal hydride-based system is considerably more expensive than sensible heat or phase change material storage. The advantages of a metal hydride-based system are discussed, and technological advances are suggested which would make such a system economically competitive with other thermal storage systems.

INTRODUCTION

At present, there are three types of materials being considered for storage of solar thermal energy, (1) sensible heat materials, (2) phase change materials, and (3) chemical systems. Sensible heat storage materials are those with high heat capacities which are capable of retaining heat for long periods of time, such as water or rock beds. Phase change materials utilize the enthalpy of a physical change, such as melting, to store heat, while chemical systems utilize the heat of a chemical reaction for the same purpose. This paper deals with a particular chemical system, the interaction of a metal and hydrogen gas.

METAL-HYDROGEN SYSTEMS

Because of their high hydrogen densities [1], metal hydrides are being widely investigated[2,3] for the storage of hydrogen as a fuel. For this application, hydrides with low heats of formation are desirable in order to minimize energy requirements for hydrogen release. It is known, however, that many metal hydrides have high heats of formation and consequently Libowitz[4] has proposed that they be used for thermal energy storage. Later Blank and Libowitz[5] suggested they also be used for solar cooling.

A large number of metals and alloys react reversibly with hydrogen as follows:

$$\frac{2}{x} M + H_2 \rightleftharpoons \frac{2}{x} MH_x \qquad (1)$$

where the hydride, MH_x is a definite chemical compound and the value of x depends upon the particular metal or alloy, M, used. During the hydriding cycle, illustrated by the isotherm in Fig. 1, hydrogen is first absorbed in the metal or alloy to form a solid solution. When the metal becomes saturated with hydrogen at

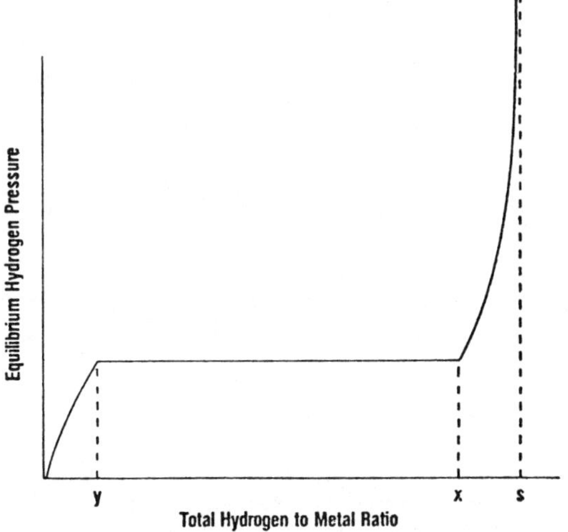

Figure 1. Typical Pressure-Composition Isotherm for a Metal-Hydrogen System

point y, a hydride phase is formed. With further addition of hydrogen, the metal continues to be converted to hydride with a corresponding evolution of heat. At a fixed temperature, the pressure remains constant across the two-phase region and this plateau pressure is referred to as the dissociation pressure of the hydride. At point x, the metal phase has been completely converted to the hydride phase (nonstoichiometric) and as the hydrogen pressure increases, the hydride approaches its stoichiometric composition, s. The assumption is made in Eq. (1) that the

363

solubility of hydrogen in the metal phase is negligible (i.e. y<<x). If this is not so, Eq. (1) should be written:

$$\frac{2}{x-y} MH_y + H_2 \rightleftarrows \frac{2}{x-y} MH_x \qquad (2)$$

This equation would also apply for the conversion of a lower hydride to a higher hydride, where y represents the solubility limit of hydrogen in the lower hydride.

During the dehydriding cycle the reverse process takes place, and heat is absorbed from the surroundings. The plateau or dissociation pressure observed during this hydrogen desorption cycle is sometimes lower than the absorption plateau pressure at the same temperature. This hysteresis effect is significant, and its existence, though not clearly understood[6], should be carefully considered whenever any specific application is proposed.

The thermal storage capacity of a metal hydride depends upon its enthalpy of formation and the amount of hydrogen which the hydride is capable of storing. Consequently a thermal figure of merit M_{th}, can be defined as follows:

$$M_{th} = \left(\frac{x-y}{2}\right)\frac{\Delta H_f}{M.W.} \qquad (3)$$

where ΔH_f is the enthalpy of formation per mole of hydrogen reacting in Eq. (2), x and y are expressed as the number of hydrogen atoms per formula unit of alloy and M.W. is the molecular weight of the hydride.

It has been experimentally observed[1] that the enthalpy of formation of a metal hydride remains essentially constant over a very wide range of temperatures (several hundred degrees C). Therefore, the dissociation pressure, P, at any temperature, can be calculated from the following equation[7]:

$$\ln P = \frac{\Delta H_f}{RT} - \frac{\Delta S_f}{R}$$

where ΔS_f is the entropy of hydride formation and R is the gas constant.

Table 1 summarizes the pertinent thermodynamic parameters for some hydrides which have dissociation pressures in the range of interest for the energy storage applications discussed in this paper. It can be seen from the data in Table 1 that because of the large differences in dissociation pressures as a function of temperature, hydrogen can be conveniently cycled between two identical or different hydrides provided the necessary heating or cooling requirements are met. It is the relative ease of dissociation and formation that make the metal hydrides so attractive for various energy related applications, including thermal energy storage.

A preliminary evaluation of this concept has been performed. The results, presented in this paper, will be compared with other thermal storage systems, and possible modifications will be discussed.

DESCRIPTION OF STORAGE SCHEMES

The concept of metal hydride thermal storage uses solar heat during sunshine periods to dissociate the hydride to metal plus hydrogen gas. The stored heat is recovered during periods of no sunshine by allowing the metal and hydrogen to recombine according to Eq. (1) and utilizing the heat of formation of the hydride.

A typical storage system, which may be employed in a future solar home, would consist of a small hydride reservoir (primary) in the basement with appropriate heat transfer surfaces through which solar heat is carried to the hydride by a heat transport fluid (e.g. air, water or water-ethylene glycol mixture) and a secondary reservoir for hydrogen storage as illustrated in Fig. 2a. As originally proposed [4], this secondary reservoir would store the hydrogen as a compressed gas under a pressure equivalent to the plateau pressure of the primary hydride at its dissociation temperature, and it could be stored underground outside of the house.

A modification of this proposal would be to store the hydrogen in a secondary less stable hydride as illustrated in Fig. 2b. Ideally, this secondary hydride should have the following properties: (a) a high hydrogen to metal ratio in order to minimize the amount of alloy required to store a given amount of hydrogen desorbed from the primary hydride, (b) a relatively low value of enthalpy of formation (as opposed to the primary hydride) to prevent overheating of the hydride as it is absorbing hydrogen; and also to avoid excessive cooling during desorption, (c) the equilibrium hydrogen pressure at ambient temperature (e.g. 21°C) should be lower than the dissociation pressure of the primary hydride at the temperature at which it is absorbing solar heat in order to have spontaneous charging of the secondary hydride. For a similar reason, it would be desirable if the equilibrium hydrogen pressure of the secondary hydride be higher than the pressure of the primary hydride at ambient temperature.

A possible system might consist of vanadium dihydride as the primary hydride and yttrium pentacobalt hydride, $YCo_5H_{2.2}$, as the secondary hydride. If a temperature of 200°F is obtained from solar radiation, it can be seen from Table I that the dissociation pressure of the primary hydride would be 39 atm.* Because this pressure is higher than the equilibrium value of 18 atm for $YCo_5H_{2.2}$ at room temperature (see Table I), the hydrogen evolved by the

---

*Since comparisons of concepts are emphasized in this paper, temperature differences across thermal resistances in each system are neglected.

Table I.   Thermodynamic Parameters of Metal Hydrides as
Related to Thermal Storage

Hydride System	$-\Delta H_f$ kJ/mole H_2	Dissociation Press.,P(atm)		M_{th} (J/g)	M_{th} (Btu/lb)	Ref.
		367°K(200°F)	294°K(70°F)			
$Mg_2NiH_{0.3} \rightarrow Mg_2NiH_4$	64	1.6×10^{-3}	9×10^{-6}	1070	460	8
$VH_{0.95} \rightarrow VH_2$	40	39	1.5	397	171	9
$FeTiH_{0.1} \rightarrow FeTiH_{1.0}$	28	34	3.5	120	52	10
$FeTiH_{1.1} \rightarrow FeTiH_{1.7}$	33	107	7.3	93	40	10
$LaNi_5H_{0.2} \rightarrow LaNi_5H_{6.7}$	30	22	2.0	220	95	11
$YCo_5H_{0.3} \rightarrow YCo_5H_{2.2}$	32	236	18	79	34	12

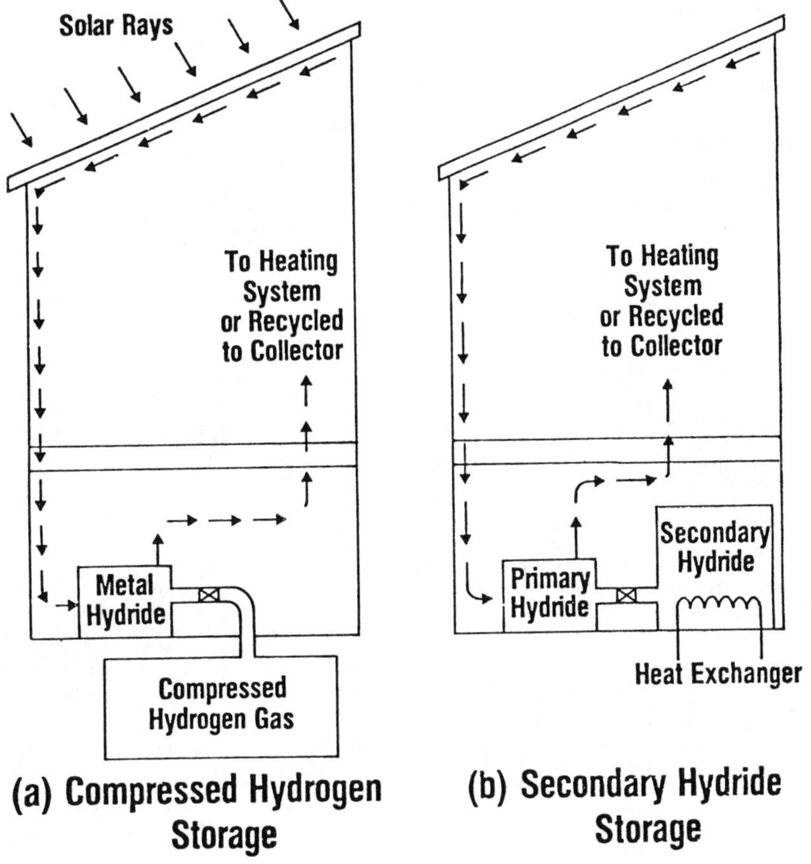

Fig. 2 - Systems for Storing Solar Thermal Energy Using Metal Hydrides

primary hydride will be absorbed by the storage (secondary) hydride. When recovery of the stored heat is desired, the hydrogen in the storage hydride (at 18 atm) is bled back to the primary hydride, which has an equilibrium pressure of 1.5 atm at room temperature. It is, of course, necessary to have efficient heat exchange between the secondary hydride and the ambient in order to maintain the pressure at 18 atm. The primary hydride will attain a temperature of 160°F[9] before the pressures become equalized. This, then, will be the temperature at which the heat will be recovered.

If the equilibrium hydrogen pressure of the storage hydride is lower than that of the primary hydride at room temperature, an auxiliary compressor may be used to pump the hydrogen back into the primary hydride during the heat recovery step.

ECONOMIC EVALUATION OF THE SYSTEM

In order to evaluate metal hydride thermal storage systems relative to sensible heat and phase change systems, it will be assumed that a total storage capacity of 10^9J (10^6BTU) (equivalent to about 2 days consumption in northeast U.S.) will be needed. It also will be assumed that the maximum attainable temperature from solar radiation is 367°K (200°F), although a new concept in solar concentration[13] should easily permit temperatures of 420°K (300°F) to be achieved. Four separate configurations will be discussed in evaluating the use of metal hydrides.

Configuration I is the original scheme[4] whereby the hydrogen evolved from the primary hydride is stored as compressed gas in a high pressure container. From Table I it can be seen that VH_2 appears to have the most desirable thermodynamic properties for application as the primary hydride. Using the values shown, it can be calculated that about 2520kg (5550 lbs) of VH_2 would be necessary to store 10^9J of heat. At a price of $5.00 per kg ($2.25 per lb.) for vanadium metal[14] and $1.75/kg ($0.80 per lb.) of hydrogen, the total cost of VH_2 (containing 100kg of H_2) would be about $13,000.

A storage tank capable of storing the hydrogen gas at 39 atm as released by the hydride would be about 15.4 m^3. The cost of such a container (~6 m long by 1.8 m diameter) constructed from stainless steel 316L would be about $13,000. At the pressures under consideration ($\leq$40 atm), an aluminum tank would also suffice. However, the increased cost in fabrication would probably offset the savings in materials costs.

Assuming a packing density of 50%, the volume required to store the hydride would be about 1.1 m^3 (40 ft^3) (using a value of 4.6 g/cc for the density of the hydride [9]). In order to obtain good heat transfer, the hydride could be stored in about

115 aluminum tubes (Schedule 40), 10 cm (4 in.) in diameter and 1.2 m (4 ft.) long. The cost[15] of each such tube with end caps would be about $60 for a total hydride storage cost of $7,000. Consequently, the total capital cost of Configuration I using vanadium hydride would be about $33,000.

Configuration IIa is the one described earlier in this paper in which a secondary hydride (e.g. $YCo_5H_{2.2}$) is used to store the hydrogen. However, the cost of YCo_5 alloy is prohibitively high and a lower cost storage hydride must be considered. At present, the most inexpensive hydrides are the iron-titanium hydrides. If the higher hydride, $FeTiH_{1.7}$, were used, the vanadium dihydride would be re-formed at a pressure of 7.3 atm (see Table I). The temperature of the VH-VH_2 isotherm[9] having this plateau pressure is 325°K(125°F). Therefore, the stored heat would be recovered at 325°K at the vanadium hydride heat exchanger. Consequently, an increased heat transfer surface would be necessary over that required if the recovery temperature were higher.

The amount of storage hydride needed depends upon the amount of hydrogen desorbed from the primary hydride. For the case of vanadium dihydride, which supplied 40 kJ/mole H_2, 26,000 moles of H_2 gas are desorbed in order to store 10^9J. Consequently, 86,000 moles of storage hydride are necessary according to the reaction (see Eq. 2 and Table I):

$$3.3 \; FeTiH_{1.1} + H_2 \rightarrow 3.3 \; FeTiH_{1.7}$$

This corresponds to 9000 kg (20,000 lbs) of iron titanium hydride. The present price of iron titanium hydride is about $4.40/kg ($2.00 per lb)[16]; however, if this hydride is used extensively to store hydrogen as a fuel, it is believed[16] that the price would drop to about $1.10 per kg., corresponding to a cost of $10,000 for the required amount of hydride.

The cost of the container also must be considered*. The 9000 kg of $FeTiH_{1.7}$ would occupy a volume of 3.3 m^3 (d = 5.5 g/cc); 50% packing density. At a cost of $7,000 per cubic meter for the Al tubes(see above) this would be about $21,000. Therefore, the total cost of Configuration IIa is $51,000.

Less iron titanium alloy would be required if both the lower and higher hydride could be used to store the hydrogen. However, the pressure-temperature relationships of the lower hydride are such that too low a

---

*In any detailed design of a system, the container configuration should be such that it will provide adequate heat transfer areas. The configuration discussed here (4 ft long tubes, 4" in diameter) are approximate and only used for purposes of comparison.

temperature would be generated on re-forma-tion of the primary hydride; i.e. a pres-sure of 3.5 atm (see Table I) corresponds to a VH-VH$_2$ plateau pressure of only 309°K (97°F). This leads to the third con-figuration to be discussed.

<u>Configuration III</u>: In Configuration III a method of pumping the hydrogen from the storage reservoir to the primary hydride is used. One way of accomplishing this is by means of a hydrogen compressor between the two hydride reservoirs. Because of the low density of hydrogen, large volumes of gas have to be handled and this re-quires oversized compressors. In addition, the low viscosity of hydrogen gas leads to increasing leakage. Thus, for the scheme described here, the cost of a hydrogen gas compressor would be at least $5,000 and probably much higher. The energy needed to run the compressor also would have to be taken into consideration.

If both the lower and higher hydrides of FeTi were used, the overall reaction would be:

$$1.25 FeTiH_{0.1} + H_2 \rightarrow 1.25 FeTiH_{1.7} \qquad (4)$$

Therefore, only 33,000 moles or 3450 kg would be required, which at $1.10 per kg amounts to $3,800. The cost of the Al storage hydride container also should be reduced a corresponding amount to $8,000, so that the total cost of the system would be $38,000 plus the compressor.

An alternative method of compressing hydro-gen into the primary hydride is to heat the lower hydride of iron titanium alloy such that the equilibrium pressure becomes equal to that of the higher hydride (7.3 atm) at room temperature. This would re-quire a temperature of 314°K (106°F). The total heat needed to raise the temperature of the hydride is $(n)(C_p)(\Delta T)$ where n is the number of moles of hydride, C_p is the molar specific heat of the hydride and T is the change in temperature, 20°K(314-294) in this case. Using a specific heat of 61 J/mole-°K[17], the heat required to raise the temperature of the hydride to 314°K is $(61)(20) = 1.2$ kJ per mole. In addition, the enthalpy of dissociation of the lower hydride, 28 kJ/mole H$_2$ (see Table I), must be supplied. According to the reaction:

$$2.22 \ FeTiH_{1.0} \rightarrow 2.22 \ FeTiH_{0.1} + H_2$$

12.7 kJ are required to dissociate one mole of hydride. Therefore, 13.9 kJ per mole of hydride are needed to recover the stores hydrogen. This energy may be obtained from the VH-VH$_2$ primary hydride reaction requir-ing an additional amount of VH$_2$ over that needed to store 10^{9}J. It can be shown (see Appendix) that in order to use this scheme, 36,600 additional moles (1930 kg) of VH$_2$ (to yield a total of 4450 kg of VH$_2$) and 55,300 moles of FeTiH$_{1.7}$ are re-quired. The corresponding costs would be

$23,000 for VH$_2$, $6400 for FeTiH$_{1.7}$, and $27,000 for the primary and secondary hy-dride containers for a total of $56,000 which is about the same total cost as Con-figuration IIa. The saving over Configura-tion IIa in the amount of secondary hydride is cancelled by the requirement of addi-tional VH$_2$ to supply energy for dissoci-ation of the storage hydride. However, it can be seen that a secondary hydride having a dissociation pressure lower than the pri-mary hydride may be utilized, if it also has a low enthalpy of formation so that not too much energy is used in recovering the hydrogen.

<u>Configuration IIb</u>: A factor in the expense of these systems is the cost of the storage containers for the hydrides. If the work-ing pressure of hydrogen could be reduced to about 10 atm, a low carbon steel tubing may be used instead of aluminum. In Con-figuration IIa, if the solar collector is adjusted such that the vanadium hydride temperature does not exceed 330°K (135°F), corresponding to a pressure of 10 atm, low carbon steel may be used for the storage containers with a corresponding decrease in costs. Four inch (10cm) diameter Schedule 80 A53 low carbon steel tubing is $13.00 per meter ($4.00/ft). Using 1.2m length tubes with end caps at $6 each, the storage cost per cubic meter would be $3200. Therefore, hydride storage would be reduced from $28,000 to $14,000.

<u>Configuration IV</u>: The fourth configuration to be considered is one proposed by Gruen and Sheft[18] whereby the primary hydride at elevated temperatures has a lower dis-sociation pressure than the storage hydride at room temperature, thus requiring com-pression to drive the hydrogen into the storage hydride. The hydrides chosen were Mg$_2$NiH$_4$ for the primary hydride because of its high enthalpy of formation (see Table I) and La Ni$_5$H$_{6.7}$ as the storage hydride. The weight of Mg$_2$NiH$_4$ necessary to store 10^9 J of energy is about 935 kg as calcu-lated from M$_{th}$ in Table I. With a cost of $3.30/kg ($1.50 per lb.)[15] for magnesium nickel hydride, the total cost of the pri-mary hydride would be $3,100. The amount of hydrogen to be stored according to the reaction:

$$0.54 \ Mg_2NiH_{0.3} + H_2 \rightleftarrows 0.54 \ Mg_2NiH_4$$

is 8800 moles which would require 1200 kg of LaNi$_5$ as calculated from the reaction:

$$0.308 \ LaNi_5H_{0.2} + H_2 \rightarrow 0.308 \ LaNi_5H_{6.7}$$

At $35/kg ($16/lb), the cost of the storage hydride would be about $42,000. The mag-nesium nickel hydride (d = 2.57 g/cc) oc-cupies a volume of 0.76 m^3 at 50% packing density and the lanthanum pentanickel hy-dride (d = 6.59 g/cc) requires a volume of 0.37 m^3. Since this is a low pressure (<10 atm) system, low carbon steel tubing can be used at a cost of $3600 for both containers.

Although Gruen and Sheft's concept utilizes the higher temperature Winston collector[13] permitting dissociation of the primary hydride at 425°K, the equilibrium pressure attained by the magnesium nickel hydride at this temperature, 0.03 atm, is considerably less than the 2 atm equilibrium pressure of the lanthanum nickel hydride, thereby requiring the use of compressors to drive the hydrogen into the storage hydride. Consequently, the cost of a compressor (>$5,000) must be added to the cost of this configuration.

COMPARISON WITH OTHER THERMAL STORAGE SYSTEMS

As mentioned in the introduction, the other types of thermal storage systems now under consideration are sensible heat storage materials (water and rock beds) and phase change materials.

Water Tanks: The price of a water storage tank has been estimated[19] at $130-200 per m^3 ($0.50 to $0.75 per gallon) including installation but excluding insulation and auxiliary components. Assuming a temperature rise of 22°K (40°F), 11 m^3 of water would be required to store 10^9 J at a tank cost of $1500 to $2300. This is in agreement with a separate estimate[20] of $1200 for 5×10^8 J (5×10^5 Btu). Assuming a cylindrical tank with a height to radius ratio of 4:1, its surface area would be about 30 m^2. At a typical price range of $22-32 per m^2 ($2-3 per square foot) of installed insulation (poly-urethane, polystyrene, etc.) the additional cost would be $800, yielding a total cost of about $3000.

Rock Bed Systems: For a temperature rise of 22°K an energy storage value of about 18 J/g (based on a specific heat of 0.8 J/g°K) is assumed. Hence, to store 10^9 J, a total of about 54,000 kg, equal to a volume of about 70 m^3 (assuming a packing density value of about 0.3) would be necessary. The cost of constructing such a system at an average rate of $14-16/man-hr, and a total of 120 man-hrs. (including container and flow distribution vanes, rock transportation and laying) would be about $1,800.

The cost of insulating a rock bed is cheaper than insulating water because heat transfer in rock beds where air is stagnant is worse than in water tanks. Assuming an approximate rectangular configuration of 7.5 × 3. × 3 m to contain the rock bed, the additional cost for insulation at $22/m^2$ ($2/ft^2$) would be $2400 to yield a total cost of about $4200.

Phase Change Materials: Phase change materials[20,21] which are candidates for thermal storage include low melting salt eutectics and waxes whose melting points are in the range 300-425°K. Lorsch[20] performed a cost analysis on two representative materials, sodium thiosulfate pentahydrate, $Na_2S_2O_3.5H_2O$ and P116 Wax (Sunoco) each having an energy storage value of 210 J/g. Assuming a price range of $0.22-0.28 per kg ($0.10-$0.40/lb) for $Na_2S_2O_3 \cdot 5H_2O$ and $0.11-0.33/kg ($.05-$.15/lb) for P-116, the cost for a heat storage system ranged from $3000 to $6400 for 10^9 J. The container and heat exchange materials were chosen on the basis of compatability with the storage materials, and in every case the least expensive combination of materials was used. Adding insulation would raise the cost approximately $500.

CONCLUSIONS

The capital costs of the various hydride schemes are summarized in Table II along

Table II. Estimated Relative Capital Costs of Thermal Storage Systems

System	Capital Cost ($)
Metal Hydride Configurations	
I - VH_2 + compressed gas storage	33,000
IIa- VH_2 + $FeTiH_2$ storage	51,000
IIb- same at lower pressures	37,000
III - VH_2 + FeTiH + $FeTiH_2$ storage with compressor	>43,000
raising temp. of storage hydride	56,000
IV - Mg_2NiH_4 + $LaNi_5H_6$ + compressor	>54,000
Water Storage	3,000
Rock Beds	4,000
Phase Change Materials	5,000

with those for other types of thermal storage. It can be seen that, with present technology, the cost of hydride storage systems would be about an order of magnitude higher than the other thermal storage systems. There are, however, several advantages to metal hydride thermal storage: (1) The rate of heat recovery can be easily controlled, and since no heat is evolved until the hydrogen is re-combined with the primary alloy, thermal energy can be stored for indefinite periods of time, if necessary, as opposed to sensible heat and phase change materials which evolve their heat spontaneously; this suggests the possibility of ultimately storing summer heat for winter use; (2) Because there is no spontaneous evolution of heat, no insulation is required; (3) Since most metal hydrides are metallic they have high thermal conductivities[22] which permit efficient heat transfer during use; (4) Metal hydrogen systems can undergo indefinite cycling with no chemical degradation, which is a problem when phase change materials are used.

Despite these advantages, the cost of a metal-hydrogen system must be brought down into the vicinity of other thermal storage systems before it would become an attractive alternative. There are several approaches which can be taken to solve this problem.

The costs of the hydride containers are approximately 50% of the total cost of the higher pressure configurations (I, IIa, and III). Consequently, the development of lower cost structural materials resistant to hydrogen embrittlement at elevated pressures would be of value. At present there is a great deal of research being pursued in this area[23].

In order to maintain a sufficient rate of mass transfer of hydrogen from one hydride bed to another, the dissociation pressures of the hydrides should be in the vicinity of one atm (or greater). There are presently many lower-cost known metal or alloy hydrides whose dissociation pressures are too low at 365°K (200°F) to be used in this application. Therefore, the development of higher temperature solar collectors would assist in lowering the cost of metal hydride storage systems. For example, a collector temperature of 525°K would raise the dissociation pressure of Mg_2NiH_4 to one atm. Also, in Configuration I, higher temperatures would permit the storage of compressed hydrogen at higher pressures and consequently, in smaller volume containers.

The most significant approach to lowering the costs of these hydride storage systems is the development of new alloy hydrides having high heats of formation and high hydrogen-to-metal ratios. Less hydride would then be required to store 10^9 J of energy, thereby reducing the hydride contribution to the total cost as well as the size of the container. If the hydride could be made from a low cost metal, further reduction in cost would be achieved. One possibility would be to find an appropriate alloy of calcium which would increase the dissociation pressure of calcium hydride[24] (10^{-19} atm at 367°K) without significantly reducing the heat of formation[25] (188 kJ/mole). The cost of calcium metal is $2.75/kg ($1.25/lb). In order to store 10^9 J only 120 kg (265 lbs) would be required using 0.3 m^3 of container volume. This would reduce the cost of the primary hydride part of the system to less than $1300.

Similarly, low cost alloys capable of forming hydrides with high hydrogen-to-metal ratios would be desirable as storage hydrides.

It can be concluded, therefore, that in order for metal hydride thermal storage to become economically competitive with other thermal storage systems, some combination of the following three requirements must be met:

(1) Lower cost container materials for storing hydrogen should be developed.

(2) New low cost alloy hydrides having high enthalpies of formation and/or high hydrogen retentive capacities must be discovered[26] which are capable of operating at reasonable pressure ranges (~1 atm).

(3) The development of higher temperature solar collectors would be of value in meeting the second requirement.

APPENDIX

As shown earlier in the paper 2520 kg, or 47,600 moles, of VH_2 are needed to store 10^9 J of heat energy. If x is the number of moles of VH_2 needed to supply the heat to dissociate the lower hydride of FeTi, the total number of moles of VH_2 needed is 47,600 + x. According to the VH_2 dissociation reaction [see Eq (2) and Table I], this corresponds to (1.05/2)(47,600 + x) moles of H_2. From Eq. (4), the number of moles of $FeTiH_{1.7}$ needed to store this hydrogen is (1.25)(1.05/2)(47,600 + x).

From Table I, the energy value of VH_2 is 397 J/g or 21.0 kJ/mole VH_2. Since 13.9 kJ/mole of $FeTiH_{1.7}$ is required to recover the hydrogen we can write 21.0 x = (13.9) x (1.25)(1.05/2)(47,600 + x). Solution of this equation yields x = 36,600 moles = 1930 kg, and the total number of moles of $FeTiH_{1.7}$ is 55,300.

ACKNOWLEDGEMENT

The authors would like to thank R. W. Armbrust and S. Levinson for helpful discussions on the economic aspects of this evaluation.

REFERENCES

1. Libowitz, G. G., *Solid State Chemistry of Binary Metal Hydrides*, W. A. Benjamin, N. Y., 1965, pp. 46-47.

2. Proc. of the First World Hydrogen Conf., Univ. of Miami, 1976, Session 8B.

3. Libowitz, G. G. and Vichr, M., "Alloy-Hydrogen Reactions and Their Application to Hydrogen Storage", 146th Meeting of the Electrochem. Soc., Abs. No. 238, New York, Oct. 1974.

4. Libowitz, G. G., "Metal Hydrides for Thermal Energy Storage", Paper #749025, Proc. of the 9th Intersoc. Energy Conv. Eng. Conf., San Francisco, 1974, pp. 322-325.

5. Z. Blank and G. G. Libowitz, "Solar Cooling Using Metal Hydrides", to be published.

6. Ref. 1, pp. 83-86.

7. Ref. 1, pp. 54-55.

8. Reilly, J. J. and Wiswall, Jr., R. H., "The Reaction of Hydrogen with Alloys of Magnesium and Nickel and the Formation of Mg_2NiH_4, Inorg. Chem. <u>7</u>, 2254 (1968).

9. Reilly, J. J. and Wiswall, Jr., R. H., "The Higher Hydrides of Vanadium and Niobium", Inorg. Chem. <u>9</u>, 1678 (1970).

10. Reilly, J. J. and Wiswall, Jr., R. H., "Formation and Properties of Iron-Titanium Hydride", Inorg. Chem. <u>13</u>, 218 (1974).

11. Kuijpers, F. A. and Van Mal, H. H., "Sorption Hysteresis in the $LaNi_5$-H and $SmCo_5H$ Systems", J. Less Common Metals, <u>23</u>, 395 (1971).

12. Takeshita, T., Wallace, W. E. and Craig, R. S., "Hydrogen Solubility in 1:5 Compounds Between Y or Th and Ni or Co", Inorg. Chem. <u>13</u>, 2282 (1974).

13. Winston, R., "Principles of Solar Concentrators of a Novel Design", Solar Energy <u>16</u>, 89 (1974).

14. Iron Age, May 12, 1975, p. 62.

15. Based on data from Popper, H., <u>Modern Cost Engineering Techniques</u>, McGraw Hill Co., N. Y. (1970); and M & S cost index of 437 in May 1975.

16. Salzano, F. J., ed., "Hydrogen Storage and Production in Utility Systems", Report BNL 19249, Brookhaven Natl. Lab. 1974.

17. Yu, W. S., <u>et al</u>, "Modelling Studies of Fixed-Bed Metal-Hydride Storage Systems", Proc. of the Hydrogen Econ. Miami Energy Conf., School of Eng. & Environmental Design, Univ. of Miami, 1974, pp. S4-22 to S4-35.

18. Gruen, D. M. And Sheft, I., "Metal Hydride Systems for Solar Energy Storage and Conversion", Proc. of the NSF-ERDA Workshop on Solar Energy Storage Systems, Charlottsville, Va., April 1975.

19. Obtained from discussions at the NSF-ERDA Workshop on Solar Energy (see Ref. 18).

20. Lorsch, H. G., Thermal Energy Storage Devices Suitable for Solar Heating, Paper #749083, Proc. of the 9th Intersoc. Energy Conv. Eng. Conf., San Francisco, Cal., 1974, pp. 572-577.

21. Hale, D. V., Hoover, M. J. and O'Neill, M. J., "Phase Change Materials Handbook" NASA report #CR-61363, 1971.

22. Mueller, W. M., Blackledge, J. P., and Libowitz, G. G., <u>Metal Hydrides</u>, Academic Press, N. Y. 1968, p. 4.

23. Proc. Conf. on Effect of Hydrogen on Behavior of Matls., Moran, Wyoming, 1975.

24. Curtis, R. W. and Chiotti, P., "Thermodynamic Properties of Calcium Hydride", J. Phys. Chem. <u>67</u>, 1061 (1963).

25. Ref. 22, pp. 201 and 204.

26. Libowitz, G. G. and Blank, Z., "Properties of Solid Metal Hydrides as Related to Their Applications in Solar Heating and Cooling", paper presented at Am. Chem. Soc. Meeting, New York City, April, 1976; to be published in Advances in Chemistry Series.

AUTHOR CITATION INDEX

Adelman, G., 215
Adt, R. R., 178
Advanced Transport Technology Staff, 209
Allen, J. W., 285
Allen, K., 83
Almquist, 182
American Petroleum Institute, 215
Anderson, J. E., 215
Andrews, D. G., 215
Angerhofer, P. E., 165
Armagnac, A. P., 178
Arntzen, J. D., 285

Bacon, F., 178
Balcomb, D., 322
Bartel, J. J., 323
Bartholme, L. G., 285
Bartlit, J. R., 172, 215
Baudendistel, L., 215
Benton, A. F., 179, 180, 183
Berman, S., 24
Bernstein, I. M., 164
Bersticker, A. C., 83
Best, J. S., 323
Billings, R. E., 215
Blackledge, J. P., 370
Blank, Z., 370
Bockris, J. O'M., 164, 178
Böhm, H., 215
Bohn, H. L., 215
Bolt, J. A., 215
Bonner, T. F., Jr., 208
Borucka, A., 322
Bramlette, T. T., 323
Brandt, A., 237, 285
Bray, A. N. G., 42
Brennon, F. L., 62
Bridgers, F. H., 323
Bronson, J. C., 172
Brown, H. L., 24
Brown, J. T., 285
Brumleve, T. D., 323

Brunshteyn, B., 164
Bryer, F., 24
Burns, 180
Burth, P. O., 178
Bush, J. B., 83
Bush, J. B., Jr., 286

Cairns, E. J., 165, 285
California Institute of Technology, 25
Chalmers, B., 24
Chang, G. C., 237
Chernishev, P. S., 157
Cherry, W. R., 215
Childs, C. E., 165
Chiotti, P., 370
Chopey, N. P., 178
Clarke, E. C., 323
Clerk, R. C., 237
Coats, K. H., 84
Convers, C. C., 178
Coogen, A. H., 83
Cornell, D., 84
Cosci, C., 164
Council, M. E., 158
Cox, K. E., 164
Craig, R. S., 370
Crittenden, 182
Cronin, J. H., 285
Curtis, R. W., 370

Daily, J. W., 286
Decker, G. L., 331
DeHouck, G., 42
Devoto, R. S., 215
Dickson, E. M., 164
Dieges, P., 178
Dobner, S. I., 215
Donner, G. A., 164
Dooker, J., 24
Dugger, G. L., 237, 285
Dulles, J. F., 323
DuPont, M., 42

Van Mal, H. H., 370
Vary, J. A., 84
Vichr, M., 369
Vielstich, W., 165
Von Ottingen, W., 165

Wagner, T. O., 178
Wallace, W. A., 178
Wallace, W. E., 370
Walsh, E., 25
Walsh, W. J., 285
Weber, O., 130
Webster, D. S., 285
Wei, J., 172
Weil, K. H., 178
Weinaug, C. E., 84
Weiss, R. O., 237, 285
Well, K. H., 215
Wentworth, T. O., 215
Wessling, F. C., Jr., 323
Westerman, A. B., 178

White, T. A., 179, 183
Whitehouse, G. D., 158
Whitlow, J. B., Jr., 209
Wilden, M. W., 323
Wilhoit, R., 165
Wilks, W., 178
Williams, A. F., 165
Williams, D. A., 165
Williams, L. O., 172, 178
Williamson, 179
Williamson, K. D., Jr., 164, 172, 178, 215
Winsche, W. E., 178, 215
Winston, R., 370
Wiswall, R. H., Jr., 178, 215, 370
Wurm, J., 178, 215

Yao, N. P., 285
Yu, W. S., 370

Zener, C., 178
Zwolinski, B., 165

SUBJECT INDEX

About the Editor

WILLIAM V. HASSENZAHL is a senior scientist at the Lawrence Berkeley Laboratory. He received the B.S. in physics from the California Institute of Technology in 1962 and the Ph.D in elementary particle physics from the University of Illinois in 1967.

During 1967 to 1980 Dr. Hassenzahl was a physicist at the Los Alamos Scientific Laboratory, where he was manager of the superconducting magnetic energy storage program from its inception in 1972 to 1978. After a sabbatical at the Centre d'Etudes Nucleaires de Saclay he returned briefly to Los Alamos before transferring to the Lawrence Berkeley Laboratory in 1980.

Dr. Hassenzahl's principal research interest has been in the development of superconducting magnets including investigations of static and transient heat transfer to liquid helium, conductor stability, and magnet performance in superfluid helium. He has authored or edited 40 papers on superconducting magnets.

He is a member of the American Physical Society and a consultant to the Electric Power Research Institute on superconducting magnetic energy storage.